Дар от Mirna
Džamonja
Спасибо Мирна!
September 2008

quaderni di matematica

volume 17

edited by

Dipartimento di Matematica
Seconda Università di Napoli

Published with the support of
Seconda Università di Napoli

quaderni di matematica

Published volumes

1 - Classical Problems in Mechanics (R. Russo ed.)

2 - Recent Developments in Partial Differential Equations (V. A. Solonnikov ed.)

3 - Recent Progress in Function Spaces (G. Di Maio and Ľ. Holá eds.)

4 - Advances in Fluid Dynamics (P. Maremonti ed.)

5 - Methods of Discrete Mathematics (S. Löwe, F. Mazzocca, N. Melone and U. Ott eds.)

6 - Connections between Model Theory and Algebraic and Analytic Geometry (A. Macintyre ed.)

7 - Homage to Gaetano Fichera (A. Cialdea ed.)

8 - Topics in Infinite Groups (M. Curzio and F. de Giovanni eds.)

9 - Selected Topics in Cauchy-Riemann Geometry (S. Dragomir ed.)

10 - Topics in Mathematical Fluids Mechanics (G.P. Galdi and R. Rannacher eds.)

11 - Model Theory and Applications (L. Belair, Z. Chatzidakis et al. eds.)

12 - Topics in Diagram Geometry (A. Pasini ed.)

13 - Complexity of Computations and Proofs (J. Krajíček ed.)

14 - Calculus of Variations: Topics from the Mathematical Heritage of E. De Giorgi (D. Pallara ed.)

15 - Dispersive Nonlinear Problems in Mathematical Physics (P. D'Ancona and V. Georgev eds.)

16 - Kinetic Methods for Nonconservative and Reacting Systems (G. Toscani ed.)

17 - Set Theory: Recent Trends and Applications (A. Andretta ed.)

Next issues

Trends in Incidence and Galois Geometries: a Tribute to Giuseppe Tallini (F. Mazzocca, N. Melone and D. Olanda eds.)

Selection Principles and Covering Properties in Topology (Lj.D.R. Kočinac)

Mathematical Modelling of Bodies with Complicated Bulk and Boundary Behavior (M. Šilhavý ed.)

Vector Bundles and Low Codimensional Subvarieties: State of the Art and Recent Developments (G. Casnati, F. Catanese, and R. Notari eds.)

Numerical Methods for Balance Laws (G. Puppo and G. Russo eds.)

Set Theory: Recent Trends and Applications

edited by Alessandro Andretta

ARACNE

Authors' addresses:

Alessandro Andretta
Dipartimento di Matematica
Università degli Studi di Torino
Via Carlo Alberto, 10
10123 Torino, Italy
email: alessandro.andretta@unito.it

Andreas Blass
Mathematics Department
University of Michigan
Ann Arbor, MI 48109–1043, U.S.A.
email: ablass@umich.edu

Mirna Džamonja
University of East Anglia
Norwich, NR4 7TJ, United Kingdom
email: M.Dzamonja@uea.ac.uk

Ilijas Farah
Dept. of Mathematics and Statistics
York University
4700 Keele Street
Toronto, M3J 1P3, Canada
email: ifarah@mathstat.yorku.ca

Sy-David Friedman
Kurt Gödel Research Center
for Mathematical Logic
Universität Wien
Währinger Str. 25
A-1090 Wien, Austria
email: sdf@logic.univie.ac.at

Juliette Kennedy
Department of Mathematics
and Statistics
University of Helsinki
P.O. Box 68 (Gustaf Hällströmin katu 2b)
FIN-00014 Helsinki, Finland
email: juliette.kennedy@helsinki.fi

Paul B. Larson
Department of Mathematics
and Statistics
Miami University
Oxford, Ohio 45056, USA
email: larsonpb@muohio.edu

Benedikt Löwe
Institute for Logic,
Language and Computation
Universiteit van Amsterdam
Plantage Muidergracht 24
1018 TV Amsterdam, The Netherlands
email: bloewe@science.uva.nl

Heike Mildenberger
Kurt Gödel Research Center
for Mathematical Logic
Universität Wien
Währinger Str. 25
A-1090 Wien, Austria
email: heike@logic.univie.ac.at

Andrzej Rosłanowski
Department of Mathematics
University of Nebraska at Omaha
Omaha, NE 68182-0243, USA
email: roslanow@member.ams.org

Saharon Shelah
Einstein Institute of Mathematics
Edmond J. Safra Campus, Givat Ram
The Hebrew University of Jerusalem
Jerusalem, 91904, Israel
email: shelah@math.huji.ac.il

Katherine Thompson
Kurt Gödel Research Center
for Mathematical Logic
Universität Wien
Währinger Str. 25
A-1090 Wien, Austria
email: thompson@logic.univie.ac.at

Jouko Väänänen
Department of Mathematics
and Statistics
University of Helsinki
Gustaf Hällströmin katu 2b, PL 68
FIN-00014 Helsinki, Finland
email: jouko.vaananen@helsinki.fi

Received June, 2006

Photocomposed copy prepared from a TEX file.

ISBN 978-88-548-0982-6

Contents

Mathematics subject classification

Disjoint Non-Free Subgroups of Abelian Groups: **20K20**, 20K27, 03E05, 03E75.

Global Complexity Results: **03E05**, 06A07, 03E35.

Absoluteness for Universally Baire Sets and ... : **03E47**, 03E40, 03E55.

Large Cardinals and L-like Universes: **03E45**, 03E55.

Applications of Regular Filters and ... : **03C55**, 03E05.

Set Theory of Infinite Imperfect Information Games: **03E60**, 03E15, 91A05.

Forcing with F_σ- and with Summable Filters: **03E05**, 03E35, 06E05, 40A05.

Reasonably Complete Forcing Notions: **03E40**, 03E35, 03E17.

Preface

Set theory started with a paper of Georg Cantor published on C
in 1874, and was developed in the following decades by Cant
many other mathematicians, including Dedekind, Gödel, vor
Zermelo. By the middle of the twentieth century, set theory l
common language in which all mathematics is formulated, but
Paul Cohen's work on the Continuum Hypothesis in 1963 t
reached full maturity. After Cohen's seminal work, a verita
results followed, and now, some forty years after, there is n
down, with a seeming inexhaustible slew of new technique
unexpected applications to other areas of mathematics.

Due to the extreme richness of the subject, it is simply n
a balanced overview of the feverish set theoretic activity in
collection of papers. The present volume, a collection of
some of the foremost set theorists, presents a snapshot of s
exciting trends and developments in the subject, making i
the beginning graduate student as well as for the experienc

All of the papers have been refereed according to the
international math journals, and I would like to thank al
their contributions and all of the Referees for their excelle

Disjoint Non-Free Subgroups of Abelian Groups

Andreas Blass and Saharon Shelah

Contents

Contents

Mathematics subject classification

Disjoint Non-Free Subgroups of Abelian Groups: **20K20**, 20K27, 03E05, 03E75.

Global Complexity Results: **03E05**, 06A07, 03E35.

Absoluteness for Universally Baire Sets and ... : **03E47**, 03E40, 03E55.

Large Cardinals and L-like Universes: **03E45**, 03E55.

Applications of Regular Filters and ... : **03C55**, 03E05.

Set Theory of Infinite Imperfect Information Games: **03E60**, 03E15, 91A05.

Forcing with F_σ- and with Summable Filters: **03E05**, 03E35, 06E05, 40A05.

Reasonably Complete Forcing Notions: **03E40**, 03E35, 03E17.

Preface

Set theory started with a paper of Georg Cantor published on Crelle's Journal in 1874, and was developed in the following decades by Cantor himself and many other mathematicians, including Dedekind, Gödel, von Neuman, and Zermelo. By the middle of the twentieth century, set theory had become the common language in which all mathematics is formulated, but it was only after Paul Cohen's work on the Continuum Hypothesis in 1963 that the subject reached full maturity. After Cohen's seminal work, a veritable explosion of results followed, and now, some forty years after, there is no sign of slowing down, with a seeming inexhaustible slew of new techniques and ideas, and unexpected applications to other areas of mathematics.

Due to the extreme richness of the subject, it is simply not possible to give a balanced overview of the feverish set theoretic activity in a reasonably sized collection of papers. The present volume, a collection of papers written by some of the foremost set theorists, presents a snapshot of some of the current exciting trends and developments in the subject, making it valuable both for the beginning graduate student as well as for the experienced researcher.

All of the papers have been refereed according to the usual standards for international math journals, and I would like to thank all of the Authors for their contributions and all of the Referees for their excellent job.

Alessandro Andretta

1. Introduction

In a discussion between the first author and John Irwin, the question arose whether every non-free, separable, torsion-free abelian group has two disjoint non-free subgroups. (Of course, in the context of subgroups, "disjoint" means that the intersection is $\{0\}$.) The main result of this paper is a strong affirmative answer. To state the result in appropriate generality, we need some terminology.

Convention 1.1. *All groups in this paper are understood to be abelian and torsion-free. In particular, "free group" means "free abelian group".*

Definition 1.1. *The* non-free rank *of a group G, written* $\operatorname{nfrk}(G)$, *is the smallest cardinal κ such that G can be split as the direct sum of a group of rank $\leq \kappa$ and a free group.*

Theorem 1.1. *If* $\operatorname{nfrk}(G)$ *is uncountable, then G has* $\operatorname{nfrk}(G)$ *pairwise disjoint, non-free subgroups.*

Recall that any countable, separable group is free. It follows that, if G is separable and not free, then $\operatorname{nfrk}(G)$ is necessarily uncountable, so the theorem applies to G. It gives not only two disjoint non-free subgroups as in the original question but $\operatorname{nfrk}(G) \geq \aleph_1$ of them.

The number of disjoint, non-free subgroups obtained in the theorem is the most one could hope for. Indeed, if $G \cong H \oplus F$ where F is free and H has rank and therefore cardinality equal to the infinite cardinal $\operatorname{nfrk}(G)$, then any non-free subgroup S of G must have a non-zero intersection with H. Otherwise the projection to F would map S one-to-one into the free group F and it would follow that S is free. Therefore, disjoint non-free subgroups of G must intersect $H - \{0\}$ in disjoint non-empty sets. So there cannot be more such subgroups than $|H| = \operatorname{nfrk}(G)$.

Although the theorem gives an optimal result for separable groups, the fact that it does not explicitly mention separability raises another question:

*Blass was partially supported by the United States National Science Foundation, grant DMS–0070723.

*Shelah was partially supported by the US-Israel Binational Science Foundation, grant 2002323. This is paper 870 in his list of publications.

Is the uncountability hypothesis really needed? We shall answer this question affirmatively by exhibiting a (necessarily non-separable) non-free group G such that $\mathrm{nfrk}(G) = \aleph_0$ and G does not have two disjoint non-free subgroups, let alone $\aleph_0$ of them.

This paper contains, in addition to this introduction and a section of known preliminary results, three sections. The first two are devoted to the proof of Theorem 1.1. Section 3 contains the proof for the case that $\mathrm{nfrk}(G)$ is (uncountable and) regular. Section 4 contains the additional arguments needed to extend the result to the case of singular $\mathrm{nfrk}(G)$. Finally, Section 5 presents our counterexample for the case of countable non-free rank.

2. Preliminaries

In this section, we collect for reference some conventions, definitions, and known results that will be needed in our proofs. The book [1] of Eklof and Mekler serves as a standard reference for this material.

Convention 2.1. *Group operations will be written additively. For $n \in \mathbb{Z}$ and x a group element, the notation nx means the sum of n copies of x if $n > 0$, it means $(-n)(-x)$ if $n < 0$, and it means 0 if $n = 0$.*

Definition 2.1. *A subgroup H of G is* pure *in G if, whenever $x \in G$ and $nx \in H$ for some non-zero integer n, then $x \in H$. If H is an arbitrary subgroup of G, then we write H_* for the* purification *of H, the smallest pure subgroup of G that includes H.*

If X is any subset of a group G, we write $\langle X \rangle$ for the subgroup of G generated by X. Its purification $\langle X \rangle_*$ is called the pure subgroup generated by *X.*

For a prime number p, the field of p-adic numbers will be denoted by $\mathbb{Q}_p$, and the subring of p-adic integers will be denoted by $\mathbb{Z}_p$.

Convention 2.2. *When we refer to the numerators or denominators of rational numbers, we always mean that the rational numbers are regarded as fractions in reduced form.*

Recall that the rational field $\mathbb{Q}$ is a subfield of $\mathbb{Q}_p$ and that the intersection $\mathbb{Q} \cap \mathbb{Z}_p$ consists of those rational numbers whose denominators are not divisible by p. A useful consequence is that a rational number is in $\mathbb{Z}_p$ for all primes p if and only if it is an integer. Recall also that $p\mathbb{Z}_p$ is the unique maximal ideal of $\mathbb{Z}_p$ and that the quotient $\mathbb{Z}_p/p\mathbb{Z}_p$ is isomorphic to the p-element field $\mathbb{Z}/p\mathbb{Z}$, the isomorphism being induced by the inclusion of $\mathbb{Z}$ in $\mathbb{Z}_p$.

Let λ be an uncountable regular cardinal and G a group of cardinality λ. A filtration of G is an increasing sequence $(G_\alpha : \alpha < \lambda)$ of subgroups of G, each of cardinality $< \lambda$, continuous at limit ordinals $\beta < \lambda$ (i.e., $\bigcup_{\alpha<\beta} G_\alpha = G_\beta$), and with $\bigcup_{\alpha<\lambda} G_\alpha = G$. Although there are many filtrations of G, any two of them, say (G_α) and (G'_α), agree almost everywhere in the sense that the set of agreement $\{\alpha < \lambda : G_\alpha = G'_\alpha\}$ includes (in fact is) a closed unbounded set (club) in λ.

Recall that a subset of λ is stationary if it intersects every club and that the intersection of any fewer than λ clubs is again a club. It follows that a stationary set cannot be the union of fewer than λ non-stationary sets. We shall also need Fodor's theorem and a variant of it. Fodor's theorem says that, if S is a stationary subset of λ and $f : S \to \lambda$ is a regressive function (i.e., $f(\alpha) < \alpha$ for all $\alpha \in S$), then f is constant on a stationary subset of S. The variant that we shall need is the following.

Lemma 2.1. *Let (G_α) be a filtration of G, let $S \subseteq \lambda$ be stationary, and let $f : S \to G$ be a function such that $f(\alpha) \in G_\alpha$ for all $\alpha \in S$. Then f is constant on some stationary subset of S.*

PROOF – Replacing S by its intersection with the club of all limit ordinals, we may assume that every $\alpha \in S$ is a limit ordinal. For such α, continuity of the filtration tells us that $f(\alpha) \in G_{g(\alpha)}$ for some $g(\alpha) < \alpha$. By Fodor's theorem, the regressive function g is constant, say with value β, on some stationary subset T of S. Since $|G_\beta| < \lambda$, we have a decomposition of T into fewer than λ sets $T_x = \{\alpha \in T : f(\alpha) = x\}$ for $x \in G_\beta$. So one of the pieces T_x must be stationary, and the lemma is established. □

The connection between filtrations and freeness is given by the following definition and lemma.

Definition 2.2. *The* Gamma invariant *of G is*

$$\Gamma(G) = \{\alpha < \lambda : G/G_\alpha \text{ has a non-free subgroup of cardinality } < \lambda\}.$$

This definition seems to depend on the choice of filtration, but in fact it doesn't, modulo restriction to a club, because any two filtrations agree on a club. In particular, the statement "$\Gamma(G)$ is stationary" has the same truth value for all choices of the filtration.

Lemma 2.2. *$\Gamma(G)$ is stationary if and only if* $\mathrm{nfrk}(G) = \lambda$.

PROOF – Suppose first that $\Gamma(G)$ is not stationary and is therefore disjoint from some club C. Let $c : \lambda \to C$ be the function enumerating C in increasing order. We construct, by induction on α, a free basis B_α for $G_{c(\alpha)}/G_{c(0)}$ such that $B_\alpha \subseteq B_\beta$ whenever $\alpha < \beta$. This is trivial for $\alpha = 0$. Continuity of c (because C is closed) lets us simply take unions at limit stages. At a successor step, say from α to $\alpha+1$, remember that $c(\alpha) \in C \subseteq \lambda - \Gamma(G)$, so $G_{c(\alpha+1)}/G_{c(\alpha)}$ has a free basis. Pick one representative in $G_{c(\alpha+1)}$ for each member of this basis, and adjoin the chosen representatives to B_α to get $B_{\alpha+1}$. At the end of the induction, $\bigcup_{\alpha<\lambda} B_\alpha$ is a free basis for $G/G_{c(0)}$. Since homomorphisms onto free groups split, G is isomorphic to the direct sum of $G_{c(0)}$ and a free group. So $\mathrm{nfrk}(G) \leq |G_{c(0)}| < \lambda$.

For the converse, suppose $G = H \oplus F$ where $|H| < \lambda$ and F is free. Fix a free basis for F, necessarily of cardinality λ, and enumerate it in a sequence of order-type λ. Then G has a filtration whose α^{th} element G_α is the subgroup generated by H and the first α elements in the enumeration of the basis of F. Then each of the quotients G/G_α is free, with a basis represented by all but the first α elements of the basis of F. Therefore, all subgroups of G/G_α are also free. So $\Gamma(G)$ is empty for this filtration, and therefore non-stationary for all filtrations. □

It will be convenient to use filtrations normalized as follows.

Lemma 2.3. *G has a filtration $(G_\alpha : \alpha < \lambda)$ such that*

– *each G_α is a pure subgroup of G, and*

– *whenever G/G_α has a non-free subgroup of cardinality $< \lambda$ (i.e., whenever $\alpha \in \Gamma(G)$ as calculated with this filtration), then $G_{\alpha+1}/G_\alpha$ is such a subgroup.*

PROOF – Starting with any filtration (H_α) of G, we produce a new filtration with the desired additional properties by defining G_α inductively. Start with $G_0 = (H_0)_*$, and at limit ordinals take unions (as demanded by the definition of filtration). At a successor step from α to $\alpha + 1$, first choose a subgroup K of G, of cardinality $< \lambda$, such that $G_\alpha \subsetneq K$ and such that, if possible, K/G_α is not free. Then let $G_{\alpha+1} = (H_{\alpha+1} + K)_*$. □

3. Proof for Regular Non-Free Rank

In this section, we shall establish Theorem 1.1 in the case that $\mathrm{nfrk}(G)$ is an uncountable, regular cardinal λ. We may assume, without loss of generality, that $|G| = \lambda$. Indeed, if G were larger, we could use, in place of G, the summand H in the decomposition $G \cong H \oplus F$ given by the definition of $\mathrm{nfrk}(G)$. This H has rank, cardinality, and non-free rank all equal to λ, and of course if we find λ pairwise disjoint non-free subgroups in H then these will serve in G as well. From now on, assume $|G| = \lambda$.

For the rest of this section, we fix a filtration $(G_\alpha : \alpha < \lambda)$ with the properties in Lemma 2.3. We define $\Gamma(G)$ using this filtration in Definition 2.2, and we note that, by Lemma 2.2, $\Gamma(G)$ is a stationary subset of λ.

For each $\alpha \in \Gamma(G)$, the properties of (G_α) in Lemma 2.3 imply that $G_{\alpha+1}/G_\alpha$ is a torsion-free, non-free group. Let Y_α be a maximal linearly independent subset of this group. By linear independence, the subgroup $\langle Y_\alpha \rangle$ generated by Y_α is free, and by maximality, its purification $\langle Y_\alpha \rangle_*$ is all of $G_{\alpha+1}/G_\alpha$. Choose, for each element of Y_α, a representative in $G_{\alpha+1}$, and let X_α be the set of these chosen representatives. Thus, the projection from $G_{\alpha+1}$ to its quotient modulo G_α maps X_α one-to-one onto Y_α.

Expressing the properties of Y_α in the quotient group as properties "modulo G_α" of X_α in the group $G_{\alpha+1}$ we obtain the following.

Lemma 3.1.

- *X_α is linearly independent modulo G_α. That is, if G_α contains a linear combination, with integer coefficients, of members of X_α, then all the coefficients are zero.*

- $(\langle X_\alpha\rangle + G_\alpha)_* = G_{\alpha+1}$.

Proof – For the first assertion, notice that such a linear combination, when projected to the quotient modulo G_α, becomes a linear combination of members of Y_α that equals zero. So the linear independence of Y_α in the quotient group gives the required conclusion.

For the second assertion, consider an arbitrary element $a \in G_{\alpha+1}$. Its image $\bar{a}$ in $G_{\alpha+1}/G_\alpha$ has a multiple $n\bar{a} \in \langle Y_\alpha\rangle$ for some non-zero $n \in \mathbb{Z}$. Since the projection maps X_α onto Y_α and thus maps $\langle X_\alpha\rangle$ onto $\langle Y_\alpha\rangle$, we have an element $z \in \langle X_\alpha\rangle$ projecting to $n\bar{a}$. So na and z project to the same element, which means $na = z + g$ for some $g \in G_\alpha$. This equation establishes that a is in the purification of $\langle X_\alpha\rangle + G_\alpha$, as required. □

Temporarily fix an arbitrary stationary subset T of $\Gamma(G)$. Using T we define a subgroup L of G by

$$L = \Big\langle \bigcup_{\beta\in T} X_\beta \Big\rangle_* .$$

Our immediate objective, and indeed the main part of our argument for Theorem 1.1 when nfrk(G) is uncountable, is to show that L is not free.

For this purpose, we shall use Lemma 2.2 and show that $\Gamma(L)$ is stationary. (So we shall have not only that L is not free but that nfrk$(L) = \lambda$.) In fact, we shall prove more, namely that $\Gamma(L)$ contains all of the stationary set T except for some non-stationary subset.

To begin the analysis of $\Gamma(L)$, we must choose a filtration of L to use in the definition of $\Gamma(L)$, and two natural choices present themselves. One is the restriction to L of the filtration we already have for G, i.e., $(G_\alpha \cap L : \alpha < \lambda)$. The other is obtained from the way L is generated by the X_α's; this filtration is $(L_\alpha : \alpha < \lambda)$, where

$$L_\alpha = \Big\langle \bigcup_{\substack{\beta\in T\\ \beta<\alpha}} X_\beta \Big\rangle_* .$$

We shall want to use each of these occasionally. Fortunately, as mentioned earlier, these two filtrations (like any two filtrations of the same group) agree on a club. Let T_1 be the intersection of T with such a club. To show that $\Gamma(L)$ contains all but a non-stationary part of T, it suffices, since $T - T_1$ is non-stationary, to prove that $\Gamma(L)$ contains all but a non-stationary part of T_1.

So our objective is now to show that the "exceptional" set

$$W = \{\alpha \in T_1 : \alpha \notin \Gamma(L)\}$$

is not stationary. Suppose, toward a contradiction, that W is stationary.

For each $\alpha \in W$, we have the following two facts:

- $L_{\alpha+1}/L_\alpha$ is free.
- $L_\alpha = L_{\alpha+1} \cap G_\alpha$.

The first of these is immediate because $\alpha \notin \Gamma(L)$. The second follows from

$$L_\alpha \subseteq L_{\alpha+1} \cap G_\alpha \subseteq L \cap G_\alpha = L_\alpha,$$

where the first inclusion uses the fact that L_α is the pure subgroup of G generated by a subset $\bigcup_{\beta \in T,\ \beta < \alpha} X_\beta$ of G_α and G_α is a pure subgroup of G. The second inclusion is trivial, and the final equality uses the fact that $\alpha \in T_1$ so the two filtrations agree at α.

Combining the two facts just established, we have, for each $\alpha \in W$, that the following group is free:

$$\frac{L_{\alpha+1}}{L_{\alpha+1} \cap G_\alpha} \cong \frac{L_{\alpha+1} + G_\alpha}{G_\alpha}.$$

Contrast this with the fact that, since $\alpha \in T \subseteq \Gamma(G)$, the following group is not free:

$$\frac{G_{\alpha+1}}{G_\alpha} = \frac{(\langle X_\alpha \rangle + G_\alpha)_*}{G_\alpha}.$$

(Here we use that our filtration was chosen to satisfy the second conclusion of Lemma 2.3.) Since subgroups of free groups are free, we conclude that

$$(\langle X_\alpha \rangle + G_\alpha)_* \not\subseteq L_{\alpha+1} + G_\alpha.$$

Choose, for each $\alpha \in W$, some element $g_\alpha \in (\langle X_\alpha \rangle + G_\alpha)_*$ that is not in $L_{\alpha+1} + G_\alpha$. By definition of purification, we have some $n_\alpha \in \mathbb{Z} - \{0\}$ such that $n_\alpha g_\alpha \in \langle X_\alpha \rangle + G_\alpha$. That is, $n_\alpha g_\alpha = c_\alpha + h_\alpha$ where c_α is a linear combination, with integer coefficients, of elements of X_α and where $h_\alpha \in G_\alpha$.

Because we assumed, toward a contradiction, that W is stationary, and because $h_\alpha \in G_\alpha$ for all $\alpha \in W$, Lemma 2.1 gives us a stationary set $W_1 \subseteq W$ such that h_α is the same element h for all $\alpha \in W_1$. Furthermore, since the values of n_α all lie in the countable set $\mathbb{Z}$, there is a stationary $W_2 \subseteq W_1$ such that n_α has the same value n for all $\alpha \in W_2$.

Let $\alpha < \beta$ be any two elements of W_2. Then we have

$$\begin{aligned} ng_\alpha &= c_\alpha + h \\ ng_\beta &= c_\beta + h. \end{aligned}$$

Subtract to get

$$n(g_\beta - g_\alpha) = c_\beta - c_\alpha.$$

Both c_α and c_β are linear combinations of elements of $X_\alpha \cup X_\beta \subseteq L_{\beta+1}$. So $L_{\beta+1}$ contains the right side of the last equation. As $L_{\beta+1}$ is pure in G, it follows that $g_\beta - g_\alpha \in L_{\beta+1}$ and so $g_\beta \in g_\alpha + L_{\beta+1}$.

Notice that G_β includes G_α (because $\alpha < \beta$), and X_α (because $X_\alpha \subseteq G_{\alpha+1}$ and $\alpha + 1 \leq \beta$), and therefore $\langle X_\alpha \rangle + G_\alpha$, and therefore $(\langle X_\alpha \rangle + G_\alpha)_*$ (because, according to the first conclusion of Lemma 2.3, G_β is a pure subgroup of G). We chose g_α from this last group, so we have $g_\alpha \in G_\beta$. But then the result from the preceding paragraph gives us that $g_\beta \in G_\beta + L_{\beta+1}$, which contradicts our choice of g_β.

This contradiction completes the proof that W cannot be stationary and therefore L is not free.

Recall that L was defined in terms of an arbitrary but fixed stationary $T \subseteq \Gamma(G)$. We shall now need to vary T, so, to indicate the dependence of L on T, we write $L(T)$ for what was previously called simply L.

Lemma 3.2. *If T_1 and T_2 are disjoint stationary subsets of $\Gamma(G)$, then the subgroups $L(T_1)$ and $L(T_2)$ are disjoint.*

PROOF – Notice first that, if two subgroups H_1 and H_2 are disjoint, then so are their purifications. Indeed, if the purifications had a common non-zero element x, then x would have non-zero multiples $n_1x \in H_1$ and $n_2x \in H_2$. But then n_1n_2x would be a non-zero element in $H_1 \cap H_2$ contrary to hypothesis.

So to prove the lemma, it suffices to prove that the subgroups generated by $\bigcup_{\beta\in T_1} X_\beta$ and $\bigcup_{\beta\in T_2} X_\beta$ are disjoint. Suppose, toward a contradiction, that they are not disjoint, choose a non-zero element in their intersection, and write it first as a linear combination c_1 of elements of $\bigcup_{\beta\in T_1} X_\beta$ and second as a linear combination c_2 of elements of $\bigcup_{\beta\in T_2} X_\beta$. Let β be the largest of the finitely many ordinals such that elements of X_β occur in these linear combinations with non-zero coefficients. Since T_1 and T_2 are disjoint, β is in only one of them, say T_1. Split c_1 as $c_1' + c_1''$ where $c_1' \neq 0$ contains the terms from X_β and c_1'' contains the terms from X_α's with $\alpha < \beta$. Then $c_1' = c_2 - c_1'' \in G_\beta$. This contradicts the linear independence of X_β modulo G_β in Lemma 3.1. □

Because of this lemma and the non-freeness of $L(T)$ for all stationary $T \subseteq \Gamma(G)$, we can get as many pairwise disjoint, non-free subgroups of G as we can get pairwise disjoint, stationary subsets of $\Gamma(G)$. It remains only to quote Solovay's famous theorem [6, Theorem 9] that, for any uncountable regular cardinal λ, every stationary subset of λ can be partitioned into λ pairwise disjoint, stationary subsets.

4. Proof for Singular Non-Free Rank

In this section, we complete the proof of Theorem 1.1 by treating the case of singular $\mathrm{nfrk}(G)$. An important ingredient of the proof is the following consequence of the second author's singular compactness theorem.

Lemma 4.1. *Assume that $\kappa < \lambda$ are infinite cardinals, that λ is singular, that G is a group of cardinality λ, and that every subgroup H of G with $|H| < \lambda$ has $\mathrm{nfrk}(H) \leq \kappa$. Then $\mathrm{nfrk}(G) \leq \kappa$.*

PROOF – This follows from the singular compactness theorem of [5]. It is explicitly stated in [2, Theorem 2]. (To avoid possible confusion, note that in

[2] "κ-generated" means generated by strictly fewer than κ elements.) For an easier proof, see [3]. □

Our use of this lemma will be via the following consequence.

Lemma 4.2. *Let G be a group whose non-free rank is a singular cardinal λ. For any infinite cardinal $\kappa < \lambda$, there exists a subgroup H of G whose non-free rank is regular and satisfies $\kappa < \mathrm{nfrk}(H) = |H| < \lambda$.*

PROOF – Given G, λ, and κ as in the statement of the lemma, let H be a subgroup of G with $\mathrm{nfrk}(H)$ as small as possible subject to the constraint that $\mathrm{nfrk}(H) > \kappa$. Such an H certainly exists, since $\mathrm{nfrk}(G) > \kappa$. By definition of $\mathrm{nfrk}(H)$, we can split H as the direct sum of a free group and a group H_1 of cardinality $\mathrm{nfrk}(H)$. Then $|H_1| = \mathrm{nfrk}(H) = \mathrm{nfrk}(H_1)$. Replacing H with H_1 if necessary, we assume from now on that $|H| = \mathrm{nfrk}(H)$.

It remains only to show that $\mathrm{nfrk}(H)$ is a regular cardinal and that $\mathrm{nfrk}(H) < \lambda$. The latter follows from the former, because λ is singular, and so we need only prove the regularity of $\mathrm{nfrk}(H)$. Suppose, toward a contradiction, that $\mathrm{nfrk}(H)$ is a singular cardinal. Notice that, for every subgroup K of H with cardinality smaller than $|H|$, we have $\mathrm{nfrk}(K) \leq |K| < |H| = \mathrm{nfrk}(H)$, and so, by minimality of $\mathrm{nfrk}(H)$, we must have $\mathrm{nfrk}(K) \leq \kappa$. By Lemma 4.1, $\mathrm{nfrk}(H) \leq \kappa$ also, but this contradicts the choice of H. □

Using the lemma, we prove the singular case of Theorem 1.1 as follows. Let G be a group with $\mathrm{nfrk}(G) = \lambda$ singular, let $\mu < \lambda$ be the cofinality of λ, and let $(\kappa_\xi : \xi < \mu)$ be a strictly increasing μ-sequence of cardinals with supremum λ. Inductively define a μ-sequence of subgroups G_ξ of G, with the following properties for all $\xi < \mu$:

- $|G_\xi| = \mathrm{nfrk}(G_\xi)$.
- $\mathrm{nfrk}(G_\xi)$ is a regular cardinal.
- $\kappa_\xi < \mathrm{nfrk}(G_\xi) < \lambda$.
- $\sum_{\eta<\xi} |G_\eta| < \mathrm{nfrk}(G_\xi)$.

Once G_η is defined and has the desired properties for all $\eta < \xi$, we obtain G_ξ by applying Lemma 4.2 with κ equal to the larger of κ_ξ and $\sum_{\eta<\xi} |G_\eta|$. Notice that this sum is strictly smaller than λ because $|G_\eta| < \lambda$ for all $\eta < \xi$ by induction hypothesis and because $\xi < \mu = \mathrm{cf}(\lambda)$. Thus, our κ is smaller than λ, and the H provided by Lemma 4.2 serves as the required G_ξ.

The regular case of Theorem 1.1, proved in the previous section, applies to each G_ξ and provides a family $\mathcal{D}_\xi$ of $\mathrm{nfrk}(G_\xi)$ pairwise disjoint non-free subgroups of G_ξ.

Although the groups in any $\mathcal{D}_\xi$ are pairwise disjoint, there may be a non-zero intersection between a group $H \in \mathcal{D}_\xi$ and a group $K \in \mathcal{D}_\eta$ for some $\eta < \xi$. But this does not happen too often. Specifically, for fixed η, K, and ξ, the number of such H's is at most $|K| \leq |G_\eta|$ because different H's from $\mathcal{D}_\xi$ would meet $K - \{0\}$ in disjoint sets. Therefore, if we keep η and ξ fixed but let K vary through all elements of $\mathcal{D}_\eta$, then the number of $H \in \mathcal{D}_\xi$ that have non-zero intersection with some such K is at most $|G_\eta| \cdot |\mathcal{D}_\eta|$. Since $\mathcal{D}_\eta$ is a family of disjoint subgroups of G_η, its cardinality is at most $|G_\eta|$, and so the product $|G_\eta| \cdot |\mathcal{D}_\eta|$ is also at most $|G_\eta|$. Now keeping ξ fixed but letting η vary through all ordinals $< \xi$, we find that the number of $H \in \mathcal{D}_\xi$ that have non-zero intersection with some K in some earlier $\mathcal{D}_\eta$ is at most

$$\sum_{\eta<\xi} |G_\eta| < \mathrm{nfrk}(G_\xi).$$

Therefore, if we discard these H's from $\mathcal{D}_\xi$, what remains is a family $\mathcal{D}'_\xi$, still of cardinality $\mathrm{nfrk}(G_\xi)$, still consisting of pairwise disjoint, non-free subgroups of G_ξ, but enjoying the additional property that all its members are disjoint from all members of earlier $\mathcal{D}_\eta$'s and, a fortiori, from all members of earlier $\mathcal{D}'_\eta$'s.

Therefore, $\mathcal{D}' = \bigcup_{\xi<\mu} \mathcal{D}'_\xi$ is a family of pairwise disjoint non-free subgroups of G. Its cardinality satisfies

$$\begin{aligned} |\mathcal{D}'| &= \sup\{|\mathcal{D}'_\xi| : \xi < \mu\} \\ &= \sup\{\mathrm{nfrk}(G_\xi) : \xi < \mu\} \\ &\geq \sup\{\kappa_\xi : \xi < \mu\} \\ &= \lambda, \end{aligned}$$

so the theorem is proved.

5. Counterexample for Countable Non-Free Rank

In this section, we construct an example of a group G with $|G| = \mathrm{nfrk}(G) = \aleph_0$ and with no two disjoint non-free subgroups. This shows that the uncountability assumption in Theorem 1.1 cannot be removed.

The group G will be a subgroup of the direct sum of $\aleph_0$ copies of the additive group $\mathbb{Q}$ of rational numbers. G will be defined as the set of solutions, in this direct sum, of infinitely many p-adic conditions for all primes p. The construction of G will proceed in three phases. The first will set up some conventions and bookkeeping. The second will define, for each prime p, a certain set of vectors over $\mathbb{Z}/p$ and a lifting of these vectors to $\mathbb{Z}_p$. The third will use these vectors to define the group G. After the construction is complete, we shall verify first that $\mathrm{nfrk}(G) = \aleph_0$ and second that G does not have two disjoint, non-free subgroups.

5.1. Conventions and bookkeeping

By a vector over a ring R, we shall usually mean an infinite sequence of elements of R, the sequence being indexed by the set $\mathbb{N}$ of positive integers. (We do not include 0 in the index set, as we shall have a separate, special use for 0 later.) The rings relevant to our work will include $\mathbb{Z}$, $\mathbb{Q}$, $\mathbb{Z}/p$, $\mathbb{Z}_p$, and $\mathbb{Q}_p$ for prime numbers p.

Occasionally, we shall need to refer to vectors of finite length l, with components indexed by $\{1, 2, \dots, l\}$, but this finiteness (and the value of l) will always be explicitly stated. Furthermore, it would do no harm to identify any finitely long vector with the infinite vector obtained by appending a sequence of zeros. We use the notation $\vec{x} \restriction l$ for the vector of length l consisting of just the first l components of $\vec{x}$; under the identification in the preceding sentence, $\vec{x} \restriction l$ can also be considered as obtained from $\vec{x}$ by replacing all components beyond the first l by 0.

We call a vector $\vec{x}$ finitely supported if the set $\{i \in \mathbb{N} : x_i \neq 0\}$, called the support of $\vec{x}$, is finite. The inner product of two vectors is defined as

$$\langle \vec{x}, \vec{y} \rangle = \sum_{i \in \mathbb{N}} x_i y_i,$$

provided at least one of the vectors is finitely supported, so that the infinite sum makes sense.

Partition the set P of prime numbers into infinitely many infinite pieces, and label the pieces as $P_{\vec{x}}$, where $\vec{x}$ ranges over the non-zero, finitely supported, infinite vectors over $\mathbb{Z}$. (The number of such vectors is $\aleph_0$, so the indexing makes sense.) Fix this indexed partition of P for the rest of the proof.

Also fix an enumeration, as $(\vec{\lambda}_i : i \in \mathbb{N})$, of all the finitely supported vectors over $\mathbb{Q}$.

Recall that $\mathbb{Q}$ is canonically identified with a subfield of $\mathbb{Q}_p$, and that under this identification $\mathbb{Z}$ becomes a subring of $\mathbb{Z}_p$. This inclusion of $\mathbb{Z}$ in $\mathbb{Z}_p$ induces an isomorphism between the quotients modulo p, $\mathbb{Z}/p \cong \mathbb{Z}_p/p\mathbb{Z}_p$. We write $[x]_p$, or just $[x]$ when p is clear from the context, for the equivalence class of x modulo p; here x is in $\mathbb{Z}$ or $\mathbb{Z}_p$, and $[x]$ is in $\mathbb{Z}/p$. We refer to x as a representative of $[x]$. We use the same notation for vectors; $[\vec{x}]_p$ is obtained from $\vec{x}$ by reducing all components modulo p.

5.2. Useful sets of vectors

For this subsection, let p be a fixed prime. Later, the work we do here will be applied to all primes, but it is notationally and conceptually easier to begin with just one p. Let $\vec{x}$ be the unique vector such that $p \in P_{\vec{x}}$. Recall that this $\vec{x}$ is a non-zero, finitely supported vector over $\mathbb{Z}$. Thus, $[\vec{x}]$ is a finitely supported (but possibly zero) vector over $\mathbb{Z}/p$.

We define a finite set M of vectors over $\mathbb{Z}/p$ as follows. Choose an integer l larger than p and all elements of the support of $\vec{x}$. M will be described as a set of finite vectors, in $(\mathbb{Z}/p)^l$, but we really mean the infinite vectors obtained by appending a sequence of zeros.

If $[\vec{x}] = \vec{0}$, then M consists of all the vectors in $(\mathbb{Z}/p)^l$. If, on the other hand, $[\vec{x}] \neq \vec{0}$, then we proceed as follows. Call an index i or the corresponding vector $\vec{\lambda}_i$ (from the enumeration fixed above) relevant if $i < p-1$ and p does not divide the denominator of any component of $\vec{\lambda}_i$. For each relevant i, the components of $\vec{\lambda}_i$ are in $\mathbb{Z}_p$, so it makes sense to reduce them modulo p, obtaining a vector $[\vec{\lambda}_i]$ over $\mathbb{Z}/p$. Choose a non-zero $a \in \mathbb{Z}/p$ that is distinct from $\langle [-\vec{\lambda}_i], [\vec{x}] \rangle$ for all relevant i. The requirement, in the definition of relevance, that $i < p-1$ means that at most $p-2$ indices are relevant, so a suitable a exists. Now let M consist of all those $\vec{m} \in (\mathbb{Z}/p)^l$ such that $\langle \vec{m}, [\vec{x}] \rangle = a$.

Before proceeding further, we summarize the properties of M that we shall need later.

Lemma 5.1. *The set $M \subseteq (\mathbb{Z}/p)^l$ defined here has the following properties.*

1. *For each $k \leq l$, the truncations to k of the vectors in M span $(\mathbb{Z}/p)^k$.*

2. *The same holds for the truncations to k of the translates $[\vec{\lambda}_i] + M$ for each relevant $\vec{\lambda}_i$.*

3. *As $\vec{m}$ varies through M, all the inner products $\langle \vec{m}, [\vec{x}] \rangle$ have the same value a.*

4. *This value a is non-zero and different from $\langle [-\vec{\lambda}_i], [\vec{x}] \rangle$ for all relevant i provided $[\vec{x}] \neq \vec{0}$.*

Proof – If $[\vec{x}] = \vec{0}$ then item 3 is obvious with $a = 0$ and item 4 doesn't apply. If $[\vec{x}] \neq \vec{0}$ then items 3 and 4 are explicitly in the definition of M.

As for items 1 and 2, it suffices to prove these for $k = l$, since truncation to smaller k's commutes with linear combinations. So assume $k = l$. If $[\vec{x}] = \vec{0}$ then both 1 and 2 are obvious as M is all of $(\mathbb{Z}/p)^l$. So assume $[\vec{x}] \neq \vec{0}$. Then M is defined as a certain affine hyperplane in $(\mathbb{Z}/p)^l$. Since $a \neq 0$, this hyperplane does not pass through the origin, and therefore it spans $(\mathbb{Z}/p)^l$. This proves 1. For 2, observe that $[\vec{\lambda}_i] + M$ is another affine hyperplane, a translate of M. It does not pass through the origin either, because $a \neq \langle [-\vec{\lambda}_i], [\vec{x}] \rangle$. So it, too, spans $(\mathbb{Z}/p)^l$, and 2 is proved. □

The next step is to lift M to an infinite set of vectors over $\mathbb{Z}_p$. For this, we regard the vectors in M as having infinite length (rather than l), by appending zeros. Arbitrarily choose, for each $\vec{m} \in M$, a countable infinity of vectors $\vec{\psi}$

over $\mathbb{Z}_p$ with $[\vec{\psi}] = \vec{m}$. Let Ψ be the set of all the resulting vectors, for all $\vec{m}$ together.

Partition the set Ψ into infinitely many pieces Ψ_k $(k \in \mathbb{N})$ with $|\Psi_k| = k+1$. For each k, we shall modify the $k+1$ vectors in Ψ_k so that their truncations to k become affinely independent over $\mathbb{Q}_p$. Affine independence means that not only do these $k+1$ vectors span $\mathbb{Q}_p^k$ but they continue to span if any fixed vector is added to all of them. Any set of $k+1$ vectors in a k-dimensional vector space over $\mathbb{Q}_p$ (or over any valued field with a non-trivial valuation) can be made affinely independent by an arbitrarily small perturbation. For our modification of Ψ_k, we use a perturbation that is small in the p-adic norm, so that each vector $\vec{\psi}$ is modified by adding something divisible in $\mathbb{Z}_p$ by p. Thus, the reduction modulo p, $[\vec{\psi}]$, is unaffected. Let the perturbed vectors constitute Φ_k, and let $\Phi = \bigcup_{k\in\mathbb{N}} \Phi_k$. Note that all the vectors in Φ have, like those in Ψ, all their components in $\mathbb{Z}_p$.

Lemma 5.2. *The set Φ of vectors over $\mathbb{Z}_p$ constructed here has the following properties.*

1. *For each $k \leq p$, the vectors $[\vec{\varphi}] \upharpoonright k$ for $\vec{\varphi} \in \Phi$ span $(\mathbb{Z}/p)^k$.*

2. *The same holds for the translates $[\vec{\lambda}_i] + [\vec{\varphi}]$, truncated at k, for each relevant $\vec{\lambda}_i$.*

3. *As $\vec{\varphi}$ varies through Φ, all the inner products $\langle [\vec{\varphi}], [\vec{x}] \rangle$ have the same value a.*

4. *This value a is non-zero and different from $\langle [-\vec{\lambda}_i], [\vec{x}] \rangle$ for all relevant i provided $[\vec{x}] \neq \vec{0}$.*

5. *For every k, there are $k+1$ vectors $\vec{\varphi}_0, \vec{\varphi}_1, \ldots, \vec{\varphi}_k \in \Phi$ such that, for any vector λ over $\mathbb{Q}$, the truncated vectors $(\vec{\varphi}_i + \vec{\lambda}) \upharpoonright k$ span $\mathbb{Q}_p^k$.*

PROOF – By construction, the reductions modulo p of the vectors in Φ are the same as those of the vectors in Ψ, namely the vectors in M (each repeated $\aleph_0$ times). Therefore, the first four conclusions of the present lemma follow immediately from the corresponding items in Lemma 5.1. (For the first item, remember that the l in Lemma 5.1 was $\geq p$.)

The final conclusion of the present lemma is the result of our modification of Ψ to obtain Φ. The $k+1$ vectors in Φ_k are affinely independent, and therefore their translates by any vector $\vec{\lambda}$ over $\mathbb{Q}_p$ are linearly independent. This applies in particular to vectors $\vec{\lambda}$ over $\mathbb{Q} \subseteq \mathbb{Q}_p$. □

5.3. The counterexample group

In the preceding subsection, we worked with a fixed prime p; now we let p vary. The set Φ constructed above will now be called $\Phi(p)$.

Our group G will be a subgroup of the additive group $\mathbb{Q} \times (\mathbb{Q})^{(\mathbb{N})}$, which consists of pairs $(x_0, \vec{x})$ where x_0 is a rational number and $\vec{x}$ is a finitely supported vector of rational numbers. Recall that the components of $\vec{x}$ are indexed as x_i for positive integers i, so our use of the notation x_0 causes no conflict. We define

$$G = \{(x_0, \vec{x}) \in \mathbb{Q} \times (\mathbb{Q})^{(\mathbb{N})} : (\forall p \in P)(\forall \vec{\varphi} \in \Phi(p))\ x_0 + \langle \vec{\varphi}, \vec{x} \rangle \in \mathbb{Z}_p\}.$$

Notice that the inner product in the definition makes sense because $\vec{x}$ is finitely supported, the components of $\vec{x}$, being in $\mathbb{Q}$, can be regarded as elements of $\mathbb{Q}_p$, and the components of $\vec{\varphi}$, being in $\mathbb{Z}_p$, are also in $\mathbb{Q}_p$. Thus, the inner product is defined in $\mathbb{Q}_p$, and the requirement is that it plus x_0 be in $\mathbb{Z}_p$.

Lemma 5.3. $\mathbb{Z} \times (\mathbb{Z})^{(\mathbb{N})} \subseteq G$.

PROOF – Remember that all components of all $\vec{\varphi}$ in $\Phi(p)$ are p-adic integers. So if the components of $\vec{x}$ are integers, and thus also p-adic integers, then the inner product $\langle \vec{\varphi}, \vec{x} \rangle$ is a p-adic integer, and so is its sum with the integer x_0. □

This lemma and the fact that $\mathbb{Q} \times (\mathbb{Q})^{(\mathbb{N})}$ is countable show that the rank and the cardinality of G are $\aleph_0$.

We close this subsection by introducing a subgroup of G that will play a central role in our proof that G has the desired properties. Let L be the subgroup of those elements of G for which at most the single component x_0 is non-zero. So

$$L = \{(x_0, \vec{0}) : (\forall p \in P)(\forall \vec{\varphi} \in \Phi(p))\ x_0 \in \mathbb{Z}_p\}.$$

Recall that a rational number is in $\mathbb{Z}_p$ for all primes p if and only if it is an integer. Therefore, $L = \mathbb{Z} \times \{0\}^{\mathbb{N}}$. Thus, L is isomorphic to $\mathbb{Z}$, and it is clearly a pure subgroup of G.

5.4. The non-free rank of G

We now prove that $\operatorname{nfrk}(G) = \aleph_0$. Since $|G| = \aleph_0$, we need only show that $\operatorname{nfrk}(G)$ is infinite. For this purpose, it is useful to consider the quotient group G/L.

Lemma 5.4. *For each element $\xi \in G/L$, there are infinitely many primes p that divide ξ, i.e., such that $\xi \in p(G/L)$.*

PROOF – Given ξ, choose a representative $(x_0, \vec{x}) \in G$ for it modulo L. The components x_i (both x_0 and the components of $\vec{x}$) are rational numbers and only finitely many are non-zero, so let d be a common denominator. Then $d\xi$ is represented by a vector $(dx_0, d\vec{x})$ of integers. If we prove the assertion of the lemma for $d\xi$ then it will follow for ξ, using the same primes except for the finitely many that divide d. (In detail, if $d\xi = p\eta$ and p doesn't divide d, then the Euclidean algorithm gives integers a, b with $ad + bp = 1$. Then $\xi = ad\xi + bp\xi = ap\eta + bp\xi$ is divisible by p.)

So, by renaming, we may assume that all the components x_i are integers. We may also assume that $\vec{x} \neq \vec{0}$, for otherwise we would have $\xi = 0$ in G/L and the conclusion of the lemma would be trivially correct. So $P_{\vec{x}}$ is an infinite set of primes. We shall show that every $p \in P_{\vec{x}}$ divides ξ, thereby completing the proof.

Since $\Phi(p)$ satisfies conclusion 3 of Lemma 5.2, we know that the inner products $\langle \vec{\varphi}, \vec{x} \rangle$ have the same value modulo p for all $\varphi \in \Phi(p)$. Write this value as $[a] \in \mathbb{Z}/p$, where $a \in \mathbb{Z}$, consider the element $(-a, \vec{x}) \in \mathbb{Z} \times (\mathbb{Z})^{(\mathbb{N})} \subseteq G$, and consider its quotient by p,

$$z = \left(-\frac{a}{p}, \frac{1}{p}\vec{x} \right) \in \mathbb{Q} \times (\mathbb{Q})^{(\mathbb{N})}.$$

Is this quotient z in G? Of the requirements for membership in G, the q-adic ones for primes q other than p are certainly satisfied, since division by p preserves the property of being a q-adic integer. There remain the p-adic requirements. But for each $\varphi \in \Phi(p)$, we have that $\langle \vec{\varphi}, \vec{x} \rangle \equiv a \pmod{p}$ and therefore

$-a+\langle\vec{\varphi},\vec{x}\rangle$ is divisible by p in $\mathbb{Z}_p$. This means that $-(a/p)+\langle\vec{\varphi},(1/p)\vec{x}\rangle\in\mathbb{Z}_p$, i.e., that z satisfies the p-adic requirement arising from φ. Since this happens for every $\varphi\in\Phi(p)$, we conclude that $z\in G$. But pz differs from $(x_0,\vec{x})$ by an element of L, namely $(x_0+a,\vec{0})$. So pz represents ξ; that is, $\xi\in p(G/L)$ as required. □

Corollary 5.1. *G/L has no non-zero, free, pure subgroup.*

PROOF – Any pure subgroup of G/L would inherit from G/L the divisibility property in the lemma and therefore cannot be free unless it is the zero group. □

Lemma 5.5. *The non-free rank of G is $\aleph_0$.*

PROOF – As remarked above, we need only prove that nfrk(G) cannot be finite. Suppose, toward a contradiction, that G can be split as $H\oplus F$, where H has finite rank and F is free. Choose a free basis B for F, and express the generator $(1,\vec{0})$ of L as a linear combination of a vector from H and a finite set B_0 of vectors from B. Then $L\subseteq H\oplus\langle B_0\rangle$. Let $B_1=B-B_0$ and notice that B_1 is infinite because G has infinite rank, H has finite rank, and B_0 is finite. Since $G=H\oplus\langle B_0\rangle\oplus\langle B_1\rangle$, we have

$$\frac{G}{L}\cong\left(\frac{H\oplus\langle B_0\rangle}{L}\right)\oplus\langle B_1\rangle.$$

This makes $\langle B_1\rangle$ a non-zero, pure, free subgroup of G/L, contrary to Corollary 5.1. □

5.5. Non-free subgroups of G

To complete the verification of our counterexample, we must show that G does not have two disjoint, non-free subgroups. The following lemma is the main ingredient in that proof.

Lemma 5.6. *If H is a subgroup of G of finite rank and $H\cap L=\{0\}$, then H is free.*

PROOF – Let H be as in the hypothesis. Because it has finite rank and because every element of G has finite support, H is a subgroup of $\mathbb{Q} \times \mathbb{Q}^k \times \{\vec{0}\}$ for some k. Whenever convenient, we shall ignore the $\{\vec{0}\}$ factor and pretend that elements of H are of the form $(x_0, \vec{x})$ with $\vec{x} \in \mathbb{Q}^k$.

Recall (for example from the proof of Lemma 3.2) that purifications of disjoint subgroups are disjoint. Since H is disjoint from L, the purification H_* of H in $\mathbb{Q} \times \mathbb{Q}^k$ is disjoint from the purification $L_* = \mathbb{Q} \times \{0\}^k$ of L. These purifications are $\mathbb{Q}$-linear subspaces of $\mathbb{Q} \times \mathbb{Q}^k$, and so we know, by linear algebra, that there is a vector $\vec{\lambda} \in \mathbb{Q}^k$ such that all elements $(x_0, \vec{x})$ of H satisfy $x_0 = \langle \vec{\lambda}, \vec{x} \rangle$. Being in G, these elements of H also satisfy

$$\langle \vec{\lambda} + \vec{\varphi}, \vec{x} \rangle = x_0 + \langle \vec{\varphi}, \vec{x} \rangle \in \mathbb{Z}_p \tag{5.1}$$

for all primes p and all $\varphi \in \Phi(p)$.

By appending zeros, we regard $\vec{\lambda}$ as an infinite vector. Being a vector over $\mathbb{Q}$ with finite support, it occurs in our enumeration as λ_i for a certain i, which we fix for the rest of this proof.

Call a prime p good if it has the following properties:

- $p \geq k$.
- $p - 1 > i$.
- p does not divide the denominator of any component of $\lambda = \lambda_i$.

We note for future reference that these conditions are satisfied by all but finitely many primes. We also note that the second and third of these conditions make λ relevant for p in the sense used in Lemmas 5.1 and 5.2.

Temporarily fix a good prime p. Assume, toward a contradiction, that p divides the denominator of some component of $\vec{x}$ for some element $(x_0, \vec{x})$ of H. Fix such an $(x_0, \vec{x})$, let p^m be the highest power of p that divides the denominator of a component of $\vec{x}$, and let $\vec{y} = p^m \vec{x}$. Then all components of $\vec{y}$ have denominators prime to p, and at least one of these components also has its numerator prime to p. Since $(x_0, \vec{x}) \in H$ and since $m > 0$, formula (5.1) above gives us, for every $\vec{\varphi} \in \Phi(p)$,

$$\langle \vec{\lambda} + \vec{\varphi}, \vec{y} \rangle = p^m \langle \vec{\lambda} + \vec{\varphi}, \vec{x} \rangle \in p\mathbb{Z}_p,$$

and so $\langle[\vec{\lambda}+\vec{\varphi}],[\vec{y}]\rangle = 0$ in $\mathbb{Z}/p$. Since this holds for all $\vec{\varphi} \in \Phi(p)$ and since, by conclusion 2 of Lemma 5.2 the corresponding vectors $[\vec{\lambda}+\vec{\varphi}] \restriction k$ span $(\mathbb{Z}/p)^k$, it follows that $[\vec{y}] \restriction k$ is the zero vector in $(\mathbb{Z}/p)^k$. The components of $\vec{y}$ beyond the first k all vanish by our choice of k. Therefore, all components of $\vec{y}$ are divisible by p as p-adic integers, which means that, as rational numbers, they have numerators divisible by p. That contradicts our choice of m and $\vec{y}$.

This contradiction shows that good primes p cannot divide denominators of components of $\vec{x}$ when $(x_0, \vec{x}) \in H$. Nor can they divide $x_0 = \langle\vec{\lambda}, \vec{x}\rangle$ because, being good, they do not divide denominators of components of $\vec{\lambda}$. Thus, good primes do not divide the denominators of any components of elements of H.

We now turn our attention to those finitely many primes p that are not good. Although it is possible for such a prime to divide the denominator of a component of an element of H, we intend to show that this divisibility cannot be with great multiplicity. That is, there is a bound $m \in \mathbb{N}$ such that no power of p higher than p^m divides the denominator of any element of H.

Fix one of these bad primes p, and choose $k+1$ vectors $\vec{\varphi}_j \in \Phi(p)$ as in assertion 5 of Lemma 5.2. Applying that assertion with our present $\vec{\lambda}$, we obtain that the $k+1$ vectors $(\vec{\varphi}_j + \vec{\lambda}) \restriction k$ span ${\mathbb{Q}_p}^k$. As ${\mathbb{Q}_p}^k$ is k-dimensional, we can select k of these vectors $\zeta_j = \vec{\varphi}_j + \vec{\lambda}$ whose truncations form a basis for ${\mathbb{Q}_p}^k$. This means that the matrix Z whose rows are these truncations is a non-singular $k \times k$ matrix over $\mathbb{Q}_p$. (We use here that the components of $\vec{\varphi}_j$ are in $\mathbb{Z}_p \subseteq \mathbb{Q}_p$ and the components of $\vec{\lambda}$ are in $\mathbb{Q} \subseteq \mathbb{Q}_p$.) Recall that any p-adic number can be written as a p-adic integer divided by a power of p. So we can choose an integer m so large that all entries of the matrix $p^m Z^{-1}$ are in $\mathbb{Z}_p$.

For any $(x_0, \vec{x}) \in H \subseteq G$, we have, by formula (5.1), that each $\langle\vec{\zeta}_j, \vec{x}\rangle \in \mathbb{Z}_p$, which means that, if we regard $\vec{x}$ as a column vector, then $Z\vec{x}$ is a vector $\vec{z}$ with components in $\mathbb{Z}_p$. Then $\vec{x} = Z^{-1}\vec{z}$ has all its components in $p^{-m}\mathbb{Z}_p$. That is, the denominator of any component of $\vec{x}$ cannot be divisible by a higher power of p than p^m.

To get the same result for x_0, we may need to increase m as follows. Since $\vec{\lambda}$ has finite support, let p^r be the highest power of p that divides the denominator of any component of $\vec{\lambda}$. Then, since $x_0 = \langle\vec{\lambda}, \vec{x}\rangle$, no power of p higher than p^{m+r} can divide the denominator of x_0.

Summarizing, we have an upper bound for the powers of p that can divide the denominator of any component of any member of H.

Let D be the product of these powers p^{m+r} for all of the finitely many bad primes. What we have shown is that every component of every element of H has, as its denominator, a divisor of D. Indeed, such a denominator cannot have good prime factors, and the remaining primes, the bad ones, cannot divide such a denominator to a higher power than they divide D. This means that H is a subgroup of

$$\frac{1}{D}\left(\mathbb{Z} \times \mathbb{Z}^k \times \{\vec{0}\}\right).$$

This group, isomorphic to $\mathbb{Z}^{k+1}$, is free, and therefore so is H. □

Corollary 5.2. *Every subgroup of G that is disjoint from L is free.*

PROOF – If H is such a subgroup, then all its finite-rank subgroups are free, by the lemma. Also, H is countable, because G is. By Pontryagin's criterion ([4, Lemma 16] or [1, Theorem IV.2.3]), it follows that H is free. □

Finally, we deduce that G does not have two disjoint non-free subgroups. Let two non-free subgroups of G be given. By the corollary just proved, each of them contains a non-zero element of L. As $L \cong \mathbb{Z}$, these elements have a common non-zero multiple, which has to be in both of the given subgroups.

References

[1] EKLOF, P. and MEKLER, A.: *Almost Free Modules. Set Theoretic Methods*, North-Holland, Amsterdam (2002), revised edition.

[2] HILLER, H., HUBER, M., and SHELAH, S.: The structure of Ext(A, Z) and $V = L$, *Math. Z.* **162** (1978), 39–50.

[3] HODGES, W.: In singular cardinality, locally free algebras are free, *Algebra Universalis* **12** (1981), 205–220.

[4] PONTRYAGIN, L.: The theory of topological commutative groups, *Ann. Math.* **35** (1934), 361–388.

[5] Shelah, S.: A compactness theorem for singular cardinals, free algebras, Whitehead problem and transversals, *Israel J. Math.* **21** (1975), 319–349.

[6] Solovay, R.: Real-valued measurable cardinals, *Axiomatic set theory*, Proc. Sympos. Pure Math. **13** Part I, Amer. Math. Soc. (1971), 397–428.

Global Complexity Results

Mirna Džamonja, Sy-David Friedman and Katherine Thompson

Contents

1. Introduction

To clarify our use of universality, we begin with the basic definitions. An embedding (also called weak embedding) for ordered sets is an injective order-preserving map. A strong embedding is an embedding which also preserves incomparability. The range of a strong embedding is an isomorphic copy of the order in the domain. For graphs, an embedding is an injective function which preserves edges and a strong embedding is an injective function whose range is an induced subgraph.

Given a set $\mathcal{A}_\lambda$ of structures each of size λ, a (strong) universal model for $\mathcal{A}_\lambda$ is an element of $\mathcal{A}_\lambda$ which (strongly) embeds all other structures in $\mathcal{A}_\lambda$. If there does not exist a universal model for $\mathcal{A}_\lambda$, then we consider its complexity, or the smallest size of a subfamily of $\mathcal{A}_\lambda$ which embeds the rest. This subfamily of structures is called a universal family. All of these notions have weak and strong counterparts depending on the type of embedding used.

In this paper, assume that λ, κ are regular, uncountable cardinals such that $\kappa \leq \lambda$.

We will now define notation for types of structures which will be discussed in this paper. As we would like to work with sets of structures rather than classes, we will assume that the structures are representatives under isomorphism. Let $\mathcal{C}(\lambda, \kappa)$ be the set of all posets of size λ which omit (i.e. do not contain) chains of size κ. Let $\mathcal{C}_a(\lambda, \kappa)$ be the set of all posets of size λ which omit antichains of size κ.

In section 3 it is shown to be consistent with ZFC that the complexity of $\mathcal{C}(\lambda, \kappa)$ and $\mathcal{C}_a(\lambda, \kappa)$ is 2^λ where the value of 2^λ can be any μ such that $\mathrm{cf}(\mu) > \lambda$. That is, μ-many posets will be added which do not embed into any other poset in the appropriate set. This will force the value for 2^λ to be μ. The posets built in this construction are well-founded, therefore, it is also consistent that the complexity for well-founded posets of size λ omitting chains or antichains of size κ is 2^λ where $2^\lambda = \mu$ as above.

*Mirna Džamonja thanks EPSRC for their support through an Advanced Fellowship and the University of Vienna and the Kurt Gödel Research Center for their kind hospitality in October 2005. The second and third authors wish to thank the Austrian Science Fund (FWF) for supporting this research.

Note that obtaining maximal weak complexity (i.e. 2^λ for structures of size λ) is optimal. That is, if there does not exist a small weakly universal family, then there certainly does not exist a strongly universal one.

In section 4, the above complexity results are shown to be consistent for all regular λ simultaneously. These use Easton products which destroy the GCH at all regular cardinals but in addition fix the complexity for posets as above.

Komjáth and Shelah in [3] give a similar localised result for graphs of size λ omitting $\aleph_1$-cliques for regular uncountable cardinals λ. Their methods were the basis of our methods for showing consistency results at one cardinal at a time, in particular the method of low sets is taken from [3]. That paper also gives another consistency result where the strong complexity for such graphs can be any $\nu \in [\lambda^+, 2^\lambda)$. The methods used for this result do not seem to apply to give strong complexity result for $\mathcal{C}(\lambda, \kappa)$ because of the difficulties imposed by transitivity.

The techniques in this paper also give a generalised statement for graphs of size λ omitting κ-cliques (for κ, λ as above). That is, we have the following results.

Theorem 1.1. *Let $\mathcal{A}$ be the set of all (representatives under isomorphism of) graphs of size λ omitting κ-cliques. Let μ be a cardinal satisfying cf(μ) $> \lambda$. Then:*

1. *(Komjáth -Shelah) for any regular $\nu \in [\lambda^+, \mu]$, there exists a cardinal and cofinality preserving forcing extension such that GCH holds below λ, $2^\lambda = \mu$ and the complexity of $\mathcal{A}$ is ν.*

2. *there are forcing extensions such that the results above hold for all λ simultaneously.*

The proof of (1) is omitted, but the proof of (2) can be found in section 4.

We will consider larger forcing conditions as the stronger ones. That is if p, p' are forcing conditions with $p \geq p'$ then we will say that p is stronger than p'. All the forcings P considered in this paper will have weakest elements, which we will call $\emptyset_P$.

The results in sections 2 and 4 are due to the second and third authors. The results in section 3 are due to the first and third authors.

2. Weak λ-closure

If a forcing notion has the property that one can guarantee an upper bound to any increasing sequence of limit length $< \lambda$ by making careful choices of upper bounds at earlier limit stages, then we say that the forcing is weakly λ-closed. More precisely:

Definition 2.1. *A forcing notion P is* weakly λ-closed, *if and only if there exists a function $F : P^{<\lambda} \to P$ such that for all limit $\tau < \lambda$ if $\langle p_\alpha : \alpha < \tau \rangle$ is a non-decreasing chain and satisfies the following:*

$$\text{for all limit } \tau' < \tau \text{ we have } p_{\tau'} \geq F(\langle p_\alpha : \alpha < \tau' \rangle)$$

then the sequence $\langle p_\alpha : \alpha < \tau \rangle$ has an upper bound.

The notion of weak λ-closure is somewhat stronger than the familiar notion of strategic λ-closure. The two notions have similar properties, but the former is easier to work with in the context of products.

Fact 2.1.

1. *If a poset is λ-closed then it is weakly λ-closed.*
2. *Weak λ-closure implies λ-distributivity. This is because a nondecreasing sequence of conditions can be built to meet any $< \lambda$ many dense subsets.*
3. *If $\theta \geq \lambda$ then weak θ-closure implies weak λ-closure.*
4. *In the definition of weak λ-closure, we may assume that if there is $\alpha < \tau$ such that $\alpha < \beta < \tau$ implies $p_\alpha = p_\beta$ then $F(\langle p_\alpha : \alpha < \tau' \rangle) = p_\alpha$ for all limit $\tau' < \tau$.*

Claim 2.1 - If P is a product forcing notion such that

$$P = \prod_{<\delta-\text{support}} \langle Q_i : i < \delta \rangle$$

for some regular $\delta \geq \lambda$ and Q_i is weakly λ-closed for all $i < \delta$, then P is weakly λ-closed.

Proof – A condition $p_\beta \in P$ can be written as $\langle q^i_\beta : i < \delta \rangle$ where $q^i_\beta \in Q_i$ for all $i < \delta$. To witness the weak λ-closure of Q_i, there exists an appropriate $F_i : Q_i^{<\lambda} \to Q_i$. Let $F_P(\langle p_\beta : \beta < \tau \rangle) = \langle F_0(\langle q^0_\beta : \beta < \tau \rangle), F_1(\langle q^1_\beta : \beta < \tau \rangle)... \rangle$. By Fact 2.1(4), this does not violate $< \delta$-support. □

Claim 2.2 - If $P = \langle (P_\alpha, \underset{\sim}{Q}_\alpha) : \alpha \in \mathrm{Ord} \rangle$ is an Easton support iteration and $\underset{\sim}{Q}_\lambda$ is weakly λ-closed for all regular λ, then for regular λ, we have that $P(\geq \lambda)$ is weakly λ-closed.

Proof – This is similar to the product, we must compose the functions from the Q_α's together. Let λ be a regular cardinal. We can write $p_\beta \in P(\geq \lambda)$ as $p_\beta = \langle \underset{\sim}{p}^i_\beta : i \in \mathrm{Ord} \rangle$ where each $p^i_\beta \in \underset{\sim}{Q}_i$. Let $F_P(\langle p_\beta : \beta < \tau \rangle) = \langle F_{\underset{\sim}{Q}_0}(\langle \underset{\sim}{p}^0_\beta : \beta < \tau \rangle), F_{\underset{\sim}{Q}_1}(\langle \underset{\sim}{p}^1_\beta : \beta < \tau \rangle)... \rangle$, again using Fact 2.1(4) to ensure Easton support. □

Claim 2.3 - If $P = \langle (P_\alpha, \underset{\sim}{Q}_\alpha) : \alpha < \mu \rangle$ is an iteration with $< \delta$-support for some regular cardinal $\delta \geq \lambda$ and cardinal μ and Q_α is weakly λ-closed for all $\alpha < \mu$, then P_α is weakly λ-closed.

Proof – Similar to the proof of 2.2. □

3. High complexity

In this section, we will show that the weak (and hence strong) complexity of $\mathcal{C}(\lambda, \kappa)$ can be 2^λ whose value will be forced to be μ for some $\mu > \lambda$.

Theorem 3.1. *Assume that $M \models V = L$ and $\mu, \lambda, \kappa > \aleph_0$ are cardinals such that $cf(\mu) > \lambda = cf(\lambda)$ and $\kappa \leq \lambda$ is regular. Then in a cardinal and cofinality preserving forcing extension M^P, GCH holds below λ and the weak (and hence strong) complexity of $\mathcal{C}(\lambda, \kappa)$ is $2^\lambda = \mu$.*

Remark 3.1 - We actually do not need the full strength of $V = L$, but only GCH and $\square_\theta$ when $\lambda = \theta^+$ and $\mathrm{cf}(\theta) < \kappa$.

Proof – First we will define and prove some lemmas about certain small subsets of λ in the ground model, called low subsets. We will use these facts in the definition of our forcing extension in order to prove that the extension preserves cofinalities.

When defining and working directly with these low sets, we will need three cases.

1. $\lambda = \kappa$
2. $\lambda > \kappa$, $\lambda = \theta^+$ and $\mathrm{cf}(\theta) < \kappa$
3. $\lambda > \kappa$ and case 2 does not apply.

For case 1 and case 3 the low subsets of λ are the nonempty bounded subsets of λ of size $< \kappa$ and of limit order type. For case 2 we will define low sets using $\square_\theta$:

Assume we are in case 2. Jensen has shown in [2] that, a $\square_\theta$-sequence exists in M. Fix a $\square_\theta$-sequence $\langle C_\delta : \delta < \lambda,\ \delta \text{ limit}\rangle$ with the following properties:

1. $C_\delta \subseteq \delta$ is a club of δ,
2. if γ is a limit point of C_δ, then $C_\gamma = \gamma \cap C_\delta$,
3. $|C_\delta| < \theta$.

Let the increasing enumeration of each C_δ be $\langle c^\delta_\xi : \xi < \mathrm{type}(C_\delta)\rangle$. Also, fix a sequence of cardinals $\{\theta_\gamma : \gamma < \mathrm{cf}(\theta)\}$ strictly increasing to θ and an injective function $\phi_{\alpha,\beta} : [\alpha, \beta) \to \theta$ for each $\alpha < \beta < \lambda$.

For case 2, the low subsets of λ are nonempty sets $A \subset \lambda$ with the following properties:

1. $|A| < \kappa$ with limit order type
2. Let $\delta = \sup(A)$. There exists a $\gamma < \mathrm{cf}(\theta)$ such that for all $\xi < \mathrm{type}(C_\delta)$ and for all $a \in A$ with $c^\delta_\xi \leq a < c^\delta_{\xi+1}$, we have $\phi_{c^\delta_\xi, c^\delta_{\xi+1}}(a) < \theta_\gamma$.

We say that A is witnessed by γ.

The next two lemmas are the reasons that we needed the existence of the $\square_\theta$-sequence. This method allows us to only have to deal with a small number of bounded subsets of λ. Consider in case 2 if we had simply taken all bounded subsets of λ of size $< \kappa$ of limit type as low. Then the number of low subsets of λ would be $\theta^{<\kappa} \geq \theta^{\mathrm{cf}(\theta)} > \theta$ by König's Lemma, so Lemma 3.1 would be contradicted when $\alpha = \theta$ and $\lambda = \theta^+$.

Also, the coherence of the $\square_\theta$-sequence will help us to shrink subsets of λ to a subset whose intersection with each of its limit point is low. This will be important as the low sets will be used to control the growth of chains when building sequences of increasing posets.

Lemma 3.1. *Fix an ordinal $\alpha < \lambda$. For $\lambda > \kappa$, the number of low subsets of λ which are subsets of α is less than λ.*

PROOF – For case 2, let $\lambda = \theta^+$ and let $\mathcal{A}_\gamma$ be the set of low subsets of λ contained in α which are witnessed by γ. The number of low subsets of λ contained in α is then $\sup\{|\mathcal{A}_\gamma| : \gamma < \mathrm{cf}(\theta)\}$. We will fix a $\gamma < \mathrm{cf}(\theta)$ and determine the size of $\mathcal{A}_\gamma$.

Let A be a set of size $< \kappa$ of limit type with $\delta = \sup(A)$ such that $\delta \leq \alpha$ and γ witnesses that A is low. The number of possibilities for A is the number of possible subsets of $\phi^{-1}_{c^\delta_\xi, c^\delta_{\xi+1}}(\theta_\gamma)$ of size $< \kappa$ times the number of intervals $[c^\delta_\xi, c^\delta_{\xi+1})$. We will calculate these values below.

For each $\xi < \mathrm{type}(C_\delta)$, let $X_\xi = \phi^{-1}_{c^\delta_\xi, c^\delta_{\xi+1}}[\theta_\gamma]$. Since $\phi_{c^\delta_\xi, c^\delta_{\xi+1}}$ is injective, we have $|X_\xi| \leq \theta_\gamma$ possibilities for $a \in X_\xi$. A low subset is a subset of $\bigcup_{\xi < \mathrm{type}(C_\delta)} Y_\xi$ of size $< \kappa$, where each $Y_\xi \subseteq X_\xi$. For each ξ, X_ξ has at most $2^{\theta_\gamma} \leq \theta_{\gamma+1}$ subsets, as GCH holds. Since $|C_\delta| < \theta$, certainly $\mathrm{type}(C_\delta) < \theta$ so there at most $\theta_{\gamma+1} \cdot \mathrm{type}(C_\delta) < \theta$ sets in $\mathcal{A}_\gamma$.

Hence, over all $\gamma < \mathrm{cf}(\theta)$, there are at most θ low subsets of α, which is less than λ.

One can see that for case 3, the number of low subsets of any $\alpha < \lambda$ is less than λ as the number of subsets of α of size $< \kappa$ of limit type is $|\alpha|^{<\kappa} < \lambda$. $\square$

Lemma 3.2. *If $\lambda > \kappa$ and $B \subseteq \lambda$ has order type κ, then there exists a cofinal subset $B' \subseteq B$ such that if $\gamma < \sup(B')$ is a limit point of B' then $B' \cap \gamma$ is low.*

PROOF – For case 2, let λ and B be as above and let $\delta = \sup(B)$. Let $\langle C_\delta : \delta < \lambda \text{ limit}\rangle$ be defined as above. Choose a cofinal subset $B' \subseteq B$ such that if $a, b \in B'$ then there exists c^δ_ξ such that $a \leq c^\delta_\xi < b$. Such a cofinal sequence exists as both B and C_δ are unbounded in δ. For all ξ such that there is $b \in B'$ with $c^\delta_\xi \leq b < c^\delta_{\xi+1}$, there is some $\alpha_\xi < \mathrm{cf}(\theta)$ such that $\phi_{c^\delta_\xi, c^\delta_{\xi+1}}(b) < \theta_{\alpha_\xi}$, since $|B'| \leq \kappa < \mathrm{cf}(\theta)$. Hence there is $\alpha < \mathrm{cf}(\theta)$ such that for all $b \in B'$ where $c^\delta_\xi \leq b < c^\delta_{\xi+1}$, we have $\phi_{c^\delta_\xi, c^\delta_{\xi+1}}(b) < \theta_\alpha$.

Since B' is cofinal in B, it is clear that type(B') is a limit, in fact, the order type of B' is κ because that is the order type of B. Because of the cofinality of B' in B, we also know that $\sup(B') = \sup(B) = \delta$. If $\gamma < \delta$ is a limit point of B' then $|B' \cap \gamma| < \kappa$ (as the order type of B' is κ). Since γ is also a limit point for C_δ, the coherence property of $\square_\theta$ holds, namely $C_\gamma = \gamma \cap C_\delta$ and thus α witnesses that $B' \cap \gamma$ is low.

For case 3, let $B' = B$. If γ is a limit point of B', then $B' \cap \gamma$ is a subset with size $< \kappa$ of limit type. □

We shall define our forcing notion $P = P(\lambda, \kappa, \mu)$ to be the $(< \lambda)$-support product of μ copies of $Q = Q(\lambda, \kappa)$, a forcing notion that will be defined below.

Let the elements of Q be $q = (\delta, X, \mathcal{A})$, where

1. $\delta < \lambda$ is an ordinal,

2. X is a poset $(X, \leq_X)$ with domain δ which omits κ-chains and in which $\alpha \leq_X \beta$ implies that $\alpha \leq \beta$,

3. the conditions on $\mathcal{A}$ are as follows:

   (a) $\mathcal{A}$ is a family of low subsets A of λ such that $A \subseteq \delta$ and the size of $\mathcal{A}$ is $< \lambda$,

   (b) if $A \in \mathcal{A}$ is such that $\sup(A) \leq x < \delta$, then $A \not\leq_X x$.

We will assign the ordering of Q to be as follows. If $q = (\delta, X, \mathcal{A})$ and $q' = (\delta', X', \mathcal{A}')$ are in Q, then $q' \geq_Q q$ (i.e. q' is stronger than q) if and only if $\delta' \geq \delta$, with X a subposet of X' such that $X = X' \restriction \delta$ and $\mathcal{A} = \mathcal{A}' \cap [\delta]^{<\kappa}$. In addition, if $A \in \mathcal{A}$ and $a \in A$ then $a \not\leq_{X'} \delta$.

Let us observe that this ordering is transitive: if $p_l = (\delta_l, X_l, \mathcal{A}_l)$ for $l < 3$ and $p_0 \leq p_1, p_1 \leq p_2$ hold, then to see that $p_0 \leq p_2$ we have to see that for every $A \in \mathcal{A}_0$ and $a \in A$ we have $a \not\leq_{X_2} \delta_0$. Since $p_0 \leq p_1$ we have $a \not\leq_{X_1} \delta_0$ but since $X_2 \restriction \delta_1 = X_1$ this implies $a \not\leq_{X_2} \delta_0$.

Next we need to show that forcing with P preserves cardinalities and cofinalities. We start by defining upper bounds for sequences in Q.

Definition 3.1. *Given $i^* < \lambda$ and $\bar{q} := \langle q_i = (\delta_i, X_i, \mathcal{A}_i) : i < i^* \rangle$ increasing in Q, we will define the* canonical upper bound *for $\bar{q}$ or $cub(\bar{q})$ such that $cub(\bar{q}) \geq q_i$ for all $i < i^*$. If i^* is a successor ordinal $j+1$, then let $cub(\bar{q}) = q_j$. Assume now that i^* is a limit ordinal.*

For $\lambda = \kappa$, let $cub(\bar{q}) = (\delta, X, \mathcal{A})$ be such that $\delta = \sup\{\delta_i : i < i^*\}$ and let $X, \mathcal{A}$ be the union over all $i < i^*$ of their respective parts in q_i, (e.g. $X = \bigcup_{i<i^*} X_i$). No extra relations are added to X.

For $\lambda > \kappa$, let $cub(\bar{q}) = (\delta, X, \mathcal{A})$ be such that δ, X are the union over all $i < i^*$ of their respective parts in q_i and let $\mathcal{A}$ be as follows. If cf$(i^*) \geq \kappa$, take $\mathcal{A} = \bigcup_{i<i^*} \mathcal{A}_i$, otherwise we also include all low subsets of λ cofinal in δ.

We will show that the cub of an increasing sequence of elements of Q is in fact a condition in Q.

Claim 3.1 - If $\alpha^* < \lambda$ and $\bar{q} := \langle q_\alpha = (\delta_\alpha, X_\alpha, \mathcal{A}_\alpha) : \alpha < \alpha^* \rangle$ is a sequence in Q such that for all limit $\alpha < \alpha^*$ we have q_α is the cub of $\langle q_\beta : \beta < \alpha \rangle$, then cub$(\bar{q}) = (\delta, X, \mathcal{A}) \in Q$ is an upper bound of the sequence $\bar{q}$.

Proof – The proof is by induction on α^*, assuming that the statement holds for all sequences of length less than α^*. In particular we may assume that $\bar{q}$ is increasing in Q.

If α^* is a successor ordinal then this is trivial, hence we shall now assume that α^* is a limit.

One can see that for all $\alpha < \alpha^*$ we have $\delta_\alpha < \lambda$ and X_α has domain δ_α, hence by the regularity of λ it follows that $\delta < \lambda$, and it is also clear that the domain of X is δ. We know that $\mathcal{A}$ is a family of low subsets of λ because all the sets that comprise each $\mathcal{A}_\alpha$ are low. If cf$(\alpha^*) \geq \kappa$ then $|\mathcal{A}| < \lambda$ because $\alpha^* < \lambda$ and $|\mathcal{A}_\alpha| < \lambda$ for all $\alpha < \alpha^*$ as $q_\alpha \in Q$. If cf$(\alpha^*) < \kappa$ then there are $< \lambda$ low sets cofinal in δ by Lemma 3.1, so we still have $|\mathcal{A}| < \lambda$.

For requirement 3b, if cf$(\alpha^*) < \kappa$ and $A \in \mathcal{A}$, then either A is cofinal in δ and so $\sup(A) = \delta$ or A is also in $\mathcal{A}_\alpha$ for some $\alpha < \alpha^*$. Thus, in the former case there is no x such that $\sup(A) \leq x < \delta$, so that the requirement is trivially met. In the latter case the same can be seen by observing that for all $\beta' \in [\beta, \alpha^*)$ we have $A \in \mathcal{A}_{\beta'}$ and using that $\delta = \sup\{\delta_\alpha : \alpha < \alpha^*\}$. If $\mathrm{cf}(\alpha^*) \geq \kappa$, then for each $A \in \mathcal{A}$ there exists an $\alpha < \alpha^*$ such that $A \in \mathcal{A}_\alpha$ so the requirement is met similarly as above.

The only remaining difficulty that might arise is if we added a κ-chain in some poset X_α. Let $\alpha \leq \alpha^*$ be the minimal ordinal such that there exists a κ-chain T of X in δ_α. Note that the domain of T is necessarily cofinal in δ_α. By shrinking T if necessary we may assume that the order type of it is exactly κ.

We must have that $\mathrm{cf}(\alpha) = \kappa$ as this is the only point where such a chain could be added. If $\lambda = \kappa$, then this cannot occur as $\alpha^* < \lambda$. Assuming then that $\lambda > \kappa$, by Lemma 3.2, there exists a cofinal $T' \subseteq T$ such that if $\gamma < \delta_\alpha$ is a limit point of T', then $T' \cap \gamma$ is low. There exists a limit ordinal $\beta < \alpha$ such that δ_β is a limit point of T' and hence $\mathrm{cf}(\beta) \neq \kappa$ (since the order type of T' is κ). Such a β can be found as both the set of limit points of an unbounded subset of δ_α (namely T') and the set of $\{\delta_\beta : \beta < \alpha\}$ form clubs in δ_α, by the assumption on $\bar{q}$. By our construction $T' \cap \delta_\beta \in \mathcal{A}_\beta$ because it is low, so $T' \cap \delta_\beta$ could not have been extended to a κ-chain later by requirement 3b on q'_β for $\beta' > \beta$.

We still need to show that $q_\alpha \leq \mathrm{cub}(\bar{\mathrm{q}})$ for $\alpha < \alpha^*$. Note that when $\mathrm{cf}(\alpha^*) \geq \kappa$ we have $\mathcal{A} \cap [\delta_\alpha]^{<\kappa} = \mathcal{A}_\alpha$ because $[\delta_\alpha]^{<\kappa}$ does not include any Y cofinal in δ. The rest is clear from the definition of $\mathrm{cub}(\bar{q})$. □

Lemma 3.3. *Q is weakly λ-closed.*

Proof – The function F witnessing the weak closure will be the cub operator defined above. Suppose that we are given a sequence $\bar{q} = \langle q_\alpha = (\delta_\alpha, X_\alpha, \mathcal{A}_\alpha) : \alpha < \alpha^* \rangle$ in Q for some $\alpha^* < \lambda$, and this sequence satisfies that for all limit $\tau < \alpha^*$ we have $q_\tau \geq \mathrm{cub}(\langle q_\alpha : \alpha < \tau \rangle)$. If α^* is a successor ordinal $\alpha + 1$ then clearly q_α is an upper bound of the sequence, so let us assume that α^* is a limit. We can then define a sequence $\bar{q}'$ of length α^* such that for $q'_0 = q_0$ and $q_{\alpha+1} = q_\alpha$, while for α limit we have q'_α is the cub of $\langle q'_\beta : \beta < \alpha \rangle$. Then $\bar{q}$ is cofinal in $\bar{q'}$ and any upper bound for $\bar{q'}$ is an upper bound for $\bar{q}$.

The existence of such an upper bound follows by Claim 3.1. □

Note that by Claim 2.1, the forcing P is also weakly λ-closed.

Lemma 3.4. *The forcing notion P is λ^+-cc.*

Proof – Suppose $\{p_\alpha : \alpha < \lambda^+\}$ is a sequence in P. We know that λ is regular and for all $\alpha < \lambda^+$ we have $|\alpha^{<\lambda}| < \lambda^+$. Also, for all $p \in P$, we know that $|\mathrm{ran}(p)| < \lambda$ since p has $< \lambda$ non-trivial elements. So, by the Δ-System Lemma, there exists an $X \subseteq \lambda^+$ such that $|X| = \lambda^+$ and $\{\mathrm{Dom}(p_\alpha) : \alpha \in X\}$ forms a Δ-system with some root r. Since $|Q| = \lambda$ and $\lambda^{<\lambda} = \lambda$, there are fewer than λ^+ possibilities for $p_\alpha \restriction r$ so there exists a $Y \subseteq X$ with $|Y| = \lambda^+$ such that for all $\alpha, \beta \in Y$, we have $p_\alpha \restriction r = p_\beta \restriction r$. Now for all $\alpha \in Y$, the p_α's are compatible, which shows that the original sequence could not have formed an antichain. □

Let G_P be P-generic over M. If in $M[G_P]$ the complexity of $\mathcal{C}(\lambda, \kappa)$ is $< \mu$, then there exists a family $\mathcal{F}$ of posets that witnesses this. This subfamily is added by a $< \mu$ sized sub-product of P, call it R. (Note that R may be a cofinal sub-product of P in case μ is singular.) By the product lemma (see e.g. [4]), if we let G_R and $G_{P/R}$ be the generics for R and P/R respectively, then $M[G_R][G_{P/R}] = M[G_{P/R}][G_R] = M[G_P]$. It is enough to show that forcing with Q over $M[G_R]$ introduces a poset Y that cannot be embedded into any poset in $M[G_R]$ of size λ which omits κ-chains. Then we can take one copy of Q from P/R and we shall have in $M[G_P]$ that $\mathcal{F}$ is not a universal family. So let us show the required property of Q.

If $G \subseteq Q$ is generic for Q then let $Y = Y_G = \bigcup\{X : (\delta, X, \mathcal{A}) \in G\}$ be such that for all $x, y \in Y$ we have $x \leq_Y y$ if and only if for some X with $(\delta, X, \mathcal{A}) \in G$ we have $x \leq_X y$.

Lemma 3.5. *For any generic G, the set $Y = Y_G$ as defined above is a poset of size λ which omits κ-chains.*

Proof – We start with a density argument to show that Y has size λ.

Claim 3.2 - For every $\gamma < \lambda$ the set $D_\gamma = \{(\delta, X, \mathcal{A}) \in Q : \delta > \gamma\}$ is dense in Q.

PROOF – Given a condition $q = (\delta, X, \mathcal{A}) \in Q$, suppose that $\delta \leq \gamma$. Let Z be the poset extending X whose universe is $\gamma+1$ such that for all ε in $[\delta, \gamma+1)$ we have that ε is not comparable to any other element of Z.

Then Z is a poset which omits κ-chains. We wish to show that $q' = (\gamma + 1, Z, \mathcal{A})$ is a condition in D_γ and that it extends q.

If $A \in \mathcal{A}$ satisfies $\sup(A) \leq x \leq \gamma$ for some $x \in Z$ then either $x \in X$ so $A \not\leq_Z x$ or $x > \delta$ and again $A \not\leq_Z x$. Also, δ is not comparable with any other element of Z so taken together these observations show that q' is as required. □

By the Claim above, we know for any generic G that Y has size λ. It is also a poset by genericity. Now we will show that Y omits κ-chains.

If $\lambda = \kappa$, we will assume that $q \in Q$ and $\underset{\sim}{T}$ are such that $q \Vdash$ "$\underset{\sim}{T}$ is an increasing κ-chain in Y". We will choose an increasing sequence $\langle q_n : n < \omega \rangle$ such that $q_0 = q$ and let $q_{n+1} = (\delta_{n+1}, X_{n+1}, \mathcal{A}_{n+1})$ be such that $q_{n+1} \Vdash$ "$t_n \in \underset{\sim}{T}$ and $\delta_n < t_n < \delta_{n+1}$". This can be done because each $r = (\delta, X, \mathcal{A}) \in Q$ satisfies $r \Vdash$ "$\underset{\sim}{Y} \restriction \delta = X$". Hence, if it is also the case that $r \geq q$, then $r \Vdash$ "$\underset{\sim}{T}$ is a κ-chain in $\underset{\sim}{Y}$" and so $r \Vdash$ "$(\exists \alpha > \delta) \alpha \in \underset{\sim}{T}$". Applying this to $r = q_n$, we may find $q_{n'} \geq q_n$ and $t_n > \delta_n$ such that $q_{n'} \Vdash$ "$t_n \in \underset{\sim}{T}$". Then by extending $q_{n'}$ to q_{n+1} if necessary, we may by Claim 3.2 assume that $\delta_{n+1} > t_n$.

Let $\bar{t} = \{t_n : n < \omega\}$. Now let $q' = (\delta, X, \mathcal{A})$ be such that $\delta = \lim\{\delta_n : n < \omega\}$ with $X = \bigcup\{X_n : n < \omega\}$ and $\mathcal{A} = \bigcup\{\mathcal{A}_n : n < \omega\} \cup \{\bar{t}\}$. As $\sup\{t_n : n < \omega\} = \sup\{\delta_n : n < \omega\}$ we know that $\sup(\bar{t}) = \delta$. Therefore, there does not exist x such that $\sup(\bar{t}) \leq_X x < \delta$ and thus requirement 3b holds for $\bar{t} \in \mathcal{A}$. The rest of the requirements hold by similar arguments to those in Claim 3.1. Thus we have that $q' \in Q$.

We Claim that $q' \Vdash$ "$\underset{\sim}{T} \subseteq \delta$". This is equivalent to the statement that for all $r \geq q'$ with $r \in Q$, we have that r does not force "$T \not\subseteq \delta$". We will suppose otherwise and arrive at a contradiction.

Let $r = (\xi, Z, \mathcal{B})$ with $r \geq q'$ and let $\beta > \delta$ be such that $r \Vdash$ "$\beta \in \underset{\sim}{T}$". Without loss of generality $\xi > \beta$. We know that $\bar{t} \in \mathcal{A}$ so it is also the case that $\bar{t} \in \mathcal{B}$. Since $\delta < \beta$, we also must have that $\bar{t} \not\leq_Z \beta$ by requirement 3b of $\mathcal{B}$ for $r \in Q$. However, since r forces $\beta \in \underset{\sim}{T}$, r must force that $\bar{t} \leq_T \beta$ and $Z \subseteq Y$.

So we proved that $q' \Vdash$ "$\underset{\sim}{T} \subseteq \delta$". (Note that this does not imply that q' forces

T to be countable as the elements of $\bar{t}$ are not necessarily the only elements of T). Since $\delta < \lambda = \kappa$, we proved that q' forces that $\underset{\sim}{T}$ has size $< \kappa$, which is a contradiction.

If $\lambda > \kappa$, then Q does not add any new κ-sequences since it is weakly λ-closed. Therefore, Q could not have added a κ-chain. □

Lemma 3.6. *The poset Y does not embed into any poset in the ground model which has size λ and omits κ-chains.*

Proof – Suppose that there exists $q \in Q$ such that $q \Vdash$ "$\underset{\sim}{f} : \underset{\sim}{Y} \to Z$ is an embedding" where Z in the ground model is a poset of size λ which omits κ-chains".

By induction on $\alpha < \kappa$, we will construct an increasing sequence $\langle q_\alpha = (\delta_\alpha, X_\alpha, \mathcal{A}_\alpha) : \alpha < \kappa \rangle$ and a function g as follows. Let $q_0 = q$ and let $q_{\alpha+1} \Vdash$ "$\underset{\sim}{f}(\delta_\alpha) = g(\alpha)$". For α a limit ordinal, define an intermediate condition $q'_\alpha = (\delta'_\alpha = \sup_{\beta<\alpha} \delta_\beta, X'_\alpha = \bigcup_{\beta<\alpha} X_\beta, \mathcal{A}'_\alpha = \bigcup_{\beta<\alpha} \mathcal{A}_\beta)$. It is easily seen that $q_\beta \leq q'_\alpha$ for all $\beta < \alpha$. Let $q_\alpha = (\delta_\alpha, X_\alpha, \mathcal{A}_\alpha)$ such that $\delta_\alpha = \delta'_\alpha + 1$ and $\mathcal{A}_\alpha = \mathcal{A}'_\alpha$. In X_α let $\delta_\beta \leq_{X_\alpha} \delta'_\alpha$ for $\beta < \alpha$, let $\delta'_\beta \leq_{X_\alpha} \delta'_\alpha$ for all $\beta < \alpha$ limit, let X_α inherit all the relations from X'_α, and finally take the transitive closure. Note that this does not affect X'_α as we are only adding pairs (x, y) in which $y = \delta'_\alpha$ and $\delta'_\alpha \notin X'_\alpha \cup \bigcup \mathcal{A}'_\alpha$.

We must see that q_α is a condition. The proof is by induction on the limit α.

It is easy to see that all the requirements for q_α to be in Q are met, except for requirement 3b which requires some argument.

Requirement 3b can fail for q_α if there exists $A \in \mathcal{A}_\alpha$ with $A \subseteq \delta_\alpha$ and $\sup(A) \leq x \leq \delta'_\alpha$ for some $x \in X_\alpha$ and $A \leq_{X_\alpha} x$. This breaks into three cases.

We cannot have that $\sup(A) \geq \delta'_\alpha$ as by the definition of q'_α for α limit, there are no sets $A \in \mathcal{A}_\alpha$ cofinal in δ'_α.

If $\sup(A) \leq x < \delta'_\alpha$ then there exists $\beta < \alpha$ such that $A \in \mathcal{A}_\beta$ and $x \in X_\beta$ as α is a limit. However, this contradicts q_β being a condition.

The last case to check is if $\sup(A) < \delta'_\alpha$ and $x = \delta'_\alpha$. We will assume this is the case and arrive at a contradiction. Nothing in X_α is connected to x except $\{\delta_\beta : \beta < \alpha\}$, $\{\delta'_\beta : \beta < \alpha \text{ limit}\}$ and points below the δ_β's which we had to connect by transitivity. If $A \leq_{X_\alpha} x$ then for every $a \in A$ there must be $\delta_{\beta_a} = \delta_\beta$ such that $a \leq_{X_\alpha} \delta_\beta$, for some $\beta < \alpha$. On the other hand, since α is a limit ordinal and $\sup(A) < \delta'_\alpha$, then by the definition of $\mathcal{A}_\alpha$ there exists $\xi < \alpha$ such that $A \in \mathcal{A}_\xi$ and hence $\sup(A) \leq \delta_\xi$. Let ξ be the first such ordinal. By the fact that $\langle q_\beta : \beta < \alpha \rangle$ is increasing and by the definition of $\leq_Q$, we must have that for every $a \in A$ the inequality $\beta_a < \xi$ holds. The first possibility is that $\sup_{a \in A} \beta_a = \xi$. If this is the case then $\xi < \alpha$ is a limit. Then by the definition of q_ξ we have that for all $a \in A$, $\delta_{\beta_a} \leq_{X_\xi} \delta'_\xi$ and in particular $A \leq_{X_\xi} \delta'_\xi$. This contradicts the inductive hypothesis that q_ξ is a condition. Hence we must have $\beta \stackrel{\text{def}}{=} \sup_{a \in A} \beta_a < \xi$. Since for every $a \in A$ we have $a \leq_{X_\alpha} \delta_{\beta_a}$ then in particular $a \leq \delta_{\beta_a}$ as ordinals, and hence $\sup(A) \leq \delta_\beta$. However, $\mathcal{A}_\xi \cap [\delta_\beta]^{<\kappa} = \mathcal{A}_\beta$, so $A \in \mathcal{A}_\beta$, contradicting the choice of ξ.

Now that we have shown that q_α is a condition it is also easy to show that $q_\beta \leq_Q q_\alpha$ for all $\beta < \alpha$. Hence we can define the sequence $\langle q_\alpha : \alpha < \kappa \rangle$, but then $\{g(\alpha) : \alpha < \kappa \text{ limit}\}$ will be a κ-chain in Z, a contradiction. □

We have now exhibited a poset of size λ which omits κ-chains in the extension and which does not embed into any such poset in $M[G_R]$. This shows that the complexity of $\mathcal{C}(\lambda, \kappa)$ is at least μ. It now only remains to show that $2^\lambda = \mu$ in M^P since 2^λ is the maximal complexity of a set of structures of size λ.

Each poset Y is a subset of λ and we introduced μ many of them. Hence in the extension $2^\lambda \geq \mu$.

Every subset of λ in M^P has a nice name, that is a name of the form $\{\langle \alpha, K_\alpha \rangle : \alpha < \lambda\}$ where each K_α is an antichain of P. By definition $|P| = \mu$ and we have shown that P is λ^+-cc, so it has at most $\mu^\lambda = \mu$ antichains. Hence, there are at most μ nice names for subsets of λ and thus $2^\lambda \leq \mu$. □

Theorem 3.2. *Assume that $M \models V = L$ and $\mu, \lambda, \kappa > \aleph_0$ are cardinals such that cf(μ) $> \lambda =$ cf(λ) and $\kappa \leq \lambda$ regular. Then in a cofinality preserving forcing extension M^P, GCH holds below λ and the weak (and hence strong) complexity of $\mathcal{C}_a(\lambda, \kappa)$ is $2^\lambda = \mu$.*

Proof – The proof of this theorem is essentially the same as the one above except for one thing: Condition 3b in the definition of Q needs to be changed.

The new version should read:

3b′. if $A \in \mathcal{A}$ and $\sup(A) \leq x < \delta$ then A is comparable to x.

Then in the lemmas (in particular Lemmas 3.3, 3.5 and 3.6 and Claim 3.2), replace $A \leq x$ with $A >< x$ and $A \not\leq x$ with "A is comparable to x". $\square$

4. Global results

Theorem 4.1. *Assume $V = L$. To each uncountable regular cardinal λ definably associate a regular uncountable cardinal $\kappa(\lambda) \leq \lambda$ and a cardinal $F(\lambda)$ such that cf$(F(\lambda)) > \lambda$ and $\lambda < \theta$ implies $F(\lambda) \leq F(\theta)$. Then there exists a definable, ZFC-preserving and cofinality preserving class forcing notion P such that in L^P the complexity of $\mathcal{C}(\lambda, \kappa(\lambda))$ is $F(\lambda) = 2^\lambda$ for each regular uncountable λ.*

Remark 4.1 - As before, we do not need the full strength of $V = L$, but only GCH and $\square_\theta$ when $\lambda = \theta^+$ and $\mathrm{cf}(\theta) < \kappa(\lambda)$.

Proof – We shall define a forcing notion $P(\lambda, F(\lambda), \kappa(\lambda))$ to be the $(< \lambda)$-support product of $F(\lambda)$ copies of $Q(\lambda, \kappa(\lambda))$, forcing notions that were defined in section 3. The relevant properties of these forcings are summarised below.

Fact 4.1.

1. *The cardinality of $Q(\lambda, \kappa(\lambda))$ is λ.*
2. *If $\lambda = \kappa(\lambda)$, then $Q(\lambda, \kappa(\lambda))$ is $(< \kappa(\lambda))$-closed. If $\lambda > \kappa(\lambda)$, then forcing with $Q(\lambda, \kappa(\lambda))$ is weakly $\kappa(\lambda)$-closed. Similar statements hold for $P(\lambda, F(\lambda), \kappa(\lambda))$.*
3. *The forcing notion $P(\lambda, F(\lambda), \kappa(\lambda))$ is λ^+-cc.*

P will be a product forcing notion of all $P(\lambda, F(\lambda), \kappa(\lambda))$ made with Easton support. Recall that Easton support means that the support is bounded at inaccessibles. We use the notation $P(< \lambda)$ to denote the part of the product made over $\theta < \lambda$ and $P(\geq \lambda)$ for the remainder of the product. Similarly, we use $G(< \lambda)$ and $G(\geq \lambda)$ for the corresponding generics.

Lemma 4.1. *$P(\geq \lambda)$ is weakly λ-closed.*

PROOF – Holds by Claim 2.2. □

Lemma 4.2. *If λ is Mahlo or the successor of a regular cardinal, then $P(< \lambda)$ is λ-cc.*

PROOF – Works by standard Δ-system arguments. See Jech [1] for details. □

Unfortunately, when λ is the successor of a singular cardinal or a non-Mahlo inaccessible, we only have that $P(< \lambda)$ is λ^+-cc.

Let G be P-generic over L.

Lemma 4.3. *$L[G]$ satisfies that the complexity of posets of size λ omitting $\kappa(\lambda)$ chains is $F(\lambda)$ for all regular uncountable λ.*

PROOF – Since P is a product, we can split $L[G] = L[G(\geq \lambda)][G(< \lambda)]$. By doing the larger forcing first, nothing is changed below λ. In particular, GCH and square still hold below λ and no new sequences of size $< \lambda$ were added, so forcing over $L[G(\geq \lambda)]$ is the same, from the point of view of the smaller forcing, as forcing over the ground model.

However, once we do the small forcing, we want to see that the complexity calculations achieved by large forcing are not changed by the small forcing. We can do this by anticipating a maximal antichain of conditions in the small forcing when we do the complexity proof for the large forcing.

If $G_Q \subseteq Q(\lambda, \kappa(\lambda))$ is generic for $Q(\lambda, \kappa(\lambda))$ then let $Y = Y(G_Q) = \bigcup\{X : (\delta, X, \mathcal{A}) \in G_Q\}$ be such that for all $x, y \in Y$ we have $x \leq_Y y$ iff for some X with $(\delta, X, \mathcal{A}) \in G_Q$ we have $x \leq_X y$. By Lemma 3.5, we know that for any generic G_Q, the set $Y = Y(G_Q)$ as defined above is a poset of size λ which omits $\kappa(\lambda)$-chains.

We also have that the poset Y does not embed into any ground model poset of size λ which omits $\kappa(\lambda)$-chains. We strengthen this argument below to show that Y does not embed into any poset of size λ which omits $\kappa(\lambda)$-chains and which exists in $L[G(< \lambda)]$. In particular, we will show that if there exists a condition in $P(\lambda, F(\lambda), \kappa(\lambda))$ which forces an embedding from

Y into a poset in $L[G(<\lambda)]$, then we can find a sequence of conditions in $P(<\lambda) \times P(\lambda, F(\lambda), \kappa(\lambda))$ which "decides" the embedding.

By the product lemma we can look at conditions from $P(<\lambda) \times Q(\lambda, \kappa(\lambda))$ instead of $P(<\lambda) \times P(\lambda, F(\lambda), \kappa(\lambda))$.

Assume that there exists $q \in Q(\lambda, \kappa(\lambda))$ such that over the ground model $L^{P(<\lambda)}$, $q \Vdash$ "$\underset{\sim}{f} : \underset{\sim}{Y} \to Z$ is an embedding of $\underset{\sim}{Y}$ into some ground model poset Z of size λ which omits $\kappa(\lambda)$-chains".

By induction on $\alpha < \kappa(\lambda)$, we will construct an increasing sequence $\langle q_\alpha = (\delta_\alpha, X_\alpha, \mathcal{A}_\alpha) : \alpha < \kappa(\lambda)\rangle$ which, together with a maximal antichain of conditions from $P(<\lambda)$, "decides" the values of the embedding. We will deal with the cases where λ is Mahlo or the successor of a regular cardinal first, as the maximal antichains of $P(<\lambda)$ have size $<\lambda$.

Let $q_0 = q$. For α limit, let $\delta_\alpha = \sup\{\delta_\beta : \beta < \alpha\}$ with $X_\alpha = \bigcup\{X_\beta : \beta < \alpha\}$ and $\mathcal{A}_\alpha = \bigcup\{\mathcal{A}_\alpha : \beta < \alpha\}$. After each limit stage α, let q'_α be an intermediate condition $(\delta_\alpha + 1, X^+_\alpha, \mathcal{A}_\alpha)$ such that $X^+_\alpha = X_\alpha \cup \{\delta_\alpha\}$ and for $\beta < \alpha$ we have $\delta_\beta \leq_{X^+_\alpha} \delta_\alpha$ and take the transitive closure. Note that this does not affect X_α as we are only adding relations to δ_α and $\delta_\alpha \notin X_\alpha$. Then we will continue constructing $q_{\alpha+1}$ by extending q'_α as below.

For $\alpha + 1$, we build the condition $q_{\alpha+1}$ by induction. Choose any $p_0 \in P(<\lambda)$ and $q^0_{\alpha+1}$ such that $(q^0_{\alpha+1}, p_0) \Vdash$"$\underset{\sim}{f}(\delta_\alpha) = g_{p_0}(\alpha)$" for some $g_{p_0}(\alpha) \in Z$. Then choose $p_1 \in P(<\lambda)$ and $q^1_{\alpha+1} \geq q^0_{\alpha+1}$ such that p_1 is incompatible with p_0 and $(q^1_{\alpha+1}, p_1) \Vdash$ "$\underset{\sim}{f}(\delta_\alpha) = g_{p_1}(\alpha)$" for some $g_{p_1}(\alpha) \in Z$. Note that by the choice of q we have that $(q, \emptyset)$ forces that $\underset{\sim}{f}$ is an embedding of the sort required, hence certainly (q^0_α, p_1) forces this. At limit ordinals γ, let $q^\gamma_{\alpha+1}$ be the $\mathrm{cub}(\langle q^\beta_{\alpha+1} : \beta < \gamma\rangle)$.

Continue choosing incompatible elements from $P(<\lambda)$ until a maximal antichain A_α is reached. Let $q_{\alpha+1} = \mathrm{cub}(\langle q^\beta_{\alpha+1} : \beta < \beta^*\rangle)$ for some $\beta^* < \lambda$, which is the length of the maximal antichain chosen.

By Lemma 3.6 all q_α are conditions in $Q(\lambda, \kappa(\lambda))$. Let $G(<\lambda)$ be $P(<\lambda)$-generic and for each α choose $p_\alpha \in G(<\lambda) \cap A_\alpha$. Then $\langle g_{p_\alpha}(\alpha) : \alpha < \kappa(\lambda)\rangle$ is a $\kappa(\lambda)$-chain through Z in $V[G(<\lambda)]$, a contradiction.

The non-Mahlo inaccessible and successor of a singular cases can be further reduced to two subcases. The first is when $\kappa(\lambda) < \lambda$. When λ is the successor of a singular, we have that $\kappa(\lambda)^{++} < \lambda$ as $\kappa(\lambda)$ is regular and $\lambda = \mu^+$ for some singular limit cardinal μ and thus $\kappa(\lambda)$ must be less than μ. If λ is inaccessible, then it is a limit cardinal, so $\kappa(\lambda)^{++} < \lambda$.

Thus, the forcing can be split somewhere between $\kappa(\lambda)$ and λ, say at $\kappa(\lambda)^{++}$. Since $\kappa(\lambda)^{++}$ is the successor of a regular cardinal, $P(< \kappa(\lambda)^{++})$ is $\kappa(\lambda)^{++}$-cc and $P(\geq \kappa(\lambda)^{++})$ is $\kappa(\lambda)^{+++}$-distributive. We may proceed as above since the conditions q_α only need to be defined for $\alpha < \kappa(\lambda)$.

If λ is the successor of a singular or a non-Mahlo inaccessible and $\kappa(\lambda) = \lambda$, then we cannot cut the forcing below λ and $P(< \lambda)$ only has the λ^+-cc. We would still like to build the conditions q_α together with a maximal antichain on $P(< \lambda)$. The idea here is to dovetail the building of conditions so that at each stage α, the value for $g_p(\alpha)$ is decided for $< \lambda$ many p in a maximal antichain and the rest of the values are decided at later stages. This in itself would be easy enough, but we want this process not only to have size λ, but also order type λ, as we only have weak closure under sequences of length $< \lambda$.

To this end, we take an elementary submodel M of some large $H(\theta) = L_\theta$ which has size λ, includes all relevant information and is closed under $< \lambda$-sequences. Enumerate the elements of M in order type λ. So we can list all the elements of $P(< \lambda) \cap M$ as $\langle p_i : i < \lambda \rangle$.

Assume that there exists $q \in Q(\lambda, \lambda)$ such that $q \Vdash$ "$\underset{\sim}{f} : \underset{\sim}{Y} \to Z$ is an embedding of $\underset{\sim}{Y}$ into some ground model poset Z of size λ which omits λ-chains". We may assume that $q \in M$.

By induction on $\alpha < \lambda$, we will construct an increasing sequence $\langle q_\alpha = (\delta_\alpha, X_\alpha, \mathcal{A}_\alpha) : \alpha < \lambda \rangle$ which, together with a maximal antichain of conditions from $P(< \lambda)$, decides the values of the embedding. The entire construction will take place inside of M. In order to accomplish the dovetailing process, let $\alpha = \langle \alpha_0, \alpha_1 \rangle$ be a pairing function on ordinals $< \lambda$.

Let $q_0 = q$ and at limit stages α, build q_α in the same way as in the case where λ is the successor of a regular cardinal, including building the intermediate condition q'_α.

For $\alpha + 1$, we will build the condition $q^i_{\alpha+1}$ by induction on $i \leq \alpha$, and set $q_{\alpha+1} = q^\alpha_{\alpha+1}$. For $i = 0$ set $q^0_{\alpha+1} = \emptyset$. At successor stages $i + 1$, extend p_i and $q^i_{\alpha+1}$ inside M to p'_{i+1} and $q^{i+1}_{\alpha+1}$ such that $(p'_{i+1}, q^{i+1}_{\alpha+1}) \Vdash$ "$\underset{\sim}{f}(\delta_{\alpha_0}) = g_{p'_{i+1}}(\alpha_0)$"

for some $g_{p'_{i+1}}(\alpha_0) \in Z$.

At limit stages i, let $q^i_{\alpha+1}$ be the cub of $\langle q^j_{\alpha+1} : j < i \rangle$.

Now we note that the conditions in M are predense in $P(< \lambda)$: Let $p^* = \langle r_\beta : \beta < \lambda \rangle$ be a condition in $P(< \lambda)$ (not necessarily in M), where we assume $r_\beta \in P(\beta)$. Let $p^* \restriction M = \langle r_\beta \restriction M : \beta \in M \cap \lambda \rangle$ where $r_\beta \restriction M = \langle s_\gamma \restriction M : \gamma \in \mathrm{supp}(r_\beta) \cap M \rangle$ and

$$\begin{aligned} s_\gamma \restriction M &= (\delta_\gamma \cap M, X_\gamma \restriction M, \mathcal{A}_\gamma \restriction M) \\ &= (\delta_\gamma, X_\gamma, \mathcal{A}_\gamma), \end{aligned}$$

well-defined as M is closed under $< \lambda$ sequences. Thus $p^* \restriction M \in P(< \lambda) \cap M$, and clearly $p^* \restriction M$ is compatible with p^*. By the same argument, any predense subset of $P(< \lambda) \cap M$ is predense in $P(< \lambda)$.

Now choose some generic G for $P(< \lambda)$. Let D_{α_0} be the set of all $p' \in M \cap P(< \lambda)$ such that (p', q_α) forces a value for $\underset{\sim}{f}(\delta_{\alpha_0})$ for some $\alpha < \lambda$. Then D_{α_0} is predense in $P(< \lambda)$ for all $\alpha_0 < \lambda$. Thus, for each α_0 we may choose $p'_{\alpha_0} \in G(< \lambda) \cap D_{\alpha_0}$. However, we have $\langle g_{p'_{\alpha_0}}(\alpha_0) : \alpha_0 < \lambda \rangle$ is a λ-chain through Z in $L[G(< \lambda)]$, a contradiction. □

We have now exhibited a poset of size λ which omits $\kappa(\lambda)$-chains in the extension which does not embed into any such poset in the ground model. This shows that the complexity of $\mathcal{C}(\lambda, \kappa(\lambda))$ is $\geq F(\lambda)$. It now remains to be seen that $2^\lambda = F(\lambda)$ in the extension. It was shown in Section 3 that in each $M^{P(\lambda, F(\lambda), \kappa(\lambda))}$ the size of 2^λ is exactly $F(\lambda)$. Since in each forcing $P(\lambda, F(\lambda), \kappa(\lambda))$ no new subsets of λ were added, this fact remains true in M^P. □

We can get a similar global result for the low complexity given in [3]. Even though the forcing at each regular cardinal is an iteration, the global forcing will still be in the form of a product.

Theorem 4.2. *Assume GCH. To each uncountable regular cardinal λ definably associate cardinals $F(\lambda), \kappa(\lambda), \nu(\lambda)$ such that $\kappa(\lambda) \leq \lambda$ is regular, $\nu(\lambda) \in [\lambda^+, F(\lambda)]$ and cf$(F(\lambda))$, cf$(\nu(\lambda)) > \lambda$ and $\lambda < \theta$ implies $F(\lambda) \leq F(\theta)$. Then there exists a ZFC-preserving, cofinality preserving class forcing notion*

P such that in V^P the strong complexity of graphs of size λ which omit cliques of size $\kappa(\lambda)$ is $\nu(\lambda)$ and $F(\lambda) = 2^\lambda$ for each regular uncountable λ.

PROOF – Let $P(\lambda, \kappa(\lambda), \nu(\lambda), F(\lambda))$ be the iterated forcing notion of length $\nu(\lambda)$ in [3, Theorem 4]. Namely, this is the forcing which adds a universal family of size $\nu(\lambda)$ for graphs of size λ omitting $\kappa(\lambda)$-cliques where 2^λ is forced to be $F(\lambda)$. The forcing notion P will be the Easton support product of P_α for all $\alpha \in \mathrm{Ord}$ such that $P_\alpha = P(\alpha, \kappa(\alpha), \nu(\alpha), F(\alpha))$ if α is a regular uncountable cardinal and trivial otherwise.

Each forcing $P(\lambda, \kappa(\lambda), \nu(\lambda), F(\lambda))$ is weakly λ-closed and satisfies the λ^+-cc.

By Claim 2.2, $P(\geq \lambda)$ is weakly λ-closed for λ regular uncountable. Also, if λ is a successor of a regular or Mahlo then $P(< \lambda)$ has the λ-cc.

The only thing left to check is that we have not changed the complexity at each λ when doing the smaller forcing. This proceeds exactly like the proof of Theorem 4.1. □

References

[1] JECH, T.: *Set Theory*, Springer Monographs in Mathematics, Springer-Verlag, Berlin Heidelberg New York (2002), 3rd millennium edition.

[2] JENSEN, R.B.: The finite structure of the constructible hierarchy, *Annals of Mathematical Logic* **4** (1972), 229–308.

[3] KOMJÁTH, P. and SHELAH, S.: Universal graphs without large cliques, *Journal of Combinatorial Theory* **63** (1) (1995), 125–135.

[4] KUNEN, K.: *Set Theory*, Studies in Logic and the Foundations of Mathematics **102**, North-Holland Publishing Co., Amsterdam (1980).

Absoluteness for Universally Baire Sets and the Uncountable I

Ilijas Farah and Paul B. Larson

Contents

1. Introduction

Cantor's Continuum Hypothesis was proved to be independent from the usual ZFC axioms of Set Theory by Gödel and Cohen. The method of forcing, developed by Cohen to this end, has lead to a profusion of independence results in the following decades. Many other statements about infinite sets, such as the Borel Conjecture, Whitehead's problem, and automatic continuity for Banach Algebras, were proved independent, perhaps leaving an impression that most nontrivial statements about infinite sets can be neither proved nor refuted in ZFC.[1] Moreover, some classical statements imply the consistency of ZFC and stronger theories, and by Gödel's incompleteness theorems the consistency of these statements with ZFC can be proved only by using strong axioms of infinity, so-called **large cardinal axioms**. A classical example is Banach's 'Lebesgue measure has a σ-additive extension to all sets of reals.' While it is fairly easy to find a model in which this is false and there is no known ZFC-proof of its negation, proving the consistency of this statement requires assuming the existence of a **measurable cardinal** ([33]).

A remarkable result was proved by Shoenfield ([32]): every statement of the form $(\exists x \in \mathbb{R})(\forall y \in \mathbb{R})\phi(x, y)$, where all quantification in ϕ is over the natural numbers and all of its parameters are real numbers, is **absolute** between models of ZFC that are transitive and contain all countable ordinals. In this form Shoenfield's theorem is best possible, as it cannot even be improved by adding one more alternation of quantifiers ranging over $\mathbb{R}$. However, a corollary that the truth of any $\mathbf{\Sigma}^1_2$ statement (i.e., one of the above syntactical form) cannot be changed by forcing turned out to be susceptible to far-reaching generalizations. One of the more striking results in modern set theory is that the existence of suitable large cardinals implies that the theory of the inner model $L(\mathbb{R})$ (the smallest inner model of ZF, the usual axioms of Set Theory without the Axiom of Choice, containing all real numbers) cannot be changed by set forcing (see

*The first author is partially supported by NSERC. The second author was supported in part by NSF grant DMS-0401603. Part of his research was conducted at the Institute for Mathematical Sciences at the National University of Singapore, under the partial support of a summer research grant from Miami University.

[1] Shelah's pcf-theory witnesses that this impression is false even in the realm of cardinal arithmetic, a subject that was considered essentially closed after Easton's work in 1970.

[13, 21]). In particular, a sentence with real parameters and any number of alterations of quantifiers ranging over $\mathbb{R}$ has a fixed truth value that cannot be changed by forcing. The impact of large cardinals on sets of reals goes well beyond $L(\mathbb{R})$ to imply absoluteness for certain canonical sets of reals, the universally Baire sets ([11]), as defined below. A remarkable consequence is that the existence of large cardinals outright implies that all sets of reals in $L(\mathbb{R})$, and indeed all universally Baire sets, share all the classical regularity properties of Borel sets such as Lebesgue measurability.

The problem of precise calibration of the large cardinal axioms required to prove absoluteness for formulas of a given complexity is currently a very active field of research. In this paper we are taking a different line by proving forcing-absoluteness of sentences that are not statements of $L(\mathbb{R})$.

Forcing-absoluteness from large cardinals does not extend to $L(\mathcal{P}(\omega_1))$ (the smallest inner model of ZF containing all subsets of the first uncountable ordinal, ω_1) or to the closely related set-model $H(\aleph_2)$ (the set of all sets whose transitive closure is of cardinality not greater than $\aleph_1$), however. Theories of these models are highly susceptible to forcing. For example, the Axiom of Choice implies that the Continuum Hypothesis can be expressed as 'there exists a well-order of all reals all of whose proper initial segments are countable,' a statement of these structures. Large cardinals imply that this reformulation of CH is false in $L(\mathbb{R})$, since such a well-order provides a non-Lebesgue measurable subset of the plane, while Cantor's original reformulation, that every infinite set of reals is either countable or equinumerous with $\mathbb{R}$, holds in $L(\mathbb{R})$. Nevertheless, large cardinals do influence the theory of the uncountable, even in contexts free of metamathematics. Todorcevic ([38]) has shown that the existence of class many Woodin cardinals influences the structure of Banach spaces satisfying Namioka's 'generic continuity principle.' Todorcevic's conclusion was previously shown to fail in L by Namioka and Pol.

Also, a number of highly nontrivial objects in $H(\aleph_2)$ have been constructed without the aid of additional set-theoretic axioms: from gaps in $\mathcal{P}(\omega)/\mathrm{Fin}$ of Hausdorff and Luzin, special Aronszajn trees, to the more recent negative partition relations ([36]) and L-spaces ([27]). Statements asserting the existence of these objects are absolute simply because they are true. Lévy's absoluteness theorem ([23]) states that all Σ_1-statements, ones of the form $(\exists X)\phi(X)$ where all quantification in ϕ is bounded, are absolute between models of ZFC that are transitive, even allowing for parameters for real numbers. One consequence is that every Σ_1-statement true in the set-theoretic universe is already witnessed by a set in $H(\aleph_1)$ (the set of all sets whose transitive closure is at most countable), thus Σ_1-statements reflect to $L(\mathbb{R})$. An absoluteness result that is genuinely about ω_1 is a consequence of Keisler's completeness theorem for the logic $L_{\omega_1\omega}(Q)$ ([19]). It has been used to prove a couple of theorems for which no proof free of metamathematics is known. Statements susceptible to Keisler's theorem are Σ_1-statements of $H(\aleph_2)$ expressed in an extension of the first-order logic that allows countably infinite conjunctions and has the quantifier 'there exists uncountably many.' These are not consequences of Lévy's theorem because the witness to the Σ_1 statement in question can be required to be uncountable. Describing the exact syntactical form of these statements is beyond this introduction (see Section 6). Keisler's theorem allows ϕ to contain predicates for arbitrary Borel sets, and a desire to add arbitrary universally Baire predicates to these absoluteness results was the starting point of our investigation. As an application, we prove that for every suitably definable partition of n-tuples of reals the statement 'there exists an uncountable homogeneous set' is forcing-absolute (Proposition 8.3) and therefore all suitably defined ccc forcings are productively ccc.

While the majority of absoluteness results depend on the syntactical form of the sentence in question, we also prove a forcing-absoluteness conditioned on a semantical assumption. Statements $(\exists X)\phi(X)$ of $H(\aleph_2)$ for which a witness $\bar{X}$ can be forced so that it is indestructible, in the sense that $\phi(\bar{X})$ holds in all further extensions in which ω_1 is not collapsed, are forcing absolute (Theorem 3.2). Such an indestructibility feature is built into classical objects of $H(\aleph_2)$ such as Hausdorff and Luzin gaps, as well as special Aronszajn trees.

Our results can be expressed in the following form: if a sufficiently large rank initial segment of the universe of sets can be forced to satisfy a sentence ϕ, then there already exists a model of ϕ which has certain correctness properties. The properties we consider in this paper are the following.

1. Containing any given $\aleph_1$ many reals.

2. Correctness about stationarity for subsets of ω_1.

3. Correctness about any given universally Baire set of reals.

The corresponding result for (1) above is essentially Woodin's Σ^2_1-absoluteness result from CH ([41], see also [21]). After completing this work we have learned that the results for parts (2) and (3) were previously known to Woodin as well. In a sequel to this paper [10] we will give some additional properties.

Let us say a word about the proofs. A Σ^2_1-*statement* is one of the form $(\exists A \subseteq \mathbb{R})\phi(A)$, where all quantifiers in ϕ range over real numbers and ϕ can have reals as parameters. Note that CH is an example of a Σ^2_1 statement. Assuming suitable large cardinals (stronger than the ones used for forcing-absoluteness of the theory of $L(\mathbb{R})$), all Σ^2_1 statements are absolute between forcing extensions satisfying CH. Moreover, every Σ^2_1 statement that holds in some forcing extension (even one in which CH is false) holds in every forcing extension satisfying CH. One of the methods used in this note is extracted from Woodin's proof; without assuming CH and applying Woodin's construction we show that certain Σ_1 statements true in the extension also hold in the ground model. We also use this method to prove an extension of Zapletal's results on isolating cardinal invariants [43], originally proved by using the Σ^2_1 absoluteness theorem. Woodin's work on the partial order $\mathbb{P}_{max}$ ([42]), and in particular the notion of an *iterable model* is used to show similar absoluteness results from a more modest large cardinal assumptions.

Organization of the paper

In Section 2 we introduce universally Baire sets and present a corollary to Woodin's construction of $\mathbb{P}_{\max}$. In Section 3 we use iterable structures to prove Σ_1-absoluteness for $H(\aleph_2)$. In Section 4 we present a proof of absoluteness for universally Baire sets between generic ultrapowers, due to John Steel. In Section 5 the proof of Σ^2_1 absoluteness is modified to give Σ_1 absoluteness for $H(\aleph_2)$ for models containing any $\aleph_1$ many reals. In Section 6 we enrich Keisler's theorem by adding predicates for universally Baire sets of reals and the nonstationary ideal. Zapletal's results on isolating cardinal invariants are extended in Section 7. Applications are given in Section 8.1 and in Section 8.2

we prove that some (comparably weak) large cardinal assumptions are necessary for some of our results. A related result of John Steel is briefly discussed in Section 8.3.

Terminology

We are using the standard set-theoretic terminology. Definitions of more advanced notions such as Woodin cardinals, $\mathbb{P}_{\max}$, or the stationary towers $\mathbb{P}_{<\delta}$ and $\mathbb{Q}_{<\delta}$ can be found e.g., in [42] or [21]. For the general theory of large cardinals see also [17].

2. Universally Baire sets and $\mathbb{P}_{\max}$

We consider the reals $\mathbb{R}$ to be ω^ω, the set of all ω-sequences of elements of ω equipped with the product topology. If γ is an ordinal, a tree on $\omega \times \gamma$ is a subset of $(\omega \times \gamma)^{<\omega}$ closed under initial segments. If T is a such a tree, the projection $p[T]$ of T is the set of $x \in \omega^\omega$ for which there exists a $z \in \gamma^\omega$ such that $(x \restriction n, z \restriction n)$ is in T for each $n \in \omega$.

For a given cardinal κ, a set of reals A is κ-universally Baire if there exist trees T and S such that $p[T] = A$, $p[S] = \omega^\omega \setminus A$ and T and S project to complements in every forcing extension by a partial order of cardinality κ (see [11]). A set is universally Baire if it is κ-universally Baire for each cardinal κ. A universally Baire set of reals then has a natural reinterpretation in every set forcing extension, the projection of T, for any T as above (it is a standard fact that this does not depend on the choice of T). This allows us to talk about properties of the set in these forcing extensions. If A is universally Baire, P is a partial order and $G \subset P$ is a V-generic filter, we denote the reinterpretation of A in $V[G]$ as A_G or A^{V^P}.

Universally Baire sets of reals, introduced in [11], tend to satisfy the standard regularity properties. For example, in the presence of a proper class of Woodin cardinals the model $L(A, \mathbb{R})$ satisfies AD^+ whenever A is universally Baire ([42]). More specifically, in the presence of a proper class of Woodin cardinals universally Baire sets are ∞-homogeneously Suslin (see [26, 35, 21]), a property from which the usual regularity properties are more easily derived.

We let Γ_{uB} denote the set of universally Baire sets of reals, and let NS_{ω_1} denote the nonstationary ideal on ω_1.

In [42] Woodin introduces the partial order $\mathbb{P}_{\max}$ and proves the following theorem (see also [22]).

Theorem 2.1. *Suppose that there exist proper class many Woodin cardinals, and let A be a universally Baire set of reals. Let ϕ be a Π_2 formula in the language of set theory with additional unary predicate symbols for NS_{ω_1} and A. Then if in some set forcing extension the structure*

$$\langle H(\aleph_2); \mathrm{NS}_{\omega_1}, A^*, \in\rangle$$

satisfies ϕ, where A^* is the induced reinterpretation of A, then the structure

$$\langle H(\aleph_2); \mathrm{NS}_{\omega_1}, A, \in\rangle$$

satisfies ϕ in the $\mathbb{P}_{\max}$ extension of $L(A, \mathbb{R})$.

The conclusion of Theorem 2.1 can be interpreted as saying that $L(A, \mathbb{R})$ satisfies the statement that the structure $\langle H(\aleph_2); \mathrm{NS}_{\omega_1}, A, \in\rangle$ of its $\mathbb{P}_{\max}$ extension satisfies ϕ. In particular, the conclusion is still meaningful if no $L(\mathbb{R})$-generic filter for $\mathbb{P}_{\max}$ exists. Since $\mathbb{P}_{\max}$ adds no reals, it is not necessary to reinterpret A in the conclusion of the theorem (i.e., we can write A instead of A^*).

Theorem 2.1 has the following immediate corollary.

Corollary 2.1. *Suppose that there exist proper class many Woodin cardinals, and let A be a universally Baire set of reals. Let ϕ be a sentence in the language of set theory with two additional unary predicate symbols for NS_{ω_1} and A, and suppose that ϕ and $\neg\phi$ are provably equivalent to Π_2 sentences over $H(\aleph_2)$ in the same expanded language. Then either ϕ holds in all set forcing extension or ϕ fails in all forcing extensions.*

Proof – Otherwise ϕ and $\neg\phi$ both hold in the $\mathbb{P}_{\max}$ extension of $L(A, \mathbb{R})$, by Theorem 2.1. □

The results mentioned here hold for the language with a symbol for each universally Baire set, since each sentence can have only finitely many such symbols, and finitely many (indeed, countably many) universally Baire sets are easily combined into one.

3. Σ_1-absoluteness for $H(\aleph_2)$

If ϕ is a Boolean combination of Σ_1 sentences, then ϕ and $\neg\phi$ are clearly both Π_2. It follows from Corollary 2.1 that in the presence of suitable large cardinals, the truth value of such ϕ cannot be changed by set forcing, even if ϕ contains additional predicates for NS_{ω_1} and any given universally Baire set. Absoluteness for Boolean combinations of Σ_1 sentences follows immediately from absoluteness for Σ_1 sentences, and, in fact, the only meaningful sentences ϕ we know of for $H(\aleph_2)$ with the property that ϕ and $\neg\phi$ are provably equivalent to Π_2 sentences are such Boolean combinations (note that while CH can be expressed as a Π_2 sentence in V, it cannot be so expressed in $H(\aleph_2)$). While the proof of Theorem 2.1 is too long to present here (and in any case is presented in full in [42, 22]), we can present a self-contained proof of Σ_1-absoluteness from a slightly weaker large cardinal assumption.

Theorem 3.1. *Suppose that there exist two Woodin cardinals $\delta < \lambda$, and let A be a λ^+-universally Baire set of reals. Let ϕ be a Σ_1 sentence in the expanded language with two additional unary predicates. Let P be a partial order in V_δ. Then ϕ holds in the structure*

$$\langle H(\aleph_2); \mathrm{NS}_{\omega_1}, A, \in\rangle$$

if and only if ϕ holds in

$$\langle H(\aleph_2)^{V^P}; \mathrm{NS}_{\omega_1}^{V^P}, A^{V^P}, \in\rangle,$$

where A^{V^P} is the induced reinterpretation of A in the P-extension.

Proof – This will follow immediately from Theorem 3.4 proved below. □

Note the similarity with Levy's absoluteness theorem. The similar-looking assertion that $H(\aleph_2) \prec_{\Sigma_1} H(\aleph_2)^P$ is much stronger (it outright implies that P preserves $\aleph_1$). By [3] (also see [34]), familiar bounded forcing axioms such as

MA or BPFA are equivalent to the assertion that $H(\aleph_2) \prec_{\Sigma_1} H(\aleph_2)^P$ for all P in the corresponding class of forcings. Reflection principles for $H(2^{\aleph_0})$ in relation to fragments of second order logic were studied already in the early 1980s, in a paper that has appeared only recently ([34]). Note that Theorem 3.1 cannot be extended to $H(\aleph_3)$ since the non-absolute '$2^{\aleph_0} \geq \aleph_2$' is a Σ_1 statement in this structure.

One way to state the idea behind the proof of Theorem 3.1 is that any sentence (with a parameter for some universally Baire set of reals) that can be forced to hold holds in some $\mathbb{P}_{\max}$ condition (Lemma 3.3) which can then be embedded elementarily into an uncountable structure which is correct about NS_{ω_1} and the universally Baire set A. Almost all of the details of the proof appear in [42, 22], in proving that certain types of $\mathbb{P}_{\max}$ conditions exist. This proof gives another result of somewhat different nature, in which the assumption on the absolute formula ϕ is semantical, rather than syntactical.

3.1. Indestructible witnesses

For certain formulas ϕ there exists a formula ψ (possibly ϕ itself) such that one can prove that whenever a set X satisfies ψ in a model N, X satisfies ϕ in any model $M \supseteq N$ such that $\mathrm{NS}_{\omega_1}^M \cap N = \mathrm{NS}_{\omega_1}^N$. In this case we say that X is an indestructible witness for ϕ. This property is at least as strong as continuing to be a witness in all NS_{ω_1}-preserving forcing extensions. Theorems proving that X is an indestructible witness essentially show that that the first-order properties of set X imply its (possibly second-order) property asserted by ϕ. Typical indestructible witnesses are Hausdorff/Luzin gaps and special Aronszajn trees. In the first case, 'zapping' between two orthogonal ω_1 towers in $\mathcal{P}(\omega)/\mathrm{Fin}$ assures that they cannot be separated by a single subset of ω (see [37, Theorem 8.6]). In the second case, an ω_1-tree is the union of countably many antichains, hence every cofinal branch defines a map from ω onto the height of the branch. Other standard constructions of objects of $H(\aleph_2)$ usually yield indestructible witnesses ([36], [27]).

Theorem 3.2. *If $(\exists x)\phi(x)$ is a formula of $H(\aleph_2)$ such that some forcing of size less than a Woodin cardinal forces an indestructible witness for it, then $(\exists x)\phi(x)$ holds in $H(\aleph_2)$.*

PROOF – Readers interested only in the proof of this theorem can skip all references to universally Baire and weakly homogeneously Suslin sets and read only the beginning of Section 3.3 and the relevant part of the proof of Lemma 3.3. By Lemma 3.3 (with $A = \emptyset$) there is an iterable model $(N, \mathrm{NS}_{\omega_1}^N)$ and $X \in H(\aleph_2)^N$ such that $N \models \phi(X)$ and X is indestructible. Let $j\colon (N, \mathrm{NS}_{\omega_1}^N) \to (N^*, \mathrm{NS}_{\omega_1}^{N^*})$ be an iteration of length ω_1 such that $NS_{\omega_1} \cap N^* = \mathrm{NS}_{\omega_1}^{N^*}$. Then $Y = j(X)$ is an indestructible witness of ϕ in $H(\aleph_2)^{N^*} = H(\aleph_2) \cap N^*$ and therefore $H(\aleph_2) \models \phi(Y)$. □

Note that an anti-large cardinal assumption such as V=L implies the conclusion of Theorem 3.2 fails. One can strengthen Theorem 3.2 in an obvious way, by allowing parameters for universally Baire sets, adding an extra Woodin cardinal and appropriately weakening the indestructibility requirement. An axiom resembling the statement of Theorem 3.2 was studied by Hamkins in [15].

3.2. Weakly homogeneously Suslin sets

We refer the reader to [17, p. 453$^+$], [42, pages 22–23], [35, Section 1], or [21, Definitions 1.3.4 and 1.3.6] for the definition of "weakly homogeneously Suslin tree/set of reals." For every λ-weakly homogeneously Suslin tree T there exists a canonical tree S on $\omega \times (2^{|T|})^+$ such that $p[S] = \omega^\omega \setminus p[T]$ (see [21, p. 23]). This is the so-called Martin–Solovay tree (see [35, Theorem 1.20]). These trees S and T project to complements in every forcing extension by a partial order of cardinality less than λ, and therefore λ-weakly homogeneously Suslin sets are λ-universally Baire. Weakly homogeneously Suslin sets will figure prominently in Section 4. The following is due to Woodin.

Theorem 3.3. *Suppose that λ is a Woodin cardinal, and that A is a λ^+-universally Baire set of reals. Then A and $\omega^\omega \setminus A$ are both $< \lambda$-weakly homogeneously Suslin.*

PROOF – This follows immediately from [21, Theorem 3.3.8]. □

The argument in the proof of the following is standard, and the theorem itself is a variation of standard facts. Since this exact statement does not seem to formally follow from the published results, we include the proof for the convenience of the reader.

Lemma 3.1. *Let θ be a regular cardinal, suppose that T is a weakly homogeneous tree on $\omega \times Z$ in $H(\theta)$, for some set Z. Let $\gamma \geq 2^\omega$ be an ordinal such that there exists a countable collection Σ of γ^+-complete measures witnessing the weak homogeneity of T. Assume that there is a measurable cardinal in the interval (γ, θ).*

Then for every elementary submodel X of $H(\theta)$ of cardinality less than γ with T, Σ, $\gamma \in X$, there is an elementary submodel Y of $H(\theta)$ containing X such that $Y \cap \theta$ is uncountable, $Y \cap \gamma = X \cap \gamma$ and $p[T \cap Y] = p[T]$.

Proof – Fixing θ, T, Σ and γ as in the statement of the theorem, the theorem follows from the following fact. Suppose that W is an elementary submodel of $H(\theta)$ of cardinality less than γ. Fix $x \in p[T]$ and a countably complete tower $\{\sigma_i : i < \omega\} \subset \Sigma$ such that for all $i < \omega$, $\{a \in Z^i : (x \restriction i, a) \in T\} \in \sigma_i$, and for each $i < \omega$, let $A_i = \bigcap(\sigma_i \cap W)$. Then since $\{\sigma_i : i < \omega\}$ is countably complete, there exists a $z \in Z^\omega$ such that for all $i < \omega$, $z \restriction i \in A_i$. Then the pair (x, z) forms a path through T, and, letting

$$W[z] = \{f(z \restriction i) \mid i < \omega \wedge f\colon Z^i \to H(\theta) \wedge f \in W\},$$

$W[z]$ is an elementary submodel of $H(\theta)$ containing W and $\{z \restriction i : i < \omega\}$, and, since each σ_i is γ^+-complete, $W \cap \gamma = W[z] \cap \gamma$.

Repeated application of this fact (starting with X) for each real in the projection of T produces a Y containing X such that $Y \cap \gamma = X \cap \gamma$, and $p[T \cap Y] = p[T]$. Furthermore, Y will have cardinality less than γ.

If $Y \cap \theta$ happens to be countable, then fix a measurable cardinal κ in Y greater than γ and a normal measure μ on κ. Then letting η be any member of $\bigcap(\mu \cap Y)$,

$$Y[\eta] = \{f(\eta) \mid f\colon \gamma \to H(\theta) \wedge f \in Y\}$$

is an elementary submodel of $H(\theta)$ end-extending Y below κ. Iterating this construction ω_1 times produces an elementary submodel Y' end-extending Y below κ with $Y' \cap \theta$ uncountable. □

The last paragraph of the preceding proof is needed only when the projection of T is analytic. Note also that the theorem implies an ostensibly stronger version where X is expanded to a Y capturing the projection of two weakly homogeneous trees in X, which in any case would be proved directly in the same way.

3.3. Iterable models

The proof of Theorem 3.4 will also use the following notion. ZFC^0 is a fragment of ZFC that holds in $H(\theta)$ for a regular $\theta \geq \aleph_2$ (the precise definition will not be needed here; see [22, Section 1]). If N_0 is a transitive model of ZFC^0 and I_0 is a normal ideal on $\omega_1^{N_0}$ in N_0, then an iteration of (N_0, I_0) of length γ is $\langle (N_\eta, I_\eta), j_{\xi\eta}, G_\eta, \xi < \eta \leq \gamma \rangle$, where $j_{\xi\eta} \colon (N_\xi, I_\xi) \to (N_\eta, I_\eta)$ is a commuting family of elementary embeddings, $G_\eta \subseteq (\mathcal{P}(\omega_1)/I_\eta)^{N_\eta}$ is a generic filter, $j_{\eta\eta+1}$ is the corresponding generic ultrapower embedding, and for a limit η and $\xi < \eta$, $j_{\xi\eta}$ and N_η are the direct limit of $j_{\xi\zeta}$ and N_ζ for $\xi < \zeta < \eta$. An iteration is well-founded if all the models occurring in it are well-founded. A pair is iterable if all of its iterations are well-founded. If $A \in N_0$ is a universally Baire set then a pair (N_0, I_0) is A-iterable if it is iterable and its iterations compute A correctly. We shall follow the standard convention and identify an iteration of length γ with the final model together with the embedding $j_{0\gamma} \colon (N_0, I_0) \to (N_\gamma, I_\gamma)$. For more information we refer the reader to [42, 22]. The term 'iterable model' is also used to describe inner models of ZFC equipped with strategies for iterating extender-based ultrapowers. This notion will not be used in the present paper.

We will also use the following fact, due to Woodin (see [12, 14, 31, 42]). For our purposes, it is only important that presaturated ideals are precipitous: the corresponding generic ultrapower is forced to be well-founded.

Lemma 3.2. *If δ is a Woodin cardinal, then after forcing with* $\mathrm{Coll}(\omega_1, {<}\delta)$ *the nonstationary ideal on ω_1 is presaturated.*

Note that the following lemma also applies in the case when ϕ has additional predicates for NS_{ω_1} and a universally Baire set. The phrase "ϕ is preserved by forcings of the form $\mathrm{Coll}(\omega_1, \gamma)$" in the statement of Lemma 3.3 means that every forcing extension of a model M of $ZFC + \phi$ by $\mathrm{Coll}(\omega_1, \gamma)^M$ for some ordinal $\gamma \in M$ satisfies ϕ.

Lemma 3.3. *Assume $\delta < \lambda$ are a Woodin and a measurable cardinal, A and $\omega^\omega \setminus A$ are δ^+-weakly homogeneously Suslin set of reals, and ϕ is a sentence* *whose truth is preserved by forcings of the form* $\mathrm{Coll}(\omega_1, \gamma)$*. If* P *is a partial order in* V_δ *that forces* ϕ*, then there exists an* A*-iterable model* $(N, \mathrm{NS}^N_{\omega_1})$ *that satisfies* ϕ*.*

Proof – Let θ be a regular cardinal greater than δ, and let X be a countable elementary submodel of $H(\theta)$ with δ, λ, A and P in X. For some fixed $\gamma \in X \setminus \delta$ there are countable sets of γ^+-complete measures in X witnessing that A and $\omega^\omega \setminus A$ are weakly homogeneously Suslin. By Lemma 3.1 there is an elementary submodel Y of $H(\theta)$ such that

- $X \subset Y$,
- $X \cap \gamma = Y \cap \gamma$,
- $Y \cap \theta$ is uncountable,
- there exist trees S, T on $\omega \times \eta$ in X, for some ordinal $\eta \in \lambda \cap X$, such that $p[S \cap Y] = A$ and $p[T \cap Y] = \omega^\omega \setminus A$.

Now let M be the transitive collapse of Y, and let $\bar{P}$, $\bar{A}$, $\bar{S}$, $\bar{T}$, $\bar{\gamma}$ be the respective images of P, A, S, T and γ under this collapse. Noting that $V^M_{\bar{\gamma}}$ is countable, let N be a forcing extension of M by the forcing $\bar{P} * \mathrm{Coll}(\omega_1, {<}\bar{\delta})$ as defined in M. Let $A_N = p[\bar{S}]$. Since $\bar{P}$ forces $\phi(A_N)$, N satisfies ϕ. By Lemma 3.2 the nonstationary ideal in N is precipitous. Furthermore, $\mathcal{P}(\mathcal{P}(\omega_1))^N$ is countable. Since $N \cap Ord$ is uncountable, [22, Lemma 1.5] implies $(N, \mathrm{NS}^N_{\omega_1})$ is iterable. Let $j\colon (N, \mathrm{NS}^N_{\omega_1}) \to (N^*, \mathrm{NS}^{N^*}_{\omega_1})$ be an iteration. The trees $\bar{S}$ and $\bar{T}$ project to complements in N (in V, in fact), and, from the point of view of V, $p[\bar{S}] \subset p[j(\bar{S})]$ and $p[\bar{T}] \subset p[j(\bar{T})]$. Since $N^* \models p[j(\bar{S})] \cap p[j(\bar{T})] = \emptyset$, and this fact is absolute, it follows that $j(A_N) = p[j(\bar{S})] \cap N^* = A \cap N^*$ and $j((\omega^\omega)^N \setminus A_N) = p[j(\bar{T})] \cap N^* = (\omega^\omega \setminus A) \cap N^*$. Therefore, N^* is correct about A, and we have proved $(N, \mathrm{NS}^N_{\omega_1})$ is A-iterable. □

It only remains to prove that the iterable structures are robust under small forcing.

Lemma 3.4. *Let* δ *be a cardinal and suppose that* A *is* δ*-universally Baire. If a pair* (N, I) *is* A*-iterable then it is* A*-iterable in any forcing extension by a partial order* P *in* V_δ*.*

PROOF – Iterability is a Π^1_2 property and it is therefore absolute (see [22, Remark 1.3]). Let S and T be trees witnessing the δ-universally Baireness of A, with $p[S] = A$. Since every real appearing in any iteration of length ω_1 appears in some countable initial segment of the iteration, it suffices to consider iterations which have countable length in some extension by a forcing in V_δ. Fix a well-founded iteration $j\colon (N, I) \to (N^*, \mathrm{NS}_{\omega_1}^{N^*})$. We need to check that $p[S] \cap j(N \cap (\omega^\omega \setminus A)) = p[T] \cap j(N \cap A) = \emptyset$. The existence of a real $z \in p[S] \cap j(N \cap (\omega^\omega \setminus A))$ is witnessed by a path x through S with first coordinate projection equal to z, a real y coding an iteration of (N, I) (of possibly illfounded ordertype) and an embedding π from the ordertype of this iteration into δ (i.e., into the ω_1 of the extension). There is a natural tree (with S as its first two coordinates and functions from ω into δ as its last coordinate) such that any such triple z, y, π corresponds to a path through the tree. Since the wellfoundedness of this tree is absolute between wellfounded models, we conclude $p[S] \cap j(N \cap (\omega^\omega \setminus A)) = \emptyset$. The argument for the nonexistence of a real $z \in p[T] \cap j(N \cap A)$, for j some wellfounded iteration of (N, I), is identical. □

Theorem 3.4. *Suppose that there exist two Woodin cardinals $\delta < \lambda$, and let A be a λ^+-universally Baire set of reals. Let ϕ be a Σ_1 sentence in the expanded language with two additional unary predicates. Then the following are equivalent.*

1. *ϕ holds in the structure $\langle H(\aleph_2); \mathrm{NS}_{\omega_1}, A, \in\rangle$.*
2. *There is an A-iterable pair $(N, \mathrm{NS}_{\omega_1}^N)$ such that $N \models \phi$.*
3. *ϕ holds in $\langle H(\aleph_2)^{V^P}; \mathrm{NS}_{\omega_1}^{V^P}, A^{V^P}, \in\rangle$ for some partial order P in V_δ.*
4. *ϕ holds in $\langle H(\aleph_2)^{V^P}; \mathrm{NS}_{\omega_1}^{V^P}, A^{V^P}, \in\rangle$ for every partial order P in V_δ.*

PROOF – Clearly, (4) implies (1) and (1) implies (3), using the trivial forcing. Assume that (3) holds, and let $P \in V_\delta$ be such that every condition in P forces ϕ to hold. By Theorem 3.3, A and $\omega^\omega \setminus A$ are $<\lambda$-weakly homogeneous, and Lemma 3.3 implies (2).

Assume (2) holds, and let $(N, \mathrm{NS}_{\omega_1}^N)$ be an A-iterable model for ϕ. There is an iteration j of $(N, \mathrm{NS}_{\omega_1}^N)$ with final model N^* such that $\mathrm{NS}_{\omega_1}^{N^*} = N^* \cap \mathrm{NS}_{\omega_1}$

(this can be assured by a straightforward bookkeeping argument, see [42] or [22]). Since the iteration is correct about A, ω_1, and NS_{ω_1}, the witness for ϕ in N^* is a witness for ϕ in V, and therefore (1) holds.

Finally, assume (2). By Lemma 3.4, an A-iterable model in which ϕ holds remains A-iterable after forcing with any $P \in V_\delta$ and therefore we may apply the implication from (2) to (1) to conclude the proof of (4). □

The following weakening of (2): 'There is an A-iterable pair (N, I) such that $N \models \phi$' does not in general imply (1). Since in every ω_1-iteration $j\colon (N, I) \to (N^*, I^*)$ every set in I^* is disjoint from the critical sequence we have $\mathrm{NS}_{\omega_1} \cap N^* \supseteq I^*$. Therefore if ϕ states that $\mathrm{NS}_{\omega_1} \neq I$ then no iteration of (N, I) can be correct about NS_{ω_1}.

Remark 3.1 - Theorem 3.4 (and other results in this paper) have natural reinterpretations in terms of Woodin's Ω-logic (see [40]). Briefly, given suitable large cardinals, there is a universally Baire set of reals A such that if M is an A-closed model (in the sense of [40]) then in M every set is a member of an inner model containing a Woodin cardinal below a measurable cardinal κ (of M), such that this inner model is iterable (in V) by some measure on κ in M. Given this fact, the arguments in this section show that if ϕ is a Σ_1 statement for $\langle H(\aleph_2), \mathrm{NS}_{\omega_1}, B, \in\rangle$ for a universally Baire set B then there is a universally Baire set A such that ϕ holds in V if and only if it holds in some A-closed model. Such an A is a proof of ϕ in Ω-logic. Similar arguments are discussed in [22].

4. Weakly homogeneously Suslin sets and stationary tower embeddings

Reader's familiarity with stationary tower forcing $\mathbb{P}_{<\delta}$ and $\mathbb{Q}_{<\delta}$ as presented in [21] is assumed in this section. We want to extend the arguments in the previous section to arguments using the stationary tower. In this case, handling the predicates for universally Baire sets requires Theorem 4.1 below (see [21, Theorem 1.7.1 and Theorem 1.7.2] for more on this result, and Section 3.2 for the definitions).

Theorem 4.1 (Steel). *Assume that T is a λ-weakly homogeneously Suslin tree for some strongly inaccessible cardinal λ, and that P is $\mathbb{Q}_{<\delta}$ or $\mathbb{P}_{<\delta}$ for some strongly inaccessible $\delta < \lambda$. If S is the Martin–Solovay tree such that $p[S] = \omega^\omega \setminus p[T]$, then the generic embedding $k\colon V \to M$ derived from P satisfies $k(S) = S$.*

This is [21, Theorem 3.3.17] and the case when T is λ-homogeneous is proved in [35, Corollary 4.6(1)]. Since the proof is only outlined in [21], we include some more details for the convenience of the reader. The proof relies on two lemmas, in which we assume λ, P, δ and k are as in Theorem 4.1.

Lemma 4.1. *Assume μ is a $|P|^+$-complete ultrafilter on $\kappa^{<\omega}$. Then P forces*

$$j_{k(\mu)} \restriction \mathrm{Ord} = j_\mu \restriction \mathrm{Ord}\,.$$

Proof – The proof closely follows the proof of [21, Lemma 1.7.3]. Let $M \subseteq V[G]$ be an inner model such that $k\colon V \to M$. Let

$$X = \{f\colon k(\kappa^{<\omega}) \to \mathrm{Ord} \,|\, f \in M\}$$
$$Y = \{g\colon \kappa^{<\omega} \to \mathrm{Ord} \,|\, g \in V\}.$$

Define $\pi\colon X \to Y$ via $(\pi(f))(s) = f(k(s))$. We claim the following.

1. For each $f \in \mathrm{dom}(\pi)$ there exists $g\colon \kappa^{<\omega} \to \mathrm{Ord}$ in V and $A \in \mu$ such that $(\pi(f))(s) = g(s)$ for all $s \in A$.

2. For each $g\colon \kappa^{<\omega} \to \mathrm{Ord}$ in V there exists an $A \in \mu$ and a $f \in \mathrm{dom}(\pi)$ such that $(\pi(f))(s) = g(s)$ for all $s \in A$.

Once proved, (1) and (2) will imply that $\pi^*\colon X/k(\mu) \to Y/\mu$ defined by $\pi^*([f]_{k(\mu)}) = [\pi(f)]_\mu$ is a bijection. We will then show that it is an order-isomorphism.

We work in V below some $p \in P$ throughout. To prove (1), fix an $f\colon k(\kappa^{<\omega}) \to \mathrm{Ord}$ and a P-name τ for f. For each $s \in \kappa^{<\omega}$ fix $p_s \leq p$ and γ_s such that $p_s \Vdash (\pi(\tau))(\check{s}) = \check{\gamma}_s$. Since μ is $|P|^+$-complete there exist $A \in \mu$ and $q \leq p$ such that for all $s \in A$ we have $q = p_a$. Define $g\colon \kappa^{<\omega} \to \mathrm{Ord}$ via $g(s) = \gamma_s$ for all $s \in \kappa^{<\omega}$. Then $q \Vdash (\pi(\tau))(\check{s}) = g(\check{s})$ for all $s \in A$.

To prove (2), fix a $g\colon \kappa^{<\omega} \to \mathrm{Ord}$ in V. For each $s \in \kappa^{<\omega}$ find $p_s \le p$, $a_s \in P$ and $h_s\colon a_s \to \mathrm{Ord}$ such that $p_s \Vdash [\check{h}_s]_G = \check{g}(\check{s})$ (see [21, Section 2.2]). Since μ is $|P|^+$-complete there exist $A \in \mu$, $q \le p$ and $a \in V_\delta$ such that for all $s \in A$ we have $q = p_s$ and $a = a_s$. Define $F\colon a \to \mathrm{Ord}^{\kappa^{<\omega}}$ by letting $(F(X))(s)$ be $h_s(X)$ when $s \in A$ and 0 otherwise. We wish to see that for each $s \in A$, q forces that $\pi([F]_G)(s)$ (which is $[F]_G(k(s))$ by definition) is equal to $g(s)$. This follows from the fact that q forces that h_s represents $g(s)$ in the embedding k, the constant function $X \mapsto s$ represents $k(s)$, and $F(X)(s) = h_s(X)$ for all $X \in a$.

In order to check π^* is order-preserving, we closely follow the proof of [21, Claim 3 on page 11] to prove the following.

3. For all $A \subseteq \kappa^{<\omega}$ in V and $C \subseteq k(\kappa^{<\omega})$ in M such that $A \in \mu$ and $C \in k(\mu)$ we have $k[A] \cap C \neq \emptyset$.

Fix A and C and $p \in P$. By extending p if necessary we may find a p' below p such that $C = [h]_G$ for some $h\colon a \to \mu$ with $p' \le a$. Since μ is $|P|^+$-complete, there is an $A \in \mu$ which is contained in $h(X)$ for all $X \in a$. Then $k(s) \in C$ for all $s \in A$.

Since (3) immediately implies π^* is order-preserving, we conclude π^* is an isomorphism. Fix an ordinal γ. Then $j_\mu(\gamma) = [g]_\mu$ for a function $g\colon \kappa^{<\omega} \to \mathrm{Ord}$ in V that is constantly equal to γ. Similarly, $j_{k(\mu)}(\gamma) = [f]_{k(\mu)}$ for $f\colon \kappa^{<\omega} \to \mathrm{Ord}$ in M that is constantly equal to γ. Since $\pi(f) = g$, the conclusion follows. □

Lemma 4.2. *Assume μ_0 is a λ-complete ultrafilter on κ^{i_0} and μ_1 is a λ-complete ultrafilter on κ^{i_1} that projects to μ_0. Then P forces*

$$j_{k(\mu_0)k(\mu_1)} \restriction \mathrm{Ord} = j_{\mu_0\mu_1} \restriction \mathrm{Ord}\,.$$

Proof – The proof is very similar to the proof of Lemma 4.1 and we omit it. □

Proof of Theorem 4.1 – This is analogous to the proof of [21, Theorem 1.3.14], but using Lemma 4.2 instead of [21, Lemma 1.1.25]. As in [21, p. 23], S is the increasing union of trees S_γ ($\gamma < (2^{|T|})^+$) and each

S_γ is defined using only $j_{\mu_i\mu_j} \restriction \gamma$, where μ_i $(i < \omega)$ is an enumeration of λ-complete measures witnessing T is weakly homogeneous. By Lemma 4.2, $j_{j_E(\mu_i)j_E(\mu_j)} \restriction \mathrm{Ord} = j_{\mu_i\mu_j} \restriction \mathrm{Ord}$ for all i, j. Furthermore, S is definable from $|T|^+$ in $L(\{j_{\mu_i\mu_j} \restriction |T|^+ : i, j < \omega\})$, and so, since $j(|T|^+) = |T|^+$, $j(S) = S$. □

Theorem 4.2. *Assume there are class many Woodin cardinals. Let A be a universally Baire set in V and let $V[G]$ be a set forcing extension of V in which $j\colon V \to M$ is a V-generic elementary embedding obtained via forcing with $\mathbb{P}_{<\delta}$ or $\mathbb{Q}_{<\delta}$. Then in $V[G]$, $j(A) = A_G$.*

Proof – Fix an inaccessible λ greater than δ and the rank of the partial order for which G is a generic filter. By Theorem 3.3, A and its complement are respectively the projections of λ-weakly homogeneous trees T_0 and T_1 which project to complements in $V[G]$. By Theorem 4.1, we have trees S_0 and S_1 in V such that

- $p[S_0] = A$,
- $p[S_1] = \omega^\omega \setminus A$,
- the pairs (S_0, T_1) and (S_1, T_0) project to complements in $V[G]$,
- $j(S_0) = S_0$,
- $j(S_1) = S_1$.

By the absoluteness of well-foundedness of trees, $j(A) = (p[S_0])^M = A_G \cap M$. □

Corollary 4.1. *Assume there are class many Woodin cardinals and let A be a universally Baire set. Suppose $V[h]$ is a forcing extension of V, $j_0\colon V \to M_0$ and $j_1\colon V[h] \to M_1$ are stationary tower embeddings existing in some forcing extension of $V[h]$. Then $j_0(A) \cap M_1 = j_1(A) \cap M_0$.*

Proof – Let δ_i, G_i be a Woodin cardinal an generic corresponding to j_i for $i < 2$. Pick an inaccessible $\lambda > \max(\delta_0, \delta_1)$ and a λ-homogeneously Suslin tree S such that $A = p[S]$. By Theorem 4.2 the conclusion follows. □

Remark 4.1 - Note that Theorem 4.2 (as well as corresponding variants of Corollary 4.1) remain true if $\delta < \zeta$ for an inaccessible ζ and $j: V_\zeta \to M$. This remark will be used in the proof of Theorem 5.1.

Remark 4.2 - It is well-known that analogues of Theorem 4.2 and Corollary 4.1 hold for generic ultrapowers obtained from precipitous ideals. This is a consequence of an appropriate analogue of Theorem 4.1, also due to Steel (see [21, Theorem 1.7.1, also Theorem 1.7.3]).

5. Σ^2_1-absoluteness with predicates for universally Baire sets and NS_{ω_1}

Theorem 5.1 below subsumes Woodin's Σ^2_1-absoluteness theorem, and is proved in this section by a generalization of his proof. An 'elementary' proof (one without using the machinery of stationary tower forcing) can be obtained by modifying Todorcevic's proof of the Σ^2_1-absoluteness theorem (see [39], [8]). This proof uses the Lévy collapse followed by $\mathcal{P}(\omega_1)/\mathcal{I}$, for a saturated ideal $\mathcal{I}$ defined in the collapse. As Matt Foreman pointed out, an iteration of Lévy collapse and $\mathcal{P}(\omega_1)/\mathcal{I}$ for a saturated I is regularly embedded into the stationary tower by [12]. It is not difficult to see that Theorem 5.1 subsumes another well-known fact, that under large cardinal assumptions for every universally Baire set A the theory of $L(A, \mathbb{R})$ cannot be changed by forcing.

Fix a cardinal κ. An $N \in \mathcal{P}_{\aleph_2}(H(\kappa))$ is internally approachable if there is a continuous chain $(N_\alpha)_{\alpha<\omega_1}$ such that $N = \bigcup_{\alpha<\omega_1} N_\alpha$ and each proper initial segment of this chain belongs to N. This is a special case of [12, Definition 2.1] when the length is equal to ω_1. We shall denote the set of all internally approachable structures by IA. Internally approachable structures form a rather large stationary subset of $\mathcal{P}_{\aleph_2}(H(\kappa))$, roughly corresponding to $\{\alpha < \kappa \mid \mathrm{cf}(\alpha) = \omega_1\}$. In all of our applications, κ can be taken to be any large enough cardinal.

If a sentence ϕ of the language of Set Theory has an additional unary predicate for a set $A \subseteq \mathbb{R}$ then a model M of ϕ is A-correct if it is transitive and $A \cap M = A^M$. Similarly (assuming $\omega_1 \subseteq M$), M is NS_{ω_1}-correct if $\mathrm{NS}_{\omega_1} \cap M = \mathrm{NS}^M_{\omega_1}$.

Theorem 5.1. *Suppose that κ is a measurable Woodin cardinal. Let A be a κ-universally Baire set of reals and let ϕ be a sentence in the language of set theory with one additional unary predicate. Then the following statement is absolute between forcing extensions by posets in V_κ:*

> *The set $\mathfrak{M} = \{M \cap H(\aleph_1) \mid M$ is an A-correct and NS_{ω_1}-correct model of $\phi\}$ contains a club in $\mathcal{P}_{\aleph_2}(H(\aleph_1))$ relative to IA.*

The strength of the large cardinal assumption in Theorem 5.1 can be weakened to 'κ is a full Woodin cardinal and there is a Woodin cardinal $\lambda > \kappa$' (see Section 5.1 below for definition). The case of Theorem 5.1 when ϕ is a Σ_1 formula of $H(\aleph_2)$ implies the conclusion of Theorem 3.1. Since under CH the stationarity of $\mathfrak{M}$ is equivalent to the existence of an A-correct and NS_{ω_1}-correct model of ϕ containing all the reals, the case when ϕ is a Σ^2_1 formula implies Woodin's theorem.

We emphasize that in Theorem 5.1 ϕ is not assumed to be a statement of $H(\aleph_2)$, and this result may be considered as an extension of the fact that every sentence that can be forced is true in a transitive model. The version of Theorem 5.1 in which $\mathbb{R}$ is replaced with $\mathcal{P}(\omega_1)$ is false: CH implies there is a subset of ω_1 that does not belong to any correct model of $\neg$CH.

5.1. Robust and full Woodin cardinals

If D is a collection of stationary sets (i.e., sets in $\mathbb{P}_\infty$), then we say that a set X captures D if there exists a $d \in D \cap X$ such that $X \cap (\bigcup d) \in d$. Say that a strongly inaccessible cardinal κ is robust if the following holds in every forcing extension by a partial order in V_κ: for every $b \in \mathbb{Q}_{<\kappa}$ and every $f\colon b \to \mathcal{P}(\omega_1) \setminus \mathrm{NS}_{\omega_1}$ the set of countable $X \prec V_{\kappa+1}$ such that

- X captures every predense $D \subset \mathbb{Q}_{<\kappa}$ in X,
- $\mathrm{o.t.}(X \cap \kappa) \in f(X \cap (\bigcup b))$ (hence in particular $X \cap \bigcup b \in b$)

is stationary and (as an element of $\mathbb{P}_{<\lambda}$ or $\mathbb{Q}_{<\lambda}$ for some $\lambda > \kappa$) compatible with every member of $\mathbb{Q}_{<\kappa}$.

In the following lemma G is a canonical name for a $\mathbb{Q}_{<\lambda}$ generic filter.

Lemma 5.1. *Assume $\kappa < \lambda$ are Woodin cardinals and κ is moreover robust. Then $\mathbb{Q}_{<\kappa}$ forces that for every name $\dot{S}$ for a stationary subset of ω_1 in the generic ultrapower there is $a \in \mathbb{Q}_{<\lambda}$ that forces $\kappa \in \dot{S}$ and $G \cap \mathbb{Q}_{<\kappa}$ is $\mathbb{Q}_{<\kappa}$-generic. The same conclusion holds when $\mathbb{Q}_{<\lambda}$ is replaced with $\mathbb{P}_{<\lambda}$.*

Proof – We shall give an argument for $\mathbb{Q}_{<\lambda}$ since the proof for $\mathbb{P}_{<\lambda}$ is identical. Fix $b \in \mathbb{Q}_{<\kappa}$ and $f : b \to \mathcal{P}(\omega_1) \setminus \mathrm{NS}_{\omega_1}$ that represents $\dot{S}$. Let a be the stationary subset of $[V_{\kappa+1}]^{\aleph_0}$ as in the definition of robust Woodin cardinals. Since $\bigcup a \supseteq V_\kappa$, for every $c \leq a$ in $\mathbb{Q}_{<\lambda}$ we have $c \cap V_\kappa \subseteq a$ and every $Y \in c$ still captures all predense subsets of $\mathbb{Q}_{<\kappa}$. Therefore if $D \subseteq \mathbb{Q}_{<\kappa}$ is predense and $d \in D$ is such that $Y \cap (\cup d) \in D$ for all $Y \in c$ then d is compatible with c in $\mathbb{Q}_{<\lambda}$. Hence a forces $G \cap \mathbb{Q}_{<\kappa}$ is generic.

The function $X \mapsto \mathrm{o.t.}(X \cap \alpha)$ represents α in any $\mathbb{Q}_{<\lambda}$ or $\mathbb{P}_{<\lambda}$ embedding for $\lambda > \alpha$ (for $\mathbb{P}_{<\kappa}$ this is [21, Fact 2.2.10] and the proof for $\mathbb{Q}_{<\kappa}$ is similar). Therefore, the requirement that $\mathrm{o.t.}(X \cap \kappa) \in f(X \cap (\cup b))$ ensures that κ will be in the set represented by f in any stationary tower embedding derived from $\mathbb{Q}_{<\lambda}$ or $\mathbb{P}_{<\lambda}$ for some $\lambda > \kappa$ by a filter containing the set above. □

Lemma 5.2. *If κ is a measurable Woodin cardinal, then the set of robust Woodin cardinals is in every normal measure on κ.*

Proof – Aside from the condition that $\mathrm{o.t.}(X \cap \kappa) \in f(X \cap (\cup b))$, this follows from Lemma [21, Lemma 2.7.14]. The remaining part follows from the fact that if κ is measurable, then every countable elementary submodel of $V_{\kappa+2}$ can be end-extended (as in the last paragraph of the proof of Lemma 3.1), and so the stationarity of the desired set reflects to a member of every normal measure on κ. □

For a strongly inaccessible cardinal λ, let $a(\lambda)$ be the set of $X \prec V_{\lambda+1}$ such that

1. $\mathrm{o.t.}(X \cap \lambda) = \omega_1$,
2. X captures every predense $D \subset \mathbb{Q}_{<\lambda}$ in X,
3. letting $\bar{X}$ be the transitive collapse of X, there exists a closed unbounded set $C \subset \bar{X} \cap \omega_1$ such that

- each member of C is a robust Woodin cardinal in $\bar{X}$,
- for each limit point γ of C, if $E \in \mathcal{P}(\gamma)^{\bar{X}}$ is a club subset of γ, then E contains a tail of $C \cap \gamma$.

Let us say that λ is full if it is a Woodin cardinal and $a(\lambda)$ is stationary.

Lemma 5.3. *Every measurable Woodin cardinal κ is full. Moreover, for any normal uniform measure on κ full Woodin cardinals form a set of measure one.*

PROOF – Except for the robustness requirement, the proof adapted from Woodin's proof of Σ_1^2 absoluteness ([41]) was given in [21, Claim on page 94]. The additional requirement of robustness does not change the proof, though, as the proof used only the fact that the set of Woodin cardinals below a measurable Woodin is in every normal measure, which holds for robust Woodin cardinals by Lemma 5.2. □

Lemma 5.4. *If λ is full Woodin and $\delta \geq \lambda$ then in $\mathbb{P}_{<\delta}$ the condition $a(\lambda)$ forces $j(\omega_1) = \lambda$ and $G \cap V_\lambda$ is V-generic for $\mathbb{Q}_{<\lambda}$. Also, it forces that there is a club $C \subset \lambda$ in M such that*

(a) each member of C is a robust Woodin cardinal in V_λ,

(b) for each limit member γ of C and each club subset E of γ in V_λ, E contains a tail of $C \cap \gamma$.

PROOF – We have $j(\omega_1) = \lambda$ by (1) and [21, Corollary 2.2.11]. Condition (2) implies the $\mathbb{Q}_{<\lambda}$ genericity of G as in the proof of Lemma 5.1. Since $a(\lambda) \in G$ and $\bigcup a(\lambda) = V_{\lambda+1}$, by [21, Fact 2.2.8] we have $j[V_{\lambda+1}] \in j(a(\lambda))$. Now V_λ, being the transitive collapse of $j[V_\lambda]$, is an element of M and (a) and (b) follow by (3). □

Lemma 5.5. *Assume λ is full Woodin, $\delta > \lambda$ is Woodin, and $G \subseteq \mathbb{P}_{<\delta}$ is V-generic below $a(\lambda)$ with $j\colon V \to M$ the corresponding generic embedding. Then in M for every forcing $P \in V_\lambda$ we can find a V-generic $h * H \subseteq P * \mathbb{Q}_{<\lambda}$. Furthermore, for a fixed h the set $\{Y \cap H(\aleph_1) \mid Y$ is a $\mathbb{Q}_{<\delta}$-generic ultrapower of $V[h]$ which is correct for $\mathrm{NS}_{\omega_1}\}$ contains a club in $\mathcal{P}_{\aleph_2}(H(\aleph_1))^M$ relative to IA.*

PROOF – We shall need only the following property of internally approachable structures.

CLAIM - If $N \prec H((2^{\mathfrak{c}})^+)$ is internally approachable then there is a well-ordering of the reals of N in type ω_1 all of whose proper initial belong to N.

PROOF – Let $(N_\alpha)_{\alpha<\omega_1}$ be the continuous chain of models with N as its union such that each initial segment belongs to N. By going to a club, we may assume this is an elementary chain. For each $X \in \mathcal{P}_{\aleph_2}(H(\aleph_1))$ fix a well-ordering $<_X$ of X in order type $\leq \omega_1$ and fix a bijection $g\colon \omega_1 \to \omega_1 \times \omega_1$. By elementarity, such g and a map $X \mapsto <_X$ both exist in N_0. For $\xi < \omega_1$ let $y_\xi \in N_{g_0(\xi)}$ be the $g_1(\xi)$-th real in $<_{N_{g_0(\xi)}}$. Then $\langle y_\xi : \xi < \omega_1\rangle$ enumerates all reals of N (with repetitions) and each proper initial segment of this ordering is in N. □

Let ζ be a strongly inaccessible cardinal in V between λ and δ. Since M is closed under sequences of length less than δ in $V[G]$ ([21, Theorem 2.5.8]), V_ζ and $G \cap V_\lambda$ are in M. Note that V is not included in M, but all the dense subsets of $P * \mathbb{Q}_{<\lambda}$ in V belong to V_ζ. Hence it will suffice to build a V_ζ generic $h * H \subseteq P * \mathbb{Q}_{<\lambda}$.

Since $\mathcal{P}(\mathbb{P})^V$ is countable in $V[G \cap V_\lambda]$, by [21, Theorem A.0.9] we may fix a V-generic filter $h \subset \mathbb{P}$ in $M \cap V[G \cap V_\lambda]$. Let C be a club as in Lemma 5.4. Now, $\mathbb{P}$ is ρ-c.c. for some $\rho < \lambda$, so for every regular $\xi \geq \rho$, every club subset of ξ in $V[h]$ contains a club set in V, which shows that members of $C \setminus \rho$ (which we may assume is the same as C) retain properties (a) and (b) in $V[h]$.

In M fix a well-ordering $<_w$ of $H((2^\delta)^+)$ and an internally approachable $(Z, <_w) \prec (H((2^\delta)^+), <_w)$ of size $\aleph_1$ containing everything relevant. In order to verify that the family of sets of reals of generic ultrapowers of $V[h]$ correct about NS_{ω_1} contains a club, we shall construct a generic ultrapower of $V[h]$ whose reals are the reals of Z. In M fix an enumeration $\{x_\alpha : \alpha < \omega_1\}$ of $\mathbb{R}^Z$ in type ω_1 all of whose proper initial segments are in Z.

Fix a partition in M of ω_1^M into stationary sets A_α $(\alpha < \omega_1)$. Working in M we will produce a continuous increasing sequence of ordinals $\langle \eta_\alpha : \alpha < \omega_1\rangle \subset C$ and a sequence $\langle H_\alpha : \alpha < \omega_1\rangle$ such that (writing $(V[h])_\eta$ for $V_\eta^{V[h]}$):

(c) each H_α is a $V[h]$-generic filter contained in $\mathbb{Q}^{V[h]}_{<\eta_\alpha}$,

(d) each $x_\alpha \in (V[h][H_{\alpha+1}])_\zeta$,

(e) $\alpha' < \alpha$ implies $H_\alpha \cap (V[h])_{\eta_{\alpha'}} = H_{\alpha'}$.

(f) The pair (η_α, H_α) is the $<_w$-minimal pair satisfying (c)–(e).

Note that (d) is equivalent to $x_\alpha \in V[h][H_{\alpha+1}]$ but the latter is not a statement of Z (or of M). Note also that (f) implies that each H_α belongs to Z. Since each of $\mathbb{Q}^{V[h]}_{<\eta_\alpha}$, $(V[h])_{\eta_\alpha}$, and H_α is countable and belongs to Z, each of these sets is included in Z. Our $V[h]$-generic filter $H \subset \mathbb{Q}_{<\lambda}$ will be $\bigcup_{\eta<\omega_1} H_\eta$.

As we choose η_α and H_α, for each $b \in \mathbb{Q}^{V[h]}_{<\lambda}$ and each $\mathbb{Q}_{<\lambda}$-name $\dot{S}$ for a stationary subset of ω_1 in the generic ultrapower we will associate some A_β and ensure that $A_\beta \cap \{\eta_\alpha : \alpha < \omega_1\}$ is included in the interpretation of $\dot{S}$, which ensures that the final image model will be correct about NS_{ω_1}.

We let η_0 be the least member of C and let H_0 be any $V[h]$-generic filter contained in $\mathbb{Q}_{<\eta_0}$.

For the successor stages of the construction we work in Z. Given $H_\alpha \subset \mathbb{Q}_{<\eta_\alpha}$, if $\eta_\beta \in A_\beta$ and the function f associated to A_β has domain in $\mathbb{Q}_{<\eta_\alpha}$, then we let a_α be the stationary subset of $[V_{\eta_\alpha+1}]^{\aleph_0}$ as guaranteed by Lemma 5.1. Then for any $\gamma > \eta_\alpha$ in C, letting

$$\mathbb{Q}_{<\gamma}(a_\alpha) = \{b \in \mathbb{Q}_{<\gamma} \mid b \leq a_\alpha\},$$

the quotient $(\mathbb{Q}_{<\gamma}(a_\alpha)/\mathbb{Q}_{<\eta_\alpha})^{V[h]}$ is a $\mathbb{Q}_{<\eta_\alpha}$-name in $V[h]$ for a partial order such that, in $V[h]$,

$$\mathbb{Q}_{<\eta_\alpha} * (\mathbb{Q}_{<\gamma}(a_{\eta_\alpha})/\mathbb{Q}_{<\eta_\alpha}) \cong \mathbb{Q}_{<\gamma}(a_{\eta_\alpha\gamma}),$$

and such that η_α is forced to be in the set represented by f.

Since each set of the form $\{h, x_\alpha, H_\alpha\}$ exists in a generic extension of V by a forcing in V_λ, if x_α is not in $V[h][H_\alpha]$ then x_α is generic over $V[h][H_\alpha]$ for some forcing in $(V[h][H_\alpha])_\lambda$, and by [21, Theorem A.0.9] that forcing can be subsumed by

$$(\mathbb{Q}_{<\gamma}(a_{\eta_\alpha})/\mathbb{Q}_{<\eta_\alpha})^{V[h]}_{H_\alpha}$$

(i.e., the realization of $(\mathbb{Q}_{<\gamma}(a_{\eta_\alpha})/\mathbb{Q}_{<\eta_\alpha})^{V[h]}$ by H_α) for some $\gamma\eta_\alpha$ in C. Finally, pick the $<_w$-minimal pair $(\eta_{\alpha+1}, H_{\alpha+1})$ meeting the conditions (c)–(e). Since in this case the pair (η_α, H_α) is uniquely determined, (f) is automatic.

At limit stages α of our construction, if we let

$$\eta_\alpha = \sup\{\eta_\beta : \beta < \alpha\}$$

then η_α is a Woodin cardinal and therefore every predense set in $(\mathbb{Q}_{<\eta_\alpha})^{V[h]}$ is predense in $(\mathbb{Q}_{<\beta})^{V[h]}$ for a club set of $\beta < \eta_\alpha$. But each of these clubs intersects $\{\eta_\beta : \beta < \alpha\}$, which shows that $H_\alpha = \bigcup\{H_\beta : \beta < \alpha\}$ is $V[h]$-generic for $(\mathbb{Q}_{<\eta_\alpha})^{V[h]}$. Let $C^* = \{\eta_\alpha : \alpha < \omega_1\}$. Since C^* is a club subset of λ, $H = \bigcup\{H_\alpha : \alpha < \omega_1\}$ is $V[h]$-generic for $\mathbb{Q}_{<\lambda}$.

If $j^* \colon V[h] \to M^*$ is the generic embedding induced by H, then $\mathbb{R}^{M^*} \supseteq \mathbb{R}^Z$ by contruction. Since the generic ultrapower is closed under ω-sequences, we have $\mathbb{R}^{M^*} = \bigcup_{\alpha<\omega_1} \mathbb{R}^{V[h][H_\alpha]} \subseteq Z$, hence $\mathbb{R}^{M^*} = \mathbb{R}^Z$. Since M^* is NS_{ω_1}-correct, the proof is complete. □

Proof of Theorem 5.1 – We first prove the following case of Theorem 5.1:

> *Suppose that some partial order $P \in V_\kappa$ forces the existence of an A-correct and NS_{ω_1}-correct model N of ϕ containing the reals. Then $\mathfrak{M} = \{M \cap H(\aleph_1) \mid M$ is an A-correct and NS_{ω_1}-correct model of $\phi\}$ contains a club in $\mathcal{P}_{\aleph_2}(H(\aleph_1))$ relative to IA.*

We closely follow the proof in [21, p. 94]. Let A, κ, ϕ and P be as in the statement of the theorem. Let $\lambda < \kappa$ be a full Woodin cardinal with $P \in V_\lambda$. Since κ is a limit of Woodin limits of Woodin cardinals, we may fix a Woodin limit of Woodin cardinals δ greater than λ. Now let $G \subset \mathbb{P}_{<\delta}$ be V-generic with $a(\lambda) \in G$, and let $j \colon V \to M$ be the corresponding generic embedding. It suffices to show that the conclusion of the theorem holds in M for $j(A)$, since then it holds in V by elementarity.

By Lemma 5.5 find V-generic $h * H \subseteq P * \mathbb{Q}_{<\lambda}$ in M. If ζ is the first inaccessible above λ in V, then $V_\zeta \subseteq M$ and in M we can define $j^* \colon V_\zeta[h] \to M^*$ the generic embedding. We have $j^*(\omega_1) = \lambda$ by Lemma 5.4.

There is a model N of ϕ in M^* with the additional symbol in ϕ interpreted as $j^*(A_h)$ (where A_h is the reinterpretation of A in $V[h]$) such that N contains the reals of M^* and is correct in M^* about $j^*(A_h)$, ω_1 and NS_{ω_1}. The universally Baire set A is absolute between M and M^* by Corollary 4.1 (see also Remark 4.1), in particular $j^*(A_h) = A \cap M^*$. Also, M^* is correct in M about ω_1 and NS_{ω_1}, hence model N is correct about ω_1, NS_{ω_1} and A in M as well. The family of intersections with $H(\aleph_1)$ of such models includes a club in $\mathcal{P}_{\aleph_2}(H(\aleph_1))$ relative to IA, and the elementarity implies the claim in V.

If CH holds then $\mathfrak{M}$ contains a club relative to IA if and only if there is an A-correct, NS_{ω_1}-correct model of ϕ containing all reals. Hence we can complete the proof using a standard fact. By [21, Theorem 2.5.10] (due to Woodin), if δ is a Woodin cardinal and some sentence ψ holds in some initial segment of the universe below δ, then ψ can be forced to hold in some initial segment of the universe by a partial order of cardinality δ after forcing with any partial order in V_δ. By Theorem 4.2 and the proof of Theorem 2.5.10 in [21], this fact holds even when one adds a predicate for a given universally Baire set to the language. Thus if $\mathfrak{M}$ includes a club relative to IA in some forcing extension then in every forcing extension satisfying CH there exists an A-correct and NS_{ω_1}-correct model of ϕ containing the reals. □

6. Universally Baire sets and $L(Q)$

If Γ is a pointclass let $\mathcal{L}^\Gamma$ be an extension of the first-order logic with formulas in which we have

1. the quantifier Qx, interpreted as 'there exists uncountably many,'
2. a predicate NS for the nonstationary ideal on ω_1,
3. predicates for all sets in Γ,
4. constants for all hereditarily countable sets.

If Γ is the class of all universally Baire sets then we write $\mathcal{L}^{\mathrm{uB}}$ for $\mathcal{L}^\Gamma$.

Definition 6.1. *If ϕ is a sentence of $\mathcal{L}^{\mathrm{uB}}$ we say that a model $\mathfrak{X}$ of ϕ is* correct *if the conditions below are satisfied.*

5. *The universe X of $\mathfrak{X}$ is an element of $H(\aleph_2)$,*

6. *all universally Baire predicates and NS_{ω_1} as well as all constants occurring in ϕ are interpreted correctly,*

7. *this model is* standard: *it interprets Qx as 'there exist uncountably many.'*

For a $\mathcal{L}^\Gamma$ sentence ϕ such that Γ=Borel and Γ does not involve the predicate NS the statement 'ϕ has a correct model' is absolute between transitive models of ZFC. This folklore consequence of Keisler's completeness theorem for $L_{\omega_1\omega}(Q)$ ([19]) was stated explicitly in [9]. An apparently stronger result in which ϕ is allowed to have countable conjunctions is still true. We drop infinitary sentences since in our context where we are dealing only with ω-models they would not add to the expressive power of the language.

Note that countable sets of reals are universally Baire, so our definition of correct model allows for real parameters in ϕ. An easy Löwenheim–Skolem argument shows that every correct model for ϕ has a correct elementary submodel of cardinality $\aleph_1$. Also, if A is a universally Baire set coding all the universally Baire sets appearing in ϕ and (N, I) is an A-iterable pair such that $N \models \phi$, then as in Subsection 3.3 there is an iteration of (N, I) of length $\omega_1 + 1$ whose last model is correct for ϕ.

Theorem 6.1. *Suppose $\kappa < \delta$ are Woodin cardinals and ϕ is a sentence in $\mathcal{L}^{\mathrm{uB}}$. Then ϕ has a correct model in a forcing extension by a partial order in V_κ if and only if ϕ has a correct model in V.*

Proof – Let A code all universally Baire sets occurring in ϕ. Since 'ϕ has a correct model' is a Σ_1 sentence in $\langle H(\aleph_2); \mathrm{NS}_{\omega_1}, A, \in\rangle$, this is a consequence of Theorem 3.1. □

In the absence of large cardinals there can exist a formula ϕ with a correct model in V such that some P forces ϕ has no correct models even with $\Gamma = \Pi^1_1$ (Proposition 8.7). By augmenting the large cardinal strength we obtain a stronger transfer theorem.

Theorem 6.2. *Suppose $\kappa < \delta$ are Woodin cardinals, κ is full, and ϕ is a sentence in $\mathcal{L}^{\mathrm{uB}}$.Then the following statement is absolute between forcing extensions by posets in V_κ:*

the set $\mathfrak{M} = \{M \cap H(\aleph_1) \mid M$ *is a correct model of* $\phi\}$ *contains a club in* $\mathcal{P}_{\aleph_2}(H(\aleph_1))$ *relative to IA.*

PROOF – Let A be the universally Baire set occurring in ϕ. If N is a model of ZFC that correctly computes A, ω_1 and NS_{ω_1}, then a correct model for ϕ in N is also correct in V. Therefore this is a consequence of Theorem 5.1. By the proof of Theorem 5.1, one needs only that κ is a full Woodin cardinal below a Woodin cardinal. □

Corollary 6.1. *If there exists a proper class of Woodin cardinals and* ϕ *is a formula of* $\mathcal{L}^{\mathrm{uB}}$*, then the set of reals* x *such that* $\phi(x)$ *has a correct model is universally Baire.*

PROOF – Let X be the set of all reals r such that $\phi(r)$ has a correct model. Assume that $G \subseteq \mathbb{Q}_{<\delta}$ is a generic filter and $j\colon V \to M$ is the induced generic embedding. If $r \in \mathbb{R}^M$, then $\phi(r)$ has a correct model in M if and only if it has a correct model in $V[G]$. This follows from the fact that M and $V[G]$ have the same reals, and Theorems 3.3 and 4.1, with P in Theorem 3.3 as the trivial forcing. (The existence of a correct model in either of M and $V[G]$ gives rise to a countable A-iterable model - for the relevant universally Baire set A - which can be iterated ω_1 times as in Subsection 3.3 to produce a correct model in the other.)

By Theorem 6.1, $\phi(r)$ has a correct model in $V[G]$ if and only if it has a correct model in $V[r]$, an extension by a forcing in V_δ. By Woodin's Tree Production Lemma ([21, Theorem 3.3.15] or [35, Theorem 4.2]) applied with $\psi(x)$ being '$\phi(x)$ has a correct model,' we conclude that $X = \{r \mid \phi(r)$ has a correct model$\}$ is universally Baire. □

Corollary 6.2 (CH). *If there exists a proper class of full Woodin cardinals and* ϕ *is a formula of* $\mathcal{L}^{\mathrm{uB}}$*, then the sets of reals* x *such that* $\phi(x)$ *has a correct model containing all the reals is universally Baire.*

PROOF – Analogous to the proof of Corollary 6.1, but using Theorem 6.2. The one key difference is in showing that $V[r]$ and M agree about whether $\phi(r)$ has a correct model, in the situation where λ is a full Woodin cardinal, $j\colon V \to M'$ is the generic embedding induced by a V-generic filter $G' \subset \mathbb{Q}_{<\lambda}$

and $r \in \mathbb{R}^{M'}$. Forcing over $V[G']$, we may fix for some Woodin cardinal $\delta\lambda$ a V-generic $G \subset \mathbb{P}_{<\delta}$ with $G \cap V_\lambda = G'$ and $a(\lambda)$ (as in Section 5) in G. Then there is an elementary embedding from M' to M with critical point greater than λ which fixes the images of the universally Baire sets from V. Furthermore, M' contains all subsets of λ in $V[G']$. If there exists a correct model of ϕ in $V[r]$, then the proof of Lemma 5.5 shows that there is one in M and so there exists one in M'. If there exists a correct model N in M', then it suffices to show that N is correct in $V[G']$. Since M, M' and $V[G]$ agree about the universally Baire sets from V, the only issue here is whether $\mathrm{NS}_{\omega_1}^{V[G']} \cap N = \mathrm{NS}_{\omega_1}^N$. This follows from the fact that $\mathrm{NS}_{\omega_1}^M \cap N = \mathrm{NS}_{\omega_1}^N$ and every club subset of λ in $V[G']$ is in M. □

The following well-known fact (e.g., [21, Theorem 3.3.10]) follows easily from Corollary 6.2 by using standard arguments.

Corollary 6.3. *Assume there are class many full Woodin cardinals and CH. Then every Σ^2_1 set X is universally Baire. Moreover, we can allow arbitrary universally Baire sets as parameters in the definition of X.*

In [10] we add $\diamond$ to our assumptions and prove variant of Theorem 6.1 for the Magidor-Malitz logic: the logic with Ramseyan quantifiers asserting the existence of uncountable homogeneous sets for partitions of any $[\omega_1]^n$ (and even $[\omega_1]^{<\omega}$).

A Σ^2_1 well-ordering of the reals can be forced starting from any model of ZFC ([2]) and therefore an additional assumption such as CH is necessary in Corollary 6.3. Also, a Δ^2_2 well-ordering of the reals in type ω_1 can be forced starting from any model of ZFC ([1]) and therefore Σ^2_1 in Corollary 6.3 cannot be improved much unless an axiom stronger than CH is assumed.

7. Isolating cardinal invariants

We now give an application of results in Section 6.

Definition 7.1. *Assume ϕ is an $\mathcal{L}^{\mathrm{uB}}$-formula in a language with an additional predicate Z. Let $E \subset \mathbb{R}^2$ be universally Baire. A set $A \subset \mathbb{R}$ is (ϕ, E)-covering* *if there is a correct model $\mathfrak{X}$ for ϕ such that $Z^{\mathfrak{X}} = A$ and moreover $(\forall x \in \mathbb{R})(\exists y \in A)(x, y) \in E$.*

The minimal cardinality of a (ϕ, E)-covering set will be denoted by $\mu_{\phi,E}$. A cardinal invariant of this form is an abstract covering number.

The set A as occurring in Definition 7.1 is not assumed to be unviersally Baire.

Every tame cardinal invariant as in [43, Section B.2] is an abstract covering number. This includes a large number of standard cardinal invariants of the continuum such as $\mathfrak{a}, \mathfrak{b}, \mathfrak{c}, \mathfrak{d}, \mathfrak{e}, \mathfrak{f}, \ldots$, as well as all cardinal invariants occurring in Cichon's diagram (see [6], [5] for the theory of cardinal invariants). The definition of a tame cardinal invariant in [43] requires ϕ to be arithmetic and E to be projective. With these additional assumptions, Theorem 7.1 as well as the auxiliary lemmas below are due to Zapletal, whose proof used Σ^2_1-absoluteness and relied on [7].

Not every abstract covering number is provably equal to a covering number of some absolutely defined σ-ideal. For example, the minimal length of a nonextendible tower in $\mathcal{P}(\omega)/\mathrm{Fin}$, $\mathfrak{t}$, is an abstract covering number but not a covering number of any σ-ideal. This is because the value of $\mathfrak{t}$ can be collapsed to $\aleph_1$ without adding reals (see [43, Example B.1.8]). The standard cardinal invariants known not to be tame, such as $\mathfrak{g}$ ([43, B.26]) or $\mathfrak{h}$ ([43, B.25]) are not abstract covering numbers, as Theorem 7.1 fails for them.

Let us introduce some of the machinery from [43, Section 3]. The following definition applies to most σ-ideals used in the theory of forcing.

Definition 7.2. *A σ-ideal I on the reals such that the collection of all codes for analytic sets in I is given by a universally Baire definition is a* universally Baire ideal. *An ideal I is* iterable and almost full *if the following hold.*

1. *It is a universally Baire σ-ideal.*

2. *Every positive universally Baire set has a positive Borel subset.*

3. *The forcing* `Borel`$/I$ *is proper in all forcing extensions.*

By (2), forcings `Borel`$/I$ and $\mathcal{P}(\mathbb{R})/I$ are forcing equivalent in $L(\mathbb{R})$ and other standard inner models of AD, in the presence of a proper class of Woodin cardinals. Futhermore (in the presence of class many Woodin cardinals), Definition 7.2 describes ideals that are both iterable (see [43, 3.1.1]) and almost full ([43, 2.2.6]) in Zapletal's terminology. This follows from [43, Claim 3.1.3].

Recall that $\mathrm{cov}(I)$ is the smallest size of a subset of I whose union covers the real line. Let $\mathrm{cov}^+(I)$ be the smallest size of a subset of I whose union covers an I-positive Borel set. If the ideal I is homogeneous, in the sense that for every Borel I-positive set B there is a Borel isomorphism $f\colon B \to \mathbb{R}$ such that $f[A] \in I$ iff $A \in I$ for all $A \subseteq B$, then clearly $\mathrm{cov}(I) = \mathrm{cov}^+(I)$.

Let $\mu_{E,I}$ be an abstract covering number. A set A is (ϕ, E, I)-covering if for every I-positive Borel set B and a Borel function $f\colon B \to \mathbb{R}$ there is $x \in A$ such that

$$\{y \in B : (f(y), x) \in E\}$$

contains an I-positive Borel set. The following is the main result of this section.

Theorem 7.1. *Assume there are class many full Woodin cardinals. For an abstract covering number $\mu_{\phi,E}$ and an iterable and almost full I the following are equivalent.*

1. *Any countable support iteration of $\mathtt{Borel}/I$ of length at least ω forces $\mu_{\phi,E} \leq \aleph_1$.*
2. *In some forcing extension we have $\mu_{\phi,E} < \mathrm{cov}^+(I)$.*
3. *In some forcing extension there exists a (ϕ, E, I)-covering set.*

Along similar lines one might expect that the consistency of $\aleph_1 < \mathrm{cov}^+(I)$ implies the properness of $\mathtt{Borel}/I$, but J. Zapletal pointed out that there is a universally Baire σ-ideal I for which this is false; see [44].

If P is a forcing notion then we say A is (ϕ, E, P)-covering if for every P-name for a real τ and every $p \in P$ there is a regular subordering Q of $\{q \in P : q \leq p\}$ that is forcing-equivalent to $\mathtt{Borel}/I$ for a universally Baire ideal I such that τ is a Q-name and A is (ϕ, E, I)-covering. Typically, P will be a countable support iteration of some quotients $\mathtt{Borel}/I$.

Lemma 7.1. *Assume there are class many Woodin cardinals, P is a proper forcing notion, and A is (ϕ, E, P)-covering. Then P forces A is (ϕ, E)-covering.*

PROOF – Assume τ is a P-name for a real and $p \in P$. Find a regular suborder Q of P such that τ is a Q-name and forcing equivalent to $\mathtt{Borel}/I$ for some universally Baire ideal I below p. Since Q is proper, by [43, Lemma 2.2.1] there is a Borel function $f\colon \mathbb{R} \to \mathbb{R}$ such that Q forces $\tau = f(\dot{r})$, where $\dot{r}$ is the canonical generic real for $\mathtt{Borel}/I$. Find $y \in A$ such that $\{x \in \mathbb{R} \mid (f(x), y) \in E\}$ contains an I-positive Borel set, B. Then B forces $(\tau, y) \in E$ because the set $C = \{z \mid (f(z), y) \in E\}$ is universally Baire (this follows from our large cardinal assumption since C is projective in E, see Corollary 6.1 or [21, Exercise 3.3.11]), $B \subseteq C$ and $\dot{r}$ is forced to belong to B. □

Let I be a σ-ideal on $\mathbb{R}$ and $\alpha < \omega_1$. For $B \subseteq \mathbb{R}^\alpha$ let

$$T_B = \{s \in \textstyle\bigcup_{\beta<\alpha} \mathbb{R}^\beta : s = x \restriction \beta \text{ for some } x \in B\}.$$

A set $B \subseteq \mathbb{R}^\alpha$ is I-perfect ([43, 3.1.4, 3.2.2] if

1. For every $s \in T_B$ the set $\{r \in \mathbb{R} : s^\frown r \in T_B\}$ is I-positive, and
2. T_B is σ-closed: if s_n $(n \in \omega)$ is an increasing sequence in T_B then $\bigcup_{n\in\omega} s_n \in T_B$.

There is another natural way to define when $B \subseteq \mathbb{R}^\alpha$ is 'large' with respect to I introduced in [43, 3.2.1]. Consider a game $G(B)$ of length α in two players, Adam and Eve. Adam plays first at all limit stages and his move is always a set $A_\xi \in I$. Eve plays reals x_ξ $(\xi < \alpha)$ so that $\langle x_\xi : \xi < \eta\rangle \in T_B$ for all η and $x_\xi \notin A_\xi$ for all ξ. If at some stage Eve is unable to make a move, then Adam wins. Otherwise, after α moves Eve wins. The family of all sets for which Adam has a winning strategy is denoted by I^α. Clearly $I^1 = I$. It is not difficult to see that I^α is a σ-ideal and that all I-perfect subsets of $\mathbb{R}^\alpha$ are I^α-positive.

The following result of Zapletal originally appeared in [43, 3.4.1].

Lemma 7.2 (Zapletal). *Assume there are class many Woodin cardinals. If I is iterable and almost full then every universally Baire set in $\mathbb{R}^\alpha$ is either in I^α or it contains a Borel I-perfect subset.*

PROOF – We shall only point to the main ingredients of the proof, as the reader can find the details in [43]. First, the large cardinal assumption implies all perfect information games of fixed countable length in which the opponents are playing reals with a universally Baire payoff are determined, and moreover the winning strategy can be coded as a universally Baire set of reals ([29, Exercise 2E.11]). Adam's moves in the game $G(A)$ need not be coded as reals, but Zapletal defines an auxiliary game $H(A)$ with this property that is equivalent to $G(A)$. Therefore $A \subseteq \mathbb{R}^\alpha$ is not in I^α if and only if Eve has a winning strategy in $G(A)$ if and only if Eve has a universally Baire winning strategy in $H(A)$. Second, Eve has a universally Baire winning strategy in $H(A)$ if and only if A contains an I-perfect Borel subset. □

Lemma 7.3. *Assume I is a σ-ideal on $\mathbb{R}$ such that for every $\alpha < \omega_1$ every I^α-positive universally Baire set contains a Borel I-perfect set. Then*

1. $\mathrm{cov}^+(I) = \mathrm{cov}^+(I^\alpha)$ *for all* $\alpha < \omega_1$.
2. *If* $\mu_{\phi,E}$ *is an abstract covering number then every* (ϕ, E, I)*-covering set is* (ϕ, E, I^α)*-covering for all* $\alpha < \omega_1$.

Proof – We prove only (1) since the proof of (2) is almost identical. It is obvious that the sequence $\mathrm{cov}^+(I^\alpha)$ is nonincreasing in α. We prove $\mathrm{cov}^+(I^\alpha) \geq \mathrm{cov}^+(I)$ by induction on α. Assume $\lambda < \mathrm{cov}^+(I)$, $B \subseteq \mathbb{R}^\alpha$ is an I-perfect Borel set, and B_ξ $(\xi < \lambda)$ are Borel sets in I^α. For $\xi < \lambda$ let τ_ξ be a winning strategy for Adam in the I-perfect game for B_ξ. In order to prove $B \not\subseteq \bigcup_{\xi<\lambda} B_\xi$, we construct a sequence $\langle x_\xi : \xi < \alpha\rangle$ of Eve's simultaneous responses to all τ_ξ in $G(B)$.

Let $T_B = \{s \in \bigcup_{\beta<\alpha} \mathbb{R}^\beta : s = x \restriction \beta \text{ for some } x \in B\}$. If $s = \langle x_\eta : \eta < \xi\rangle$ has been chosen, let $Z_\eta = \bigcup_{\xi<\lambda} Z_{\eta,\xi}$, where $Z_{\eta,\xi} \in I$ is the set provided to Adam by τ_ξ as a response to Eve's moves, coded by $s \in T_B$. Note that this is not necessarily a legal move in $G(B)$ for Adam since Z_η need not be in I. The set $X = \{r : s^\frown r \in T_B\}$ is universally Baire and I-positive, and by assumption it contains a Borel I-positive set. Since each $Z_{\eta,\xi}$ is in I and $\lambda < \mathrm{cov}^+(I)$, we have that Z_η does not cover X. Eve extends s by picking $x_\xi \in X \setminus Z_\xi$. This shows that Eve can play at any successor stage of the game. Since T_B is σ-closed and B is I-perfect, Eve wins the game. The obtained sequence belongs to $B \setminus \bigcup_{\xi<\lambda} B_\xi$. Since B and B_ξ $(\xi < \lambda)$ were arbitrary, this shows that an I-perfect set cannot be covered by fewer than $\mathrm{cov}^+(I)$ Borel sets in I^α. □

Lemma 7.4. *Assume there are class many Woodin cardinals. Assume that $\mu_{\phi,E}$ is an abstract covering number and I is iterable and almost full. Then every (ϕ, E, I)-covering set is (ϕ, E, P)-covering, where P is any countable support iteration of* `Borel`$/I$.

Proof – Lemma 7.2 and Lemma 7.3 together imply that every (ϕ, E, I)-covering set is (ϕ, E, I^α)-covering for all $\alpha < \omega_1$. It is well-known (see [43, Section 3.1]) that for every condition B in a countable support iteration P_κ of `Borel`$/I$ of any length there is $\alpha < \omega_1$ and a regular subalgebra Q of P_κ forcing-equivalent to the algebra of Borel subsets of $\mathbb{R}^\alpha$ modulo I^α. It is also known that every positive set contains an I-perfect positive subset. Since the assumption on I being iterable implies the iteration is proper, Q can be chosen so that any given name for a real is a Q-name. □

Proof of Theorem 7.1 – Since the countable support iteration of `Borel`$/I$ of ω_2 forces $\mathrm{cov}^+(I) > \aleph_1$, (1) implies (2). Also, (2) readily implies (3). Assume (3). Let ψ be the $\mathcal{L}^{\mathrm{uB}}$-sentence in the language with an additional predicate A stating that A is a (ϕ, E, I)-covering. Then it is consistent that ψ has a correct model containing all the reals. Consider the extension by countable support iteration of `Borel`$/I$ of length $\geq \omega$. A countable support iteration of forcings adding reals of length ω forces CH. Using Theorem 6.2 in this intermediate model we can find a correct model $\mathfrak{X}$ for ψ containing all reals of this model. Since the factor forcing is a countable support iteration of `Borel`$/I$, Lemma 7.4 implies that the (ϕ, E, I)-covering set provided by $\mathfrak{X}$ is (ϕ, E, P)-covering. By Lemma 7.1, (1) follows. □

Given $\alpha < \omega_1$ and a sequence I_ξ $(\xi < \alpha)$ of σ-ideals one can define ideal $\prod_{\xi<\alpha} I_\xi$ analogously to the iterated power I^α. If all of these ideals are iterable and almost full and there are class many Woodin cardinals then the analogue of Lemma 7.2 holds and the proof of Lemma 7.3 shows that $\mathrm{cov}^+(\prod_{\xi<\alpha} I_\xi) = \sup_{\xi<\alpha} \mathrm{cov}^+(I_\xi)$. So we have the following extension of Theorem 7.1.

Theorem 7.2. *Assume there are class many full Woodin cardinals. Further assume $\mathfrak{Z}$ is a set of abstract covering numbers and $\mathcal{F}$ is a family of iterable and almost full ideals. Then the following are equivalent.*

1. *Every countable support iteration of of length at least* ω *consisting of partial orders of the form* $\texttt{Borel}/I$ *for some* $I \in \mathcal{F}$ *forces that* $\mu \leq \aleph_1$ *for all* $\mu \in \mathfrak{Z}$.

2. *There is a forcing extension in which* $\mu < \mathrm{cov}^+(I)$ *for all* $\mu \in \mathfrak{Z}$ *and* $I \in \mathcal{F}$.

3. *For each* $\mu \in \mathfrak{Z}$ *and* $I \in \mathcal{F}$ *there is a forcing extension in which* $\mu < \mathrm{cov}^+(I)$.

4. *For each* $\mu_{\phi,E} \in \mathcal{F}$ *and each* $I \in \mathcal{F}$ *there is a forcing extension in which a* (ϕ, E, I)*-covering set exists.*

8. Concluding remarks

8.1. Applications

In this section let Γ be a pointclass such that the conclusion of Theorem 3.1 holds for all ϕ whose predicates for universally Baire sets are restricted to sets in Γ. If $\Gamma \subseteq \texttt{Borel}$ then the results below are provable in ZFC.

For $B \subseteq \omega_1$ let $B^* = \{x \in \mathbb{R} : x \text{ codes an ordinal in } B\}$ be its code.

Proposition 8.1. *Assume* $B \subseteq \omega_1$. *If* $B^* \in \Gamma$ *then there is a club* $C \subseteq \omega_1$ *that is either contained in or disjoint from* B.

Proof – Assume B is stationary. Then in an ω_1-preserving forcing extension in which no new reals are added there is a club $D \subseteq B$. Then there is a function $f\colon \omega_1 \to \omega_1$ such that for every $\alpha < \omega_1$ closed under f we have $\alpha \in B^*$. By Theorem 3.1 such a function exists in V and therefore B contains a club. □

Proposition 8.2.

1. *If* $\Gamma \supseteq \mathcal{P}(\mathbb{R})^M$ *for some inner model* M *such that* $\omega_1^M = \omega_1$ *then* ω_1 *is measurable in* M. [2]

[2] As pointed out to us by Ralf Schindler and Alessandro Andretta, this implies that for every δ there is an inner model with δ measurable cardinals [25].

2. *If Γ contains all codes for subsets of ω_1 constructible from x then $x^{\#}$ exists.*

PROOF – (1) By Proposition 8.1 the club filter on ω_1 in M is an ultrafilter. (2) By Proposition 8.1, the club filter on ω_1 is an $L[x]$-ultrafilter. □

If $n \geq 1$ then $[X]^n = \{s \subseteq X \mid |s| = n\}$. For $K \subseteq [X]^n$ a set $Y \subseteq X$ is K-homogeneous if $[Y]^n \subseteq K$.

Proposition 8.3. *For every partition $[\mathbb{R}]^n = K_0 \cup K_1$ such that K_0 is a continuous image of a set in Γ the statement 'there is an uncountable K_0-homogeneous set' is forcing-absolute.*

PROOF – Let A be such that $f[A] = K_0$ for some continuous f. Note that the graph of f is in Γ since it is Borel. Let ϕ be a sentence 'there exist an uncountable $X \subseteq \mathbb{P}$ and $Y \subseteq A$ such that $[X]^n \subseteq f[Y]$' and apply Theorem 3.1. □

The ZFC case of Proposition 8.3, when $\Gamma =$ `Borel` was proved in [20]. By Proposition 8.7 this theorem is optimal since even the Π^1_1 case cannot be proved in ZFC. It is known (see [30]) that for a Borel $K \subseteq [\mathbb{R}]^2$ 'there exists an uncountable K-homogeneous set' is strictly weaker than 'there exists a perfect K-homogeneous set.' Therefore Proposition 8.3, as well as the following two propositions, do not follow from forcing-absoluteness for universally Baire sets of reals.

A forcing notion $\mathbb{P}$ is in Γ if $\mathbb{P} \subseteq \mathbb{R}$ and $\mathbb{P}$, $\leq_{\mathbb{P}}$ and $\perp_{\mathbb{P}}$ are all in Γ. Proposition 8.3 immediately implies the following.

Proposition 8.4. *If P is a ccc forcing in $p[\Gamma]$ (the pointclass of continuous images of sets in Γ) then it is productively ccc.*

In particular, assuming there exists a proper class of Woodin cardinal all universally Baire ccc posets are productively ccc. In [4] productiveness of various projective ccc forcings with ccc forcings in Solovay's model was proved, starting from more modest large cardinal assumptions. The case when Γ is `Borel` is a well-known result of Shelah ([16]). Apparently the only known proof of Shelah's theorem uses Keisler's completeness theorem for $L_{\omega_1\omega}(Q)$, and our proof is modeled on it.

A variant of Proposition 8.3 may be worth stating. If $A \subseteq X \times Y$ and $x \in X$ then $A_x = \{y \mid (x,y) \in A\}$. First note that if there are class many Woodin cardinals and $A \subseteq \mathbb{R}^2$ is universally Baire, then the set $B = \{x \mid A_x$ is uncountable$\}$ is universally Baire. This is because under our assumption every set projective in a universally Baire set is universally Baire [21, Exercise 3.3.11]) and $x \in B$ if and only if $(\exists y \in \mathbb{R}^{\mathbb{N}})(\forall z \in A_x)(\exists n \in \mathbb{N})z = y(n)$. A higher-dimensional variant of this fact is also true.

Proposition 8.5. *Assume there are class many measurable Woodin cardinals. If $n \geq 1$ and $A \subseteq \mathbb{R} \times [\mathbb{R}]^n$ is universally Baire, then the set $\{x \mid$ there is an uncountable $Y \subseteq \mathbb{R}$ such that $[Y]^n \subseteq A_x\}$ is universally Baire.*

Proof – With $\phi(x, A)$ being an $\mathcal{L}^{\mathrm{uB}}$-formula 'there is an uncountable A_x-homogeneous set of reals' this is an immediate consequence of Corollary 6.1. □

We leave other applications, such as those involving homogeneous sets for square bracket partition relations or homogeneous clubs, to an interested reader.

8.2. Limitations

The case of Theorem 6.2 when $\phi \in \mathcal{L}^{\texttt{Borel}}$ and does not include a predicate for NS_{ω_1} is a well-known ZFC consequence of Keisler's completeness theorem for $L_{\omega_1\omega}(Q)$ ([19]). It was apparently stated explicitly for the first time in [9]. This was proved by associating an $L_{\omega_1\omega}(Q)$ sentence ϕ^M in an expanded language to each $\mathcal{L}^{\texttt{Borel}}$ sentence ϕ such that a (reduction of a) standard model for ϕ^M is a correct model for ϕ. By Proposition 8.6 below this method cannot be applied if Γ includes all Π^1_1 sets. The proof relies on the following observation, communicated to us by Boban Velickovic and included here with his kind permission ($\mathbf{A}, \mathbf{E}$ are unary and binary relation symbols and A, E are their interpretations).

Lemma 8.1. *Assume that ϕ is an $L_{\omega_1\omega}(Q)$ sentence in a language including $\{\mathbf{A}, \mathbf{E}\}$ such that $(A, E)^{\mathfrak{X}}$ is an uncountable well-ordering in some correct* *model $\mathfrak{X}$ of ϕ. Then $(A, E)^{\mathfrak{Y}}$ is not a well-ordering in some corect model $\mathfrak{Y}$ of ϕ.*

PROOF – We assume familiarity with the theory of $L_{\omega_1\omega}(Q)$, as presented in [19]. Assume the contrary. Let $\alpha = \sup\{\text{o.t.}(A, <)^{(\mathfrak{Y},q)} | (\mathfrak{Y}, q)$ a weak model of $\phi\}$. Since the set of all weak models of ϕ is Borel, by the Boundedness Theorem ([18, Theorem 31.2]) α is countable. Let M be a countable transitive model of a large enough fragment of ZFC such that $\omega_1^M > \alpha$. By Keisler's theorem, there exists a correct model $\mathfrak{Y}$ of ϕ in M. If $q = \{B \subseteq Y | B$ is definable and uncountable in $\mathfrak{Y}\}$ then $(\mathfrak{Y}, q)$ is a weak model of ϕ and $\text{o.t.}(A, E)^{\mathfrak{Y}} > \alpha$, a contradiction. □

Proposition 8.6. *There is a Π^1_1 set A such that there is no $L_{\omega_1\omega}(Q)$ sentence ϕ in language containing a predicate* **A** *for A with the property that all standard models of ϕ are correct about A.*

PROOF – Assume the contrary, that for every Π^1_1 set A there is a sentence ϕ such that every standard model of ϕ is correct about A. Let A be the Π^1_1 set of all subsets $\mathbb{N}^2$ that code a well ordering. Let ϕ be a sentence of $L_{\omega_1\omega}(Q)$ such that each standard model of ϕ is correct for A. Let ψ be the conjunction of ϕ and the sentence '**E** is an ω_1-like ordering of **B** and every proper initial segment is order-isomorphic to some $x \in A$.' Then in every model of ψ we have that E is a well-ordering of B in type ω_1, contradicting Lemma 8.1. □

Proposition 8.6 in particular shows that the approach of [9] cannot be applied to prove the version of Theorem 6.2 for $\mathcal{L}^{\Pi^1_1}$. The following Proposition shows that the analogue of this theorem for $\mathcal{L}^{\Pi^1_1}$ cannot be proved in ZFC.

Proposition 8.7. *If the statement "ϕ has a correct model" is forcing absolute for every ϕ in $\mathcal{L}^{\Pi^1_1}$ then ω_1 is inaccessible in L.* [3]

PROOF – Assume otherwise. Let x be a real such that $\omega_1^{L[x]} = \omega_1$. Then there exists an uncountable $\Pi^1_1(x)$ set A with no perfect subset. Then 'A has

[3] Ralf Schindler pointed out that if the assumption of Proposition 8.7 is forcing absolute, then all sets have sharps. The argument uses the following standard consequence of Covering Lemma: If $x^{\#}$ does not exist for some x, then in some forcing extension there is a real r such that $L[r]$ correctly computes ω_1. This remark also applies to Proposition 8.9 below.

an uncountable subset' is a sentence of $\mathcal{L}^{\mathbf{\Pi}^1_1}$ true in V but false in its extension obtained by collapsing ω_1. □

Assume ϕ is a sentence of $L_{\omega_1\omega}(Q)$ whose language includes $\{\mathbf{X},\mathbf{E}\}$ such that the interpretation of $(\mathbf{X},\mathbf{E})$ is an ω_1-like linear order in every model of ϕ. Many arguments using $L_{\omega_1\omega}(Q)$ in set theory would be simplified if the statement

$W(\phi)$: Sentence ϕ has a model in which the interpretation of $(\mathbf{X},\mathbf{E})$ is a well-ordering

were known to be forcing-absolute.

Proposition 8.8. *Assume there are class many Woodin cardinals. Using the notation from the last paragraph, $W(\phi)$ is forcing-absolute.*

Proof – By [21, Theorem 2.5.10] it will suffice to show that if $W(\phi)$ holds in a forcing extension then it holds in V. Assume $W(\phi)$ holds in some forcing extension. By Theorem 3.4 there is a transitive model M^* of ZFC with correct ω_1 such that $W(\phi)$ holds in M^*. Then a standard well-founded model of ϕ is a standard well-founded model of ϕ in V. □

The conclusion of Proposition 8.8 has some large cardinal strength.

Proposition 8.9. *Assume $W(\phi)$ is forcing-absolute for every ϕ in $L_{\omega_1\omega}(Q)$. Then ω_1 is inaccessible in L.*

Proof – Assume $\omega_1 = \omega_1^{L(x)}$ for a real x. There exists an uncountable $\Pi^1_1(x)$ set B with no perfect subset (see e.g., [28, 5.A.8] or [24, 3.19]). Let A be the Π^1_1 set of codes for countable ordinals. Since A is Π^1_1-complete, we have $B = f^{-1}(A)$ for some continuous function f. Let ϕ be an $L_{\omega_1\omega}(Q)$ formula in the language $\{\mathbf{A},\mathbf{B},\mathbf{f},\mathbf{X},\mathbf{E}\}$ saying:

1. $\mathbf{A},\mathbf{B}$ are sets of reals,

2. each element of $\mathbf{B}$ codes a countable linear ordering,

3. $\mathbf{E}$ is an ω_1-well-ordering of a set $\mathbf{X}$,

4. each initial interval $(-\infty, x]$ of $(\mathbf{X}, \mathbf{E})$ is coded by some element of $\mathbf{B}$,

5. The interpretation of $\mathbf{f}$ is the restriction of $\mathbf{f}$ to the reals of the model,

6. $\mathbf{A}$ is the $\mathbf{f}$-preimage of $\mathbf{B}$.

If in a model of ϕ the interpretation of $(\mathbf{B}, \mathbf{E})$ is well-founded, then it is correct about B and A. But such a model contains uncountably many elements of A, therefore if $\omega_1^{L(x)}$ is collapsed then ϕ cannot have such models. □

Problem 8.1. *Using large cardinals develop a forcing-absolute syntax for $\mathcal{L}^{\mathrm{uB}}$ and prove the corresponding completeness theorem.*

A positive solution to this problem would be a substantial improvement of Theorem 6.1.

8.3. Special trees in $L(\mathbb{R})$

A consequence of Theorem 3.1 is that if $T \subseteq \mathbb{R}^{<\omega_1}$ is a tree coded by a universally Baire set, then large cardinals imply that the statement 'T has an ω_1-branch' is forcing-absolute. Such an ω_1-branch constructs an ω_1-sequence of reals and therefore cannot exist in $L(\mathbb{R}, T)$ (again, assuming sufficient large cardinals to ensure that universally Baire sets satisfy the regularity properties), but it is natural to ask whether it is possible to determine whether T has an ω_1-branch in V by working in the model $L(\mathbb{R}, T)$. It turns out that this question has already been considered.

After the first author presented some of our results at the Annual Meeting of ASL at Stanford in March 2005, John Steel communicated Theorem 8.1 to us. The following definition is also due to him.

Assume $T \subseteq \mathbb{R}^{<\omega_1}$ is coded by a universally Baire set. Write

$$T^+ = \{p \in \mathbb{R}^{<\omega_1} \mid (\forall \alpha < \mathrm{dom}(p)) p \restriction \alpha \in T\}.$$

If there is a function g such that $g(p)$ is a surjection from ω onto $\mathrm{dom}(p)$ for every $p \in T^+$ we say that T is specialized by g. Every tree that is covered by countably many antichains is specialized by a function definable from the sequence of antichains. In the presence of the Axiom of Choice every tree is specialized by a function, but the following justifies the name in the context of the inner models of determinacy. A 'universally Baire function' is a function whose graph is a universally Baire set of pairs of reals.

Lemma 8.2. *Assume there are two Woodin cardinals $\delta < \lambda$ and T is a universally Baire tree specialized by a universally Baire function g. Then T has no ω_1 branches.*

PROOF – Assume the contrary. By Theorem 3.3 both T and g are weakly homogeneously Suslin, and by Lemma 3.3 there is a T, g-iterable model $(N, \mathrm{NS}^N_{\omega_1})$ satisfying 'T has an ω_1-branch,' p. (We only need $\mathrm{NS}^N_{\omega_1}$ to be precipitous and the generic ultrapower to be correct about T and g.) Let $j\colon N \to N^*$ be the generic ultrapower embedding. Then $p = j(p) \restriction \omega_1^N$ belongs to N^*, hence by elementarity and T-correctness $p \in T$. Thus $g(p)$, a surjection from ω onto ω_1^N, is a real in N^*. If τ is a $\mathrm{NS}^+_{\omega_1}$-name in N for $g(p)$, then it is forced that $\tau = g(p)$, and therefore $g(p) \in N$, a contradiction to the assumption that p is uncountable. □

Lemma 8.2 has a converse.

Theorem 8.1 (Steel). *Suppose that there exist infinitely many Woodin cardinals below a measurable cardinal. Let $T \subset \mathbb{R}^{<\omega_1}$ be a tree in $L(\mathbb{R})$. Then exactly one of the following holds:*

- *T has an uncountable branch in V.*
- *T is specialized by a function in $L(\mathbb{R})$.*

The relevance of the theorem is that for each Σ_1 statement ϕ for $\langle H(\aleph_2); A, \in\rangle$, where A is a set of reals in $L(\mathbb{R})$ as in Theorem 3.1 there is a tree in $L(\mathbb{R})$ which has an uncountable branch in V if and only if ϕ holds (such trees are typically called the tree of attempts to produce a witness for ϕ). The second conclusion of the theorem is a statement in $L(\mathbb{R})$, and thus forcing-absolute. The theorem then gives an absolute impediment witnessing the failure of a given Σ_1 statement. At a key place in the proof of Theorem 8.1 a rather deep result from the inner model theory is employed to construct a T-iterable model (see Section 3.3) of 'T has an ω_1-branch.' As pointed out to us by Stuart Zoble, extending to arbitrary universally Baire sets necessitates a different proof since the required inner model theory result is not known to be true in this context. A version of Steel's theorem for any tree T coded by a universally Baire set is proved in [10, Section 5].

Acknowledgements. We would like to thank Richard Ketchersid for discussions on this topic and Teruyuki Yorioka for useful remarks on an early version of the present paper.

References

[1] ABRAHAM, U. and SHELAH, S.: A Δ^2_2 well-order of the reals and incompactness of $L(Q^{\rm MM})$, *Ann. Pure Appl. Logic* **59** (1) (1993), 1–32.

[2] ABRAHAM, U. and SHELAH, S.: Coding with ladders a well ordering of the reals, *J. Symbolic Logic* **67** (2) (2002), 579–597.

[3] BAGARIA, J.: Bounded forcing axioms as principles of generic absoluteness, *Arch. Math. Logic* **39** (6) (2000), 393–401.

[4] BAGARIA, J. and BOSCH, R.: Solovay models and forcing extensions, *J. Symbolic Logic* **69** (3) (2004), 742–766.

[5] BARTOSZYNSKI, T.: Invariants of measure and category, in M. Foreman and A. Kanamori, editors, *Handbook of Set Theory*, to appear.

[6] BLASS, A.: Cardinal invariants of the continuum, in M. Foreman and A. Kanamori, editors, *Handbook of Set Theory*, to appear.

[7] CIESIELSKI, K. and PAWLIKOWSKI, J.: *The covering property axiom, CPA. A combinatorial core of the iterated perfect set model*, Cambridge Tracts in Mathematics **164**, Cambridge University Press (2004).

[8] FARAH, I.: A proof of the Σ^2_1-absoluteness theorem, in S. Gao et al., editors, *Advances in Logic*, Contemporary Mathematics, Amer. Math. Soc., to appear.

[9] FARAH, I., HRUŠÁK, M., and MARTÍNEZ RANERO, C.A.: A countable dense homogeneous set of size $\aleph_1$, *Fundamenta Mathematicae* **186** (2005), 71–77.

[10] Farah, I., Ketchersid, R., Larson, P.B., and Magidor, M.: Absoluteness for universally Baire sets and the uncountable II, preprint (2005).

[11] Feng, Q., Magidor, M., and Woodin, W.H.: Universally Baire sets of reals, in H. Judah, W. Just, and W.H. Woodin, editors, *Set Theory of the Continuum*, Springer-Verlag (1992), 203–242.

[12] Foreman, M. and Magidor, M.: Large cardinals and definable counterexamples to the continuum hypothesis, *Ann. Pure Appl. Logic* **76** (1995), 47–97.

[13] Foreman, M., Magidor, M., and Shelah, S.: Martin's Maximum, saturated ideals, and nonregular ultrafilters I, *Ann. of Math.* (2) **127** (1) (1988), 1–47.

[14] Goldring, N.: Woodin cardinals and presaturated ideals, *Ann. Pure Appl. Logic* **55** (1992), 285–303.

[15] Hamkins, J.D.: A simple maximality principle, *The Journal of Symbolic Logic* **68** (2003), 527–550.

[16] Ihoda, J.I. and Shelah, S.: Souslin forcing, *J. Symbolic Logic* **53** (1988), 1188–1207.

[17] Kanamori, A.: *The higher infinite: large cardinals in set theory from their beginnings*, Perspectives in Mathematical Logic, Springer–Verlag, Berlin–Heidelberg–New York (1995).

[18] Kechris, A.S.: *Classical descriptive set theory*, Graduate texts in mathematics **156**, Springer (1995).

[19] Keisler, H.J.: Logic with the quantifier "There exists uncountably many", *Annals of Mathematical Logic* **1** (1970), 1–93.

[20] Kubiś, W. and Shelah, S.: Analytic colorings, *Ann. Pure Appl. Logic* **121** (2-3) (2003), 145–161.

[21] Larson, P.B.: *The stationary tower*, University Lecture Series **32**, American Mathematical Society, Providence, RI (2004). Notes on a course by W. Hugh Woodin.

[22] LARSON, P.B.: Forcing over models of determinacy, in M. Foreman and A. Kanamori, editors, *Handbook of Set Theory*, to appear.

[23] LÉVY, A.: A hierarchy of formulas in set theory, *Memoirs of the Amer. Math. Soc.* **57** (1965).

[24] MANSFIELD, R. and WEITKAMP, G.: *Recursive aspects of descriptive set theory*, Oxford Logic Guides **11**, The Clarendon Press Oxford University Press, New York (1985). With a chapter by Stephen Simpson.

[25] MARTIN, D.A. and MITCHELL, W.: On the ultrafilter of closed, unbounded sets, *J. Symbolic Logic* **44** (4) (1979), 503–506.

[26] MARTIN, D.A. and STEEL, J.: A proof of projective determinacy, *Journal of the American Mathematical Society* **2** (1989), 71–125.

[27] MOORE, J.T.: A solution to the L space problem, *J. Amer. Math. Soc.* **19** (3) (2006), 717–736.

[28] MOSCHOVAKIS, Y.N.: *Descriptive set theory*, Studies in logic and foundations of mathematics **100**, North–Holland (1980).

[29] NEEMAN, I.: *The determinacy of long games*, de Gruyter Series in Logic and its Applications **7**, Walter de Gruyter GmbH & Co. KG, Berlin (2004).

[30] SHELAH, S.: Borel sets with large squares, *Fund. Math.* **159** (1999), 1–50.

[31] SHELAH, S. and WOODIN, W.H.: Large cardinals imply that every reasonably definable set of reals is Lebesgue measurable, *Israel Journal of Mathematics* **70** (1990), 381–394.

[32] SHOENFIELD, J.R.: The problem of predicativity, in Y. Bar-Hillel et al., editors, *Essays on the Foundations of Mathematics*, The Magens Press, Jerusalem (1961), 132–142.

[33] SOLOVAY, R.: Real-valued measurable cardinals, in D. Scott, editor, *Axiomatic set theory*, Proceedings of symposia in pure mathematics **13** (1971), 397–428.

[34] STAVI, J. and VÄÄNÄNEN, J.: Reflection principles for the continuum, in *Logic and algebra*, Contemp. Math. **302**, Amer. Math. Soc., Providence, RI (2002), 59–84.

[35] STEEL, J.: The derived model theorem, preprint (2003).

[36] TODORCEVIC, S.: Partitioning pairs of countable ordinals, *Acta Math.* **159** (1987), 261–294.

[37] TODORCEVIC, S.: *Partition Problems in Topology*, Contemporary mathematics **84**, American Mathematical Society, Providence, Rhode Island (1989).

[38] TODORCEVIC, S.: Universally meager sets and principles of generic continuity, preprint (2004).

[39] TODORCEVIC, S.: A letter to John Steel, March 14 1991.

[40] WOODIN, W.H.: Set theory after Russell: the journey back to Eden, in *One hundred years of Russell's paradox*, de Gruyter Ser. Log. Appl. **6**, de Gruyter, Berlin (2004), 29–47.

[41] WOODIN, W.H.: Σ^2_1-absoluteness, Handwritten note (1985).

[42] WOODIN, W.H.: *The Axiom of Determinacy, forcing axioms and the nonstationary ideal*, de Gruyter Series in Logic and Its Applications **1**, de Gruyter, Berlin New York (1999).

[43] ZAPLETAL, J.: Descriptive set theory and definable forcing, *Mem. Amer. Math. Soc.* **167** (2004).

[44] ZAPLETAL, J.: *Forcing idealized*, in preparation.

Large Cardinals and L-like Universes

Sy-David Friedman

Contents

1. Introduction

There are many different ways to extend the axioms of ZFC. One way is to adjoin the axiom $V = L$, asserting that every set is constructible. This axiom has many attractive consequences, such as the generalised continuum hypothesis (GCH), the existence of a definable wellordering of the class of all sets, as well as strong combinatorial principles, such as $\diamondsuit$, $\square$ and the existence of morasses.

However $V = L$ adds no consistency strength to ZFC. As many interesting set-theoretic statements have consistency strength beyond ZFC, it is common in set theory to assume at least the existence of inner models of V which contain large cardinals.

Can we simultaneously have the advantages of both the axiom of constructibility and the existence of large cardinals? Unfortunately even rather modest large cardinal hypotheses, such as the existence of a measurable cardinal, refute $V = L$. We can however hope for the following compromise:

V is an "L-like" model containing large cardinals.

In this article we explore the possibilities for this assertion, for various notions of "L-like" and for various types of large cardinals.

There are two approaches to this problem. The first approach is via the

Inner model program. Show that any universe with large cardinals has an L-like inner model with large cardinals.

The inner model program, through use of fine structure theory and the theory of iterated ultrapowers, has succeeded in producing very L-like inner models containing many Woodin cardinals.

An alternative approach is given by the

Outer model program. Show that any universe with large cardinals has an L-like outer model with large cardinals.

We will show that L-like outer models with extremely large cardinals can be obtained using the method of iterated forcing.

*The author wishes to thank the Austrian Science Fund (FWF) for its generous support and the Centre de Recerca Matemàtica (CRM), where this article was written in the month of September 2005.

2. Large cardinals

A cardinal κ is inaccessible iff it is uncountable, regular and larger than the power set of any smaller cardinal. It is measurable iff there is a κ-complete, nonprincipal ultrafilter on κ.

Measurability is equivalent to a property expressed in terms of embeddings, and stronger large cardinal properties are also expressed in this way. As usual, V denotes the universe of all sets. Let M be an inner model, i.e., a transitive proper class that satisfies the axioms of ZFC. A class function $j : V \to M$ is an embedding iff it preserves the truth of formulas with parameters in the language of set theory and is not the identity. If j is an embedding then there is a least ordinal κ such that $j(\kappa) \neq \kappa$, called the critical point of j, which is a measurable cardinal.

For an ordinal α, $j : V \to M$ is α-strong iff V_α is contained in M. A cardinal κ is α-strong iff there is an α-strong embedding with critical point κ. Strong means α-strong for all α.

Kunen ([10]) showed that no embedding is strong. However a cardinal can be strong, as embeddings witnessing its α-strength can vary with α. Stronger properties are obtained by requiring $j : V \to M$ to have strength depending on the image under j of its critical point. For example, κ is superstrong iff there is a nontrivial elementary embedding $j : V \to M$ with critical point κ which is $j(\kappa)$-strong. An important weakening of superstrength is the property that for each $f : \kappa \to \kappa$ there is a $\bar{\kappa} < \kappa$ closed under f and a nontrivial elementary embedding $j : V \to M$ with critical point $\bar{\kappa}$ which is $j(f)(\bar{\kappa})$-strong; such κ are known as Woodin cardinals. The consistency strength of the existence of a Woodin cardinal is strictly between that of a strong cardinal and a superstrong cardinal.

We can demand more than superstrength. A cardinal κ is hyperstrong iff it is the critical point of an embedding $j : V \to M$ which is $j(\kappa) + 1$-strong. For a finite $n > 0$, n-superstrength is obtained by requiring j to be $j^n(\kappa)$-strong, where $j^1 = j$, $j^{k+1} = j \circ j^k$. Finally, κ is ω-superstrong iff it is the critical point of an embedding $j : V \to M$ which is n-superstrong for all n. Kunen's result [10] shows that no embedding j with critical point κ is $j^\omega(\kappa) + 1$-strong, where $j^\omega(\kappa)$ is the supremum of the $j^n(\kappa)$ for finite n.

3. The inner model program

If κ is inaccessible, then κ is also inaccessible in L, the most L-like model of all. This is not the case for measurability, however if κ is measurable then κ is measurable in an inner model $L[U]$, where U is an ultrafilter on κ, which has a definable wellordering and in which GCH, $\diamondsuit$, $\square$ hold and gap 1 morasses exist. For a strong cardinal κ there is a similarly L-like inner model $L[E]$ in which κ is strong, where E now is not a single ultrafilter, but rather a sequence of generalised ultrafilters, called extenders. More recent work yields similar results for Woodin cardinals, and even for Woodin limits of Woodin cardinals (see [12]).

However, progress beyond that has been impeded by the so-called iterability problem.

4. The outer model program

How can we obtain L-like outer models with large cardinals? For inaccessibles one has the following result of Jensen (see [1]):

Theorem 4.1 (L-coding). *There is a generic extension $V[G]$ of V such that*

a. *ZFC holds in $V[G]$.*

b. *$V[G] = L[R]$ for some real R.*

c. *Every inaccessible cardinal of V remains inaccessible in $V[G]$.*

There are similar $L[U]$ and $L[E]$ coding theorems (see [6] for the former), providing outer models of the form $L[U][R]$ and $L[E][R]$, R a real, which are just as L-like as $L[U]$ and $L[E]$, preserving measurability and strength, respectively.

However the approach via coding is limited in its use. Obtaining L-like outer models via coding depends on the existence of L-like inner models, such as $L[U]$ or $L[E]$, which, as we have observed, are not known to exist beyond Woodin limits of Woodin cardinals. And there are problems with the coding method itself which arise already just past a strong cardinal.

A more promising approach is to use iterated forcing. To illustrate this, consider the problem of making the GCH true in an outer model. Begin with an arbitrary model V of ZFC. Using forcing, we can add a function from $\aleph_1$ onto $2^{\aleph_0}$ without adding reals, thereby making CH true. By forcing again, we add a function from (the possibly new) $\aleph_2$ onto (the possibly new) $2^{\aleph_1}$ without adding subsets of $\aleph_1$, thereby obtaining $2^{\aleph_1} = \aleph_2$. Continue this indefinitely (via a reverse Easton iteration) and the result is a model of the GCH.

Do we preserve large cardinal properties if we make GCH true in this way? The answer is Yes.

Theorem 4.2 (Large cardinals and the GCH). *If κ is superstrong then there is an outer model in which κ is still superstrong and the GCH holds. The same holds for hyperstrong, n-superstrong for finite n and ω-superstrong.*

PROOF – First we describe in more detail the above iteration to make GCH true. By induction on α we define the iteration P_α of length α: P_0 is the trivial forcing. For limit λ, P_λ is the inverse limit of the P_α, $\alpha < \lambda$, if λ is singular and is the direct limit of the P_α, $\alpha < \lambda$, if λ is regular. For successor $\alpha + 1$, $P_{\alpha+1} = P_\alpha * Q_\alpha$, where Q_α is the forcing that collapses $2^{\aleph_\alpha}$ to $\aleph_{\alpha+1}$ using conditions of size at most $\aleph_\alpha$. The desired class forcing P is the direct limit of the P_α's. For any cardinal κ of the form $\beth_{\alpha+1}$, the entire iteration P can be factored as $P_\kappa * P^\kappa$, where P_κ has a dense subset of size κ and P^κ is κ^+-closed. In particular, strongly inaccessible cardinals remain strongly inaccessible after forcing with P.

Now suppose that κ is superstrong, witnessed by the embedding $j : V \to M$, and that G is P-generic. Let P^* denote M's version of P. To show that κ is superstrong in $V[G]$, it suffices to find a P^*-generic G^* which satisfies $G^*_{j(\kappa)} = G_{j(\kappa)} \cap V_{j(\kappa)}$ and which contains $j[G]$, the pointwise image of G under j, as a subclass. For then, we extend $j : V \to M$ to $j^* : V[G] \to M[G^*]$ by sending σ^G to $j(\sigma)^{G^*}$. The property $j[G] \subseteq G^*$ implies that j^* is well-defined and elementary. And $V_{j(\kappa)}$ of $V[G]$ is the same as $V_{j(\kappa)}[G_{j(\kappa)} \cap V_{j(\kappa)}] = V_{j(\kappa)}[G^*_{j(\kappa)}]$ and therefore belongs to $M[G^*]$, as $V_{j(\kappa)}$ belongs to M.

Now P^*_α is the same as P_α for $\alpha < j(\kappa)$, as j is a superstrong embedding. The first difference between P^* and P is at $j(\kappa)$: $P^*_{j(\kappa)}$ is the direct limit of the P_α, $\alpha < j(\kappa)$, as $j(\kappa)$ is inacessible in M; but $j(\kappa)$ is not necessarily regular in

V and therefore it is possible that $P_{j(\kappa)}$ is the inverse limit of the P_α, $\alpha < j(\kappa)$. So we cannot simply choose $G^*_{j(\kappa)}$ to be $G_{j(\kappa)}$, as the latter is generic for the wrong forcing.

But this problem is easily fixed: As $j(\kappa)$ is in fact Mahlo in M, it follows that $P^*_{j(\kappa)}$ has the $j(\kappa)$-cc in M. So any $G^*_{j(\kappa)}$ contained in $P^*_{j(\kappa)}$ whose intersection with each P_α, $\alpha < j(\kappa)$, is P_α-generic must also be $P^*_{j(\kappa)}$-generic. It follows that we can take $G^*_{j(\kappa)}$ to simply be the intersection of $G_{j(\kappa)}$ with $P^*_{j(\kappa)}$. Notice that $G^*_{j(\kappa)}$ trivially contains the pointwise image of G_κ under j as j is the identity below κ.

Finally we must define a generic $G^{*,\ j(\kappa)}$ for the "upper part" $P^{*,\ j(\kappa)}$ of the P^* iteration, which starts at $j(\kappa)$ and is defined in the ground model $M[G^*_{j(\kappa)}]$. In addition, $G^{*,\ j(\kappa)}$ must contain the pointwise image of G^κ under j^*, where j^* is the lifting of j to $V[G_\kappa]$ and G^κ is generic for P^κ, the iteration starting at κ defined over the ground model $V[G_\kappa]$.

In fact, with a harmless additional assumption about j, this latter requirement completely determines $G^{*,\ j(\kappa)}$: We assume that $j : V \to M$ is given as an extender ultrapower embedding. This means that each element of M is of the form $j(f)(a)$, where a belongs to $V^M_{j(\kappa)} = V_{j(\kappa)}$ and f is a function (in V) with domain V_κ. This assumption is harmless, as if the initial $j : V \to M$ does not satisfy it, we can replace M by the transitive collapse $\bar{M}$ of H, the elementary submodel of M consisting of all $j(f)(a)$ of the above form, and replace j by $k \circ j$, where $k : H \simeq \bar{M}$.

Lemma 4.1. $j^*[G^\kappa]$ *generates a* $P^{*,\ j(\kappa)}$*-generic over* $M[G^*_{j(\kappa)}]$*, i.e., each predense subclass of* $P^{*,\ j(\kappa)}$ *which is definable over* $M[G^*_{j(\kappa)}]$ *has an element which is extended by a condition in* $j^*[G^\kappa]$.

PROOF – We only consider predense subsets D of $P^{*,\ j(\kappa)}$ in $M[G^*_{j(\kappa)}]$; a similar argument works for predense subclasses.

As j is given as an extender ultrapower embedding, D is of the form $\sigma^{G^*_{j(\kappa)}}$ where the name σ can be written as $j(f)(a)$ with $f : V_\kappa \to V$ in V and $a \in V_{j(\kappa)}$. Now using the κ^+-closure of P^κ, choose a condition p in G^κ which extends an element of $f(\bar{a})$ whenever $\bar{a}$ belongs to V_κ and $f(\bar{a})^{G_\kappa}$ is predense on $P^{*,\ \kappa}$. Then $j^*(p)$ obviously belongs to $j^*[G^\kappa]$ and extends an element of

$j(f)(a)^{G^*_{j(\kappa)}} = \sigma^{G^*_{j(\kappa)}} = D$, as desired. □

This completes the construction of G^* and therefore the proof that P preserves superstrong cardinals.

Now suppose that κ is hyperstrong. Again we need to find a P^*-generic G^* containing $j[G]$ as a subclass. The forcings $P_{j(\kappa)+1} = P_{j(\kappa)} * Q_{j(\kappa)}$ and $P^*_{j(\kappa)+1}$ agree as $j(\kappa)$ is regular in V and M contains $V_{j(\kappa)+1}$. Also, $j^*[g_\kappa]$, where j^* is the lifting of j to $V[G_\kappa]$ and g_κ is the Q_κ-generic chosen by G at stage κ of the iteration, is a set of conditions in $Q_{j(\kappa)}$ which belongs to $M[G_{j(\kappa)}]$ and has size 2^κ there; therefore $j^*[g_\kappa]$ has a lower bound in $Q_{j(\kappa)}$. By choosing our generic G so that $g_{j(\kappa)}$ includes this lower bound, we can succeed in lifting j to $V[G_{\kappa+1}]$. We may assume that $j : V \to M$ is given by a hyperextender; this means that each element of M is of the form $j(f)(a)$ where f is a function in V with domain $V_{\kappa+1}$ and a is an element of $V_{j(\kappa)+1}$. Then we can use the argument of Lemma 4.1 to generate the entire generic G^* containing $j[G]$.

The case of n-superstrongs raises a new difficulty. We first treat the case $n = 2$. As in the superstrong case, P and P^* may take different limits at $j^2(\kappa)$, as the latter may be singular in V. As in that case, we can obtain a $P^*_{j^2(\kappa)}$-generic by intersecting $G_{j^2(\kappa)}$ with $P^*_{j^2(\kappa)}$. However we must also ensure that $G^*_{j^2(\kappa)}$ contain $j[G_{j(\kappa)}]$ as a subset. Write $P_{j(\kappa)}$ as $P_\kappa * P^\kappa_{j(\kappa)}$; it suffices to arrange that $G^{*,\ j(\kappa)}_{j^2(\kappa)}$ contain $j^*[G^\kappa_{j(\kappa)}]$ as a subset, where j^* is the lifting of j to $V[G_\kappa]$ and $G^\kappa_{j(\kappa)}$ is $P^\kappa_{j(\kappa)}$-generic over $V[G_\kappa]$.

We argue as follows. If $j[j(\kappa)]$ is bounded in $j^2(\kappa)$ then the set of conditions $j^*[G^\kappa_{j(\kappa)}]$ has a lower bound in $P^{*,\ j(\kappa)}_{j^2(\kappa)} \subseteq P^{j(\kappa)}_{j^2(\kappa)}$. (Actually, this first case can be ruled out by choosing $j^2(\kappa)$ minimally; see Lemma 4.4 below.) Otherwise $j^2(\kappa)$ is singular, so $P_{j^2(\kappa)}$ is an inverse limit and again the set of conditions $j^*[G^\kappa_{j(\kappa)}]$ has a lower bound in $P^{j(\kappa)}_{j^2(\kappa)}$. We therefore assume that our generic G has been chosen so that $G^{j(\kappa)}_{j^2(\kappa)}$ contains the greatest lower bound of $j^*[G^\kappa_{j(\kappa)}]$. Then we can take $G^*_{j^2(\kappa)}$ to be the intersection of $G_{j^2(\kappa)}$ with $P^*_{j^2(\kappa)}$ and thereby obtain $j[G_{j(\kappa)}] \subseteq G^*_{j^2(\kappa)}$. This allows us to lift j to $V[G_{j(\kappa)}]$. Then we can use the argument from Lemma 4.1 to generate the entire generic G^* containing $j[G]$.

For the case $n > 2$ the argument is similar; we must choose $G^{j^{n-1}(\kappa)}_{j^n(\kappa)}$ to contain the greatest lower bound of $j^*[G^{j^{n-2}(\kappa)}_{j^{n-1}(\kappa)}]$, where j^* is the lifting of j to the model $V[G_{j^{n-2}(\kappa)}]$.

Finally we consider ω-superstrength. Again we must choose G^* to be P^*-generic over M and to contain the pointwise image of G under j. Let $j^\omega(\kappa)$ denote the supremum of the $j^n(\kappa)$, $n \in \omega$. As before it suffices to find $G^*_{j^\omega(\kappa)}$ which is $P^*_{j^\omega(\kappa)}$-generic and contains $j[G_{j^\omega(\kappa)}]$ as a subset. Note that $j[G_\kappa] = G_\kappa$ is trivially contained in $G_{j^\omega(\kappa)}$ and $j^*[G^\kappa_{j^\omega(\kappa)}]$ has a lower bound in $P^{j(\kappa)}_{j^\omega(\kappa)}$ (as defined in $V[G_{j(\kappa)}]$); by choosing $G^{j(\kappa)}_{j^\omega(\kappa)}$ to contain this lower bound we can take $G^*_{j^\omega(\kappa)}$ to be $G_{j^\omega(\kappa)}$ and thereby obtain $j[G_{j^\omega(\kappa)}] \subseteq G^*_{j^\omega(\kappa)}$. And again we can use the argument of Lemma 4.1 to generate the entire generic G^* containing $j[G]$. So it only remains to show:

Lemma 4.2. *$G_{j^\omega(\kappa)} \cap P^*_{j^\omega(\kappa)}$ is $P^*_{j^\omega(\kappa)}$-generic over M.*

Proof – Suppose that $D \in M$ is dense on $P^*_{j^\omega(\kappa)}$ and write D as $j(f)(a)$ where f has domain $V_{j^\omega(\kappa)}$ and a belongs to $V_{j^{n+1}(\kappa)}$ for some n. (We may assume that every element of M is of this form.) Choose p in $G_{j^\omega(\kappa)}$ such that p reduces $f(\bar{a})$ below $j^n(\kappa)$ whenever $\bar{a}$ belongs to $V_{j^n(\kappa)}$ and $f(\bar{a})$ is open dense on $P_{j^\omega(\kappa)}$, in the sense that if q extends p then q can be further extended into $f(\bar{a})$ without changing q at or above $j^n(\kappa)$. Such a p exists using the $j^n(\kappa)^+$-closure of $P^{j^n(\kappa)}_{j^\omega(\kappa)}$ in $V[G_{j^n(\kappa)}]$. Then $j(p)$ belongs to $j[G_{j^\omega(\kappa)}]$ and reduces D below $j^{n+1}(\kappa)$. As $G_{j^{n+1}(\kappa)}$ is $P_{j^{n+1}(\kappa)}$-generic and P, P^* agree below $j^{n+1}(\kappa)$, it follows that $G_{j^\omega(\kappa)} \cap P^*_{j^\omega(\kappa)}$ intersects D, as desired. □

This completes the proof of Theorem 4.2. □

Another important property of L is the existence of a definable wellordering of the universe.

Theorem 4.3 (Large cardinals and definable wellorderings). *If κ is superstrong then there is an outer model in which κ is still superstrong and there is a definable wellordering of the universe. The same holds for hyperstrong, n-superstrong for finite n and ω-superstrong.*

Proof – By Theorem 4.2 we may assume the GCH. Let κ have one of the large cardinal properties mentioned in the theorem, as witnessed by the embedding $j : V \to M$. Choose λ to be a cardinal greater than $j^\omega(\kappa)$. By the method of L-coding (see Theorem 4.1), we can enlarge V without adding subsets of λ to a universe of the form $L[A]$, A a subset of λ^+. By the argument of Lemma 4.1 the embedding j lifts to $L[A]$ and therefore κ retains its large cardinal properties.

Now we introduce a definable wellordering. Perform a reverse Easton iteration of length λ^+, indexed by successor cardinals greater than λ^+, where at the i-th successor cardinal, an i^+-Cohen set is added iff i belongs to A. The result is that i belongs to A iff not every subset of the successor of the i-th successor cardinal is constructible from a subset of the i-th successor cardinal. Now the result of this iteration is a model of the form $L[B]$ where B is a subset of $\lambda^{(\lambda^+)}$, the "λ^+-th cardinal greater than λ". Repeat this to code B using the next interval of successor cardinals. Continuing this indefinitely yields a model with a wellordering definable from the parameter λ.

To eliminate the parameter λ, use a pairing function $f : \mathrm{Ord} \times \mathrm{Ord} \to \mathrm{Ord}$ on the ordinals and arrange that the universe is of the form $L[C]$ where C is a class of ordinals and for any i, i is in C iff some subset of the successor to the $f(i,j)$-th successor cardinal is not constructible from a subset of the $f(i,j)$-th successor cardinal, for all sufficiently large j. $\square$

Jensen's (global) $\square$ principle asserts the existence of a sequence $\langle C_\alpha \mid \alpha$ singular$\rangle$ such that C_α has ordertype less than α for each α and $C_{\bar{\alpha}} = C_\alpha \cap \bar{\alpha}$ whenever $\bar{\alpha} \in \mathrm{Lim}\ C_\alpha$. The following strengthens a result of Doug Burke [2]:

Theorem 4.4 (Superstrong cardinals and $\square$). *If κ is superstrong then there is an outer model in which κ is still superstrong and $\square$ holds.*

Proof – By Theorem 4.2, we may assume the GCH. Consider now the reverse Easton iteration P where at the regular stage α, Q_α is a P_α-name for the forcing which adds a $\square$-sequence on the singular limit ordinals less than α. A condition in Q_α is a sequence $\langle C_\beta \mid \beta \leq \gamma,\ \beta$ singular$\rangle$, $\gamma < \alpha$, such that C_β has ordertype less than β for each β and $C_{\bar{\beta}} = C_\beta \cap \bar{\beta}$ whenever $\bar{\beta}$ belongs to $\mathrm{Lim}\ C_\beta$.

Using the fact that P_α forces $\square$-sequences of any regular length less than α, it is easy to verify by induction that any condition in Q_α can be extended to have arbitrarily large length less than α. Also Q_α, and indeed the entire iteration from stage α on, is α-distributive.

Let P^* denote M's version of P. We want to construct G^* to be P^*-generic over M, to agree with G strictly below $j(\kappa)$ and to contain $j[G]$ as a subclass. As in earlier arguments, P and P^* agree strictly below $j(\kappa)$ but not necessarily at $j(\kappa)$, which is regular in M but may be singular in V; as before we take $G^*_{j(\kappa)}$ to be $G_{j(\kappa)} \cap P^*_{j(\kappa)}$. Our new task is to define a $Q^*_{j(\kappa)}$-generic g over $M[G^*_{j(\kappa)}]$.

Lemma 4.3. *Assume GCH and let $j : V \to M$ witness the superstrength of κ with $j(\kappa)$ minimal. Then $j(\kappa)$ has cofinality κ^+.*

Proof – Let $\langle f_i \mid i < \kappa^+ \rangle$ be a list of all functions from κ to κ. Then the sequence $\langle j(f_i) \mid i < \kappa \rangle$ belongs to M, as it equals $j(\langle f_i \mid i < \kappa^+ \rangle) \restriction \kappa$. For any ordinal $\alpha < \kappa^+$ we can use a bijection between α and κ to similarly conclude that $\langle j(f_i) \mid i < \alpha \rangle$ belongs to M.

Now for each $\alpha < \kappa^+$ let κ_α be least so that κ_α is closed under each $j(f_i)$, $i < \alpha$. Then κ_α is less than $j(\kappa)$, as $j(\kappa)$ is regular in M. Let κ^* be the supremum of the κ_α's. It suffices to show that there is a superstrong embedding j^* with critical point κ such that $j^*(\kappa) = \kappa^*$; then by the minimality of $j(\kappa)$, we must have $j(\kappa) = \kappa^*$ and therefore $j(\kappa)$ is the supremum of a non-decreasing κ^+-sequence of ordinals strictly less than $j(\kappa)$. It follows that $j(\kappa)$ has cofinality κ^+.

To obtain j^* define $H = \{j(f)(a) \mid f : V_\kappa \to V,\ a \in V_{\kappa^*}\}$. Then H is an elementary submodel of M and $H \cap j(\kappa) = \kappa^*$. Let $\pi : H \simeq M^*$; then $j^* = \pi \circ j : V \to M^*$ satisfies $j^*(\kappa) = \pi(j(\kappa)) = \kappa^*$ and witnesses the superstrength of κ, as desired. □

We can assume that j is given by an ultrapower, and therefore that j is continuous at κ^+. It follows that $(j(\kappa)^+)^M$ has cofinality κ^+ and $H(j(\kappa)^+)^M$ is closed under κ-sequences. Therefore we can write the collection of $(j(\kappa)^+)^M$-many open dense subsets of $Q^*_{j(\kappa)}$ as the union of κ^+ subcollections, each of which belongs to $M[G^*_{j(\kappa)}]$ and has size less than $j(\kappa)$ there. Now we can build g in κ^+ steps, using $j(\kappa)$-distributivity to meet fewer than $j(\kappa)$ open dense sets at each step (and defining the $\square$-sequence coherently at limit stages). We must also ensure that g extend g_κ; but this is easy to arrange as the latter is a condition in the forcing $Q^*_{j(\kappa)}$.

Finally the rest of G^* can be generated from $j[G]$ as before. □

The proof of the previous theorem does not work for hyperstrong κ, and there is a good reason for this. κ is subcompact iff for any $B \subseteq H_{\kappa^+}$ there are $\mu < \kappa$, $A \subseteq H_{\mu^+}$ and an elementary embedding $j : (H_{\mu^+}, A) \to (H_{\kappa^+}, B)$ with critical point μ. (Note that by elementarity, j must send μ to κ.)

Proposition 4.1.

a. *If κ is hyperstrong then κ is subcompact.*

b. *(Jensen) If there is a subcompact cardinal then $\Box$ (even when restricted to ordinals between κ and κ^+) fails.*

Proof – (a) Suppose that $j : V \to M$ witnesses hyperstrength. Then for all subsets B of $j(\kappa)^+$ in the range of j, j gives an elementary embedding of (H_{κ^+}, A) into $(H_{j(\kappa)^+}, B)$, where $j(A) = B$; moreover this embedding belongs to M as j is hyperstrong and $j \restriction H_{\kappa^+}$ belongs to $H_{j(\kappa)^+}$. As the range of j is an elementary submodel of M, it follows that there is an elementary embedding of some (H_{μ^+}, A) into $(H_{j(\kappa)^+}, B)$ (sending μ to $j(\kappa)$) which belongs to the range of j. So $j(\kappa)$ is subcompact in Range j and therefore by elementarity subcompact in M. As j is elementary, κ is subcompact in V.
(b) Suppose that κ is subcompact and $\vec{C} = \langle C_\alpha \mid \kappa < \alpha < \kappa^+,\ \alpha \text{ singular}\rangle$ has the properties of a $\Box$-sequence. By thinning out the C_α's we can ensure that each has ordertype at most κ. Let j be an embedding from $(H_{\mu^+}, \bar{\vec{C}})$ to $(H_{\kappa^+}, \vec{C})$, sending μ to κ. Let α be the supremum of the ordinals in the range of j. Then α has cofinality μ^+. The ordinals in the range of j form a $< \mu$-closed and therefore ω-closed unbounded subset of α. And Lim C_α is a closed unbounded subset of α. Therefore the intersection D of these two sets is unbounded in α. By the coherence property of $\vec{C}$, the ordertype of C_β for sufficiently large β in D is at least μ. But as the ordertype of C_α is at most κ (in fact less than κ), the ordertype of C_β for all β in D is strictly less than κ. Thus there are β in $D \subseteq$ Range j with C_β of ordertype not in Range j, contradicting the elementarity of j. □

Remark - Cummings-Schimmerling [4] establish the optimality of Jensen's result above by proving the consistency of $\square$ with 1-extendibility, a notion whose strength lies between that of superstrength and hyperstrength. (κ is 1-extendible iff there is an elementary embedding j with critical point κ from $V_{\kappa+1}$ into $V_{j(\kappa)+1}$). For their result, they perform the reverse Easton iteration of Theorem 4.4 and check that j lifts to $V_{\kappa+1}[G_\kappa]$, using the fact that j is the identity below κ.

For uncountable, regular κ, $\diamondsuit_\kappa$ says that there exists $\langle D_\alpha \mid \alpha < \kappa \rangle$ such that D_α is a subset of α for each α and for every subset D of κ, $\{\alpha < \kappa \mid D_\alpha = D \cap \alpha\}$ is stationary in κ. $\diamondsuit$ asserts that $\diamondsuit_\kappa$ holds for every uncountable, regular κ.

Theorem 4.5 (Large cardinals and $\diamondsuit$). *If κ is superstrong then there is an outer model in which κ is still superstrong and $\diamondsuit$ holds. The same holds for hyperstrong, n-superstrong for finite n and ω-superstrong.*

PROOF – The proof combines the proofs of Theorems 4.2 and 4.4. As in the latter proof, we use a reverse Easton iteration P where at each regular stage α, Q_α is α-distributive (in fact, in the present context the entire iteration starting with α is α-closed). A condition in Q_α is a sequence $\langle D_\beta \mid \beta < \gamma \rangle$, $\gamma < \alpha$, such that D_β is a subset of β for each $\beta < \gamma$. It is easy to show that a Q_α-generic yields a $\diamondsuit_\alpha$-sequence, using the α-closure of Q_α.

The proof in the superstrong case is just as in Theorem 4.4, where we take $G^*_{j(\kappa)}$ to be the intersection of $G_{j(\kappa)}$ with $P^*_{j(\kappa)}$ and then build a $Q^*_{j(\kappa)}$-generic containing the condition g_κ. For hyperstrong κ (witnessed by $j : V \to M$), we need only observe that $j[g_{\kappa^+}]$ has a lower bound in the forcing $Q_{j(\kappa)^+}$ and choose $g^*_{j(\kappa)^+} = g_{j(\kappa)^+}$ to contain this lower bound. (This is where the argument with $\square$ breaks down.)

For n-superstrongs, $1 < n$ finite, we can take $G^*_{j^n(\kappa)}$ to be the intersection of $G_{j^n(\kappa)}$ with $P^*_{j^n(\kappa)}$ (requiring the latter to contain the greatest lower bound of $j[G_{j^{n-1}(\kappa)}]$), but face the problem of defining a $Q^*_{j^n(\kappa)}$-generic containing the image of $g_{j^{n-1}(\kappa)}$ under (the lifting to $V[G_{j^{n-1}(\kappa)}]$ of) j. We use the following.

Lemma 4.4. *Suppose that n is greater than 1 and $j : V \to M$ witnesses the n-superstrength of κ, with $j^n(\kappa)$ chosen minimally. Then j is continuous at $j^{n-1}(\kappa)$ (i.e., the range of j is cofinal in $j^n(\kappa)$).*

PROOF – Let κ^* be the supremum of the range of j intersect $j^n(\kappa)$. It suffices to show that there is an n-superstrong embedding j^* with critical point κ such that $j^* \restriction j^{n-1}(\kappa) = j \restriction j^{n-1}(\kappa)$ and $(j^*)^n(\kappa) = \kappa^*$.

Let H consist of all elements of M of the form $j(f)(a)$, where $f : V_\kappa \to V$ and a belongs to V_{κ^*}. Then H is an elementary submodel of M: If $M \vDash \varphi(y, j(f_1)(a_1), \ldots, j(f_n)(a_n))$ for some y in M, where $f_i : V_\kappa \to V$, $a_i \in V_{\kappa^*}$ for each i, then choose $g : V_\kappa \to V$ so that $g(\langle x_1, \ldots, x_n\rangle) = y$ is a solution to $\varphi(y, f_1(x_1), \ldots, f_n(a_n))$ in V (if there is such a solution y in V). Then by elementarity, $M \vDash \varphi(y, j(f_1)(a_1), \ldots, j(f_n)(a_n))$ where y equals $j(g)(\langle a_1, \ldots, a_n\rangle)$. The latter is an element of H.

Note that $H \cap j^n(\kappa) = \kappa^*$: If $j(f)(a)$ is less than $j^n(\kappa)$, where $f : V_\kappa \to V$ and $a \in V_{\kappa^*}$, then $j(f)(a)$ is less than the supremum of $j(f)[V_\alpha] \cap j^n(\kappa)$ where $\alpha \in$ Range $(j) \cap \kappa^*$ is large enough so that V_α contains a; as the latter supremum belongs to Range (j), it follows that $j(f)(a)$ is less than κ^*.

Now let $\pi : H \simeq M^*$ be the transitive collapse of H and define $j^* = \pi j$. Then $j^* : V \to M^*$ is elementary and has critical point κ. As π sends $j^n(\kappa)$ to κ^* and is the identity on κ^*, it follows that $(j^*)^m(\kappa) = j^m(\kappa)$ for $m < n$ and $(j^*)^n(\kappa) = \kappa^*$. And j^* is n-superstrong as $V_{\kappa^*} = V^M_{\kappa^*} = V^{M^*}_{\kappa^*}$. □

We may assume that every element of M is of the form $j(f)(a)$ where $f : V_{j^{n-1}(\kappa)} \to V$ and a belongs to $V_{j^n(\kappa)}$. Now we claim that the image of $g_{j^{n-1}(\kappa)}$ under (the lifting to $V[G_{j^{n-1}(\kappa)}]$ of) j generates a generic for $Q^*_{j^n(\kappa)}$, in the sense that every dense subset of $Q^*_{j^n(\kappa)}$ which belongs to $M[G^*_{j^n(\kappa)}]$ is met by a condition in $j[g_{j^{n-1}(\kappa)}]$. For, if D is such a dense set, then D has a name of the form $j(f)(a)$ where a belongs to $V_{j(\alpha)}$ for some $\alpha < j^{n-1}(\kappa)$. By the $j^{n-1}(\kappa)$-distributivity of $Q_{j^{n-1}(\kappa)}$, there is a condition $\bar{p} \in g_{j^{n-1}(\kappa)}$ which meets all dense sets with names of the form $f(\bar{a})$, $\bar{a} \in V_\alpha$; then $j(\bar{p}) = p$ meets D.

Finally, ω-superstrength is handled just as in Theorem 4.2. □

Remark - The proof of Lemma 4.4 shows a bit more: Any n-superstrong embedding j with critical point κ has an "approximating" n-superstrong embedding j^* which agrees with j below $j^{n-1}(\kappa)$ and is continuous at $(j^*)^{n-1}(\kappa) = j^{n-1}(\kappa)$. A similar remark applies to the proof of Lemma 4.3.

The technique of the previous proof can also be used to force a weakened form of $\square$, preserving very large cardinals. $\square$ holds at small cofinalities iff the $\square$ principle holds when restricted to singular ordinals of cofinality at most the least superstrong cardinal.

Theorem 4.6 (Large cardinals and $\square$ at small cofinalities). *If κ is hyperstrong then there is an outer model in which κ is still hyperstrong and $\square$ holds at small cofinalities. The same holds for n-superstrong for finite n and ω-superstrong.*

PROOF – Perform a reverse Easton iteration where at each regular stage α, Q_α adds a $\square$-sequence on the singular ordinals less than κ which have cofinality at most the least superstrong cardinal. If $j : V \to M$ witnesses that κ is hyperstrong, then we take $G^*_{j(\kappa)^+}$ to be $G_{j(\kappa^+)}$ and observe that $j[g_{\kappa^+}]$ does have a greatest lower bound in $Q_{j(\kappa)^+}$, because its supremum is an ordinal of cofinality κ^+, greater than the least superstrong cardinal of M. By choosing $g^*_{j(\kappa)^+} = g_{j(\kappa)^+}$ to contain this greatest lower bound, we can lift j to $V[G_{\kappa^++1}]$, and then to all of $V[G]$. If $j : V \to M$ witnesses the 2-superstrength of κ then similarly we get a greatest lower bound for $j[G_{j(\kappa)}]$ in $P_{j^2(\kappa)}$ as for each regular $\alpha \in (\kappa, j(\kappa))$, the supremum of $j[\alpha]$ is an ordinal of cofinality greater than κ, which is superstrong (and more) in M. Then we use the argument of the preceding proof to lift j to all of $V[G]$. A similar argument handles ω-superstrength. $\square$

Remark - I mention some previous work on preserving weakened forms of $\square$ in the presence of very large cardinals. If κ is supercompact then Solovay proved that $\square_\lambda$ fails at all singular $\lambda > \kappa$. However Baumgartner showed that there is a forcing extension in which κ remains supercompact and $\square_\lambda$ does hold for all λ greater than κ, provided we restrict the square sequences to ordinals of cofinality less than some fixed $\mu < \kappa$. And Cummings-Foreman-Magidor [3] show that κ remains supercompact in some forcing extension in which for each singular cardinal λ of cofinality at least κ, the Weak Square principle $\square^*_\lambda$ holds (allowing λ-many clubs in each limit ordinal $\alpha < \lambda^+$ instead of just one).

Theorem 4.7 (Large cardinals and Gap 1 morasses). *If κ is superstrong then there is an outer model in which κ is still superstrong and gap 1 morasses* *exist at each regular cardinal. The same holds for hyperstrong, n-superstrong for finite n and ω-superstrong.*

Proof – For the definition of a gap 1 morass we refer the reader to [5]. Assume GCH and let κ be superstrong. We apply the reverse Easton iteration P where at each regular stage α, Q_α adds a gap 1 morass at α. A condition in Q_α is a size $< \alpha$ initial segment of a morass up to some top level, together with a map of an initial segment of this top level into α^+ which obeys the requirements of a morass map. To extend a condition, we end-extend the morass up to its top level and require that the map from the given initial segment of its top level into α^+ factor as the composition of a map into the top level of the stronger condition followed by the map given by the stronger condition into α^+. The forcing Q_α is α-closed and, using a Δ-system argument, is α^+-cc.

To obtain the desired G^*, we must build a $Q^*_{j(\kappa)}$-generic which extends the image under (the lifting to $V[G_\kappa]$ of) j of the Q_κ-generic g_κ. As in the case of $\square$ we use minimisation of $j(\kappa)$ to ensure that it has cofinality κ^+ and then build a $Q^*_{j(\kappa)}$-generic in κ^+ steps. Note that any condition in $j[g_\kappa]$ is extended by one which has top level κ and maps an initial segment of the top level into $j(\kappa)^+$ using j. Now given fewer than $j(\kappa)$ maximal antichains in $M[G^*_{j(\kappa)}]$, we can choose $\alpha < j(\kappa)^+$ of cofinality $j(\kappa)$ in M so that these maximal antichains are maximal when restricted to conditions which are "below α" in the sense that they map an initial segment of their top level into α. Moreover, there is a condition which serves as a lower bound to all conditions in $j[g_\kappa]$ which are below α in this sense. Therefore we can choose a condition below α meeting all of the given maximal antichains compatibly with the conditions in $j[g_\kappa]$ which are below α, and therefore compatibly with all conditions in $j[g_\kappa]$. Repeating this in κ^+ steps for increasingly large $\alpha < j(\kappa)^+$ of M-cofinality $j(\kappa)$ (taking unions at limit stages) yields the desired $Q^*_{j(\kappa)}$-generic. The remainder of the generic G^* can be generated using Lemma 4.1.

Now suppose that κ is hyperstrong. We must define a suitable $Q^*_{j(\kappa)^+}$-generic. We may assume that j is given by a hyperextender and therefore j is cofinal from κ^{++} into $j(\kappa)^{++}$ of M. Let S consist of those morass points at the top level (i.e., level κ^+) of g_{κ^+} which have cofinality κ^+. For each σ in S let $g_{\kappa^+} \restriction \sigma$ denote the set of conditions in g_{κ^+} which are below σ. Then $j[g_{\kappa^+} \restriction \sigma]$

has a greatest lower bound p_σ in $Q^*_{j(\kappa)^+}$.

The collection of maximal antichains of $Q^*_{j(\kappa)^+}$ which belong to $M[G^*_{j(\kappa)^+}]$ can be written as a union $\bigcup_{i<j(\kappa)^+} X_i$ where for each i and each σ in S, $X_i \restriction j(\sigma)$ (the subset of X_i consisting of those maximal antichains all of whose elements are below $j(\sigma)$) is a set of size at most $j(\kappa)$ in M. By induction on $\sigma \in S$ choose a condition q_σ extending p_σ and all q_τ, $\tau \in S \cap \sigma$, which meets all antichains in $X_0 \restriction j(\sigma)$. By hyperstrength, the sequence of $q_\tau \restriction j(\sigma)$, $\tau \in S$, has a greatest lower bound p^1_σ for each $\sigma \in S$. Now repeat this construction for $X_1, X_2, \ldots$ for $j(\kappa)^+$ steps, resulting in a set of conditions which generates a generic $g_{j(\kappa)^+}$ for $Q^*_{j(\kappa)^+}$. As before, the remainder of the generic G^* can be generated using Lemma 4.1.

The cases of n-superstrength, $2 \leq n$ finite, are handled as in the proof of Theorem 4.5. ω-superstrength is handled as in the proof of Theorem 4.2. □

Question 4.1. *It is possible to force a definable wellordering of the universe over a model of GCH preserving the superstrength of all superstrong cardinals, at the cost of some cardinal collapsing. Is it possible to do this without cardinal collapsing? Is it possible to preserve the superstrength of all superstrong cardinals while forcing not only the universe but also each $H(\kappa)$, $\kappa > \omega_1$, to have a definable wellordering?*

Question 4.2. *Is it consistent with a superstrong cardinal to have a gap 2 morass at every regular cardinal?*

Question 4.3. *To what extent are the condensation and hyperfine structural properties of L (see [9]) consistent with large cardinals? For the former, see the forthcoming [8].*

References

[1] Beller, A., Jensen, R., and Welch, P.: *Coding the Universe*, London Math Society Lecture Note Series **47**, Cambridge University Press (1982).

[2] Burke, D.: Generic embeddings and the failure of Box, *Proceedings of the American Mathematical Society* **123** (9) (1995), 2867–2871.

[3] CUMMINGS, J., FOREMAN, M., AND MAGIDOR, M.: Scales, squares and reflection, *Journal of Mathematical Logic* **1** (2001), 35–98.

[4] CUMMINGS, J. AND SCHIMMERLING, E.: Indexed squares, *Israel Journal of Mathematics* **131** (2002), 61–99.

[5] DEVLIN, K.: *Constructibility*, Springer-Verlag, Berlin (1984).

[6] FRIEDMAN, S.: Coding over a measurable cardinal, *Journal of Symbolic Logic* **54** (4) (1989), 1145–1159.

[7] FRIEDMAN, S.: *Fine structure and class forcing*, de Gruyter, Berlin New York (2000).

[8] FRIEDMAN, S.: Forcing condensation, in preparation.

[9] FRIEDMAN, S. AND KOEPKE, P.: An elementary approach to the fine structure of L, *Bulletin of Symbolic Logic* **3** (4) (1997), 453–468.

[10] KUNEN, K.: Elementary embeddings and infinitary combinatorics, *Journal of Symbolic Logic* **36** (1971), 407–413.

[11] MITCHELL, W.: An introduction to inner models and large cardinals, a chapter for the *Handbook of Set Theory*, to appear.

[12] NEEMAN, I.: Inner models in the region of a Woodin limit of Woodin cardinals, *Annals of Pure and Applied Logic* **116** (2002), 67–115.

[13] ZEMAN, M.: *Inner models and large cardinals*, de Gruyter Series in Logic and Its Applications **5**, de Gruyter, Berlin New York (2002).

Applications of Regular Filters and Square Principles in Model Theory

Juliette Kennedy and Jouko Väänänen

Contents

1. Introduction

It is well-known that $\Box_\lambda$ is useful in constructing models of size λ^+. To systematize the study of models of size λ^+ the logic $L(\exists^{\geq\lambda^+})$ was introduced. Shelah, in [15], defined square-like principles $\Box_\lambda^{a^*}$ and $\Box_\lambda^{b^*}$ that are equivalent to basic model theoretic properties of $L(\exists^{\geq\lambda^+})$ such as the Omitting Types Theorem and the Completeness Theorem. In [10] and [11] a further step was taken by establishing the equivalence of a weaker form $\Box_{\lambda,D}^{fin}$ of $\Box_\lambda^{b^*}$ with various model theoretic properties of structures of size $\geq \lambda^+$ and regular filters. In this expository paper we give an overview of these developments, emphasizing the central role of Ehrenfeucht-Fraïssé games.

2. The Back-and-Forth Method

One of the basic methods in model theory is the back and forth method. This method is perhaps best captured by the Ehrenfeucht-Fraïssé game. This game is played on two relational structures $\mathfrak{M}$ and $\mathfrak{N}$, which are assumed to have disjoint domains M and N, respectively. There are two players, called Abelard and Eloise. During round i of the game Abelard picks an element x_i from the domain of one of the models, and then Eloise picks an element y_i from the domain of the other model. The length of the game is fixed in advance. Thus after n rounds a partial mapping

$$\pi(x) = \begin{cases} y_i & \text{if } x = x_i \in M \\ x_i & \text{if } x = y_i \in M \end{cases} \tag{2.1}$$

where $i = 0, ..., n-1$, is produced, and Eloise wins if π is a partial isomorphism $\mathfrak{M} \to \mathfrak{N}$. By varying the length of the game and by letting Abelard choose more complex objects than just elements, the game can capture different applications of the back-and-forth method. Let us denote this game by $\mathrm{EF}_n(\mathfrak{M}, \mathfrak{N})$ if the number of moves is n. The number of moves may just as well be also an infinite ordinal α and then we denote the game by $\mathrm{EF}_\alpha(\mathfrak{M}, \mathfrak{N})$.

*Research partially supported by grant 40734 of the Academy of Finland.

The nice thing about $\mathrm{EF}_\kappa(\mathfrak{M},\mathfrak{N})$ is that if the models are of cardinality $\leq \kappa$ then Eloise has a winning strategy if and only if $\mathfrak{M} \cong \mathfrak{N}$. So we can consider the property of Eloise having a winning strategy in this game as a weaker form of isomorphism, graded by the length of the game.

Perhaps the most well-known of the back and forth arguments is the one demonstrating that elementarily equivalent saturated models of the same cardinality are isomorphic. In game theoretic terms this result says that, given two κ-saturated models, if Eloise has a winning strategy in the Ehrenfeucht-Fraïssé game of any finite length, she has a winning strategy in the Ehrenfeucht-Fraïssé game of length κ. Thus κ-saturation allows us to "stretch" the winning strategies of Eloise to length κ. The strategy is the following. Suppose we are in a position, let us call it P_ζ in which elements a_α, $\alpha < \zeta$, have been played in $\mathfrak{M}$ and elements b_α, $\alpha < \zeta$, in $\mathfrak{N}$ so far, and Eloise has not lost yet. We assume $\zeta < \kappa$. During the game, Eloise has been looking forward and has made sure that she can win any finite number of moves more. Suppose Abelard now plays a_ζ in $\mathfrak{M}$. We show that the type τ of a_ζ in $(\mathfrak{M}, a_\alpha)_{\alpha<\zeta}$ is finitely consistent as a type τ' in $(\mathfrak{N}, b_\alpha)_{\alpha<\zeta}$. Let $\phi(\underline{b}_{\alpha_1}, ..., \underline{b}_{\alpha_n}, x)$ be the conjunction of a finite part of τ' (we use underlining to indicate names for elements of the models). Since Eloise can still win a game of the length of the quantifier rank of $\exists x\phi(\underline{a}_{\alpha_1}, ..., \underline{a}_{\alpha_n}, x)$ starting from position P_ζ, the sentence $\exists x\phi(\underline{b}_{\alpha_1}, ..., \underline{b}_{\alpha_n}, x)$ is true in $(\mathfrak{N}, b_\alpha)_{\alpha<\zeta}$, and we have proved that τ' is finitely consistent in $(\mathfrak{N}, b_\alpha)_{\alpha<\zeta}$. By the κ-saturation of $\mathfrak{N}$, there is an element b_ζ in $\mathfrak{N}$ satisfying this type. This is the move of Eloise. The same argument applies if Abelard plays his move b_ζ in $\mathfrak{N}$. We can summarize this strategy as follows: the compactness feature of κ-saturated structures permits us to turn the ability of Eloise to win any finite game to the ability to win a game of length κ.

The above strategy works also if only one of the models is κ-saturated, but then the game can be played in one direction only, that is, Abelard is allowed to play in one model only. Let us denote this game by $\mathrm{EF}_\kappa^{\rightarrow}(\mathfrak{M},\mathfrak{N})$, if Abelard can only play in $\mathfrak{M}$. If Eloise has a winning strategy in this game and $\mathfrak{M}$ has cardinality $\leq \kappa$, then there is an embedding $\mathfrak{M} \to \mathfrak{N}$. Moreover, *any* substructure of $\mathfrak{M}$ of cardinality $\leq \lambda$ can then be embedded into $\mathfrak{N}$. In such a situation we only get a homomorphism $\mathfrak{M} \to \mathfrak{N}$, if the game is modified so that the winning condition is that the mapping π of equation (2.1) is a partial homomorphism, rather than a partial isomorphism. These remarks show the versatile nature of the game $\mathrm{EF}_\kappa(\mathfrak{M},\mathfrak{N})$. The method of Ehrenfeucht-Fraïssé games is by no means restricted to first order logic. Most logics have their own Ehrenfeucht-Fraïssé game.

3. Reduced Products

Rather than κ-saturated structures we concentrate on reduced products of structures. In fact, they are the most popular sources of κ-saturated structures, and in a sense "canonical" among them. Of course, reduced products are not always κ-saturated, but if D is an ω-incomplete κ-good ultrafilter on the index set, then $\prod \mathfrak{M}_i/D$ is indeed κ-saturated. The question of saturation[1] of ultraproducts is a well-studied area of model theory, that we will not go into, although we discuss the related concept of universality.

Intuitively, a reduced product $\prod \mathfrak{M}_n/D$ is a "limit" of the structures $\mathfrak{M}_n$. There are various ways to make this intuition precise. The idea of saturation of the reduced product is that any type that is finitely consistent in the models $\mathfrak{M}_n$, as manifested by a sequence $a_0, a_1, \ldots$ of elements satisfying more and more of the type, is actually satisfied in the reduced product by an element which is built from the very sequence $a_0, a_1, \ldots$. In this sense the concept of reduced product helps us to organize the finitely consistent types into one model, even if the model then has to consist of sequences of elements (or rather their equivalence classes). This idea is more canonical if we consider only reduced powers $\mathfrak{M}^\kappa/D$. Then the structure $\mathfrak{M}^\kappa/D$ in a sense collects together enough elements to satisfy all the types of $\mathfrak{M}$. The reduced power is a kind of completion of $\mathfrak{M}$, as reflected by many universality properties of $\mathfrak{M}^\kappa/D$. The difference between a κ-saturated structure and a reduced power $\mathfrak{M}^\kappa/D$ is that the reduced power has an explicit construction for an element of a given finitely consistent type, while in a κ-saturated structure we merely know that there *is* some element of that type.

[1]Shelah [16] has proved that $\mathfrak{M}^I/D$ is λ^+-saturated for all λ-regular filters on I if and only if $\mathfrak{M}$ does not have the Finite Cover Property.

By a judicious choice of the filter D one can control what kind of universality properties the reduced power $\mathfrak{M}^\kappa/D$ will have. We formulate these universality properties in terms of a version of the Ehrenfeucht-Fraïssé game. It turns out the the concept of (κ, λ)-regularity of the filter D plays a key role here.

4. Regular Filters

A filter D on a set I is said to be regular if there is a family $E \subseteq D$, such that $|E| = |I|$ and $\bigcap F = \emptyset$ for all infinite $F \subseteq E$. We then say that E is a regular family in D. Regular filters have given rise to hard set theoretical problems but at the same time they are very useful in model theory. The regularity criterion is reminiscent of compactness: all finite intersections are non-empty but all infinite intersections are empty. Indeed, regularity has a close connection to both the Compactness Theorem of model theory and to saturated structures.

Regularity is the ultimate denial of countable completeness of the filter: not only is *some* infinite intersection of filter-elements empty, but *every* infinite intersection of elements of the subset E, which is as big as the domain of the filter itself, is empty. On a countable set regularity and countable completeness are the only possibilities for an ultrafilter. On uncountable sets the situation is more complicated. A regular (proper) filter D on an infinite set is always uniform: if a small set X is in the filter, it can meet only a small number of sets in the regular family E, but on the other hand X should meet *every* member of E, as D is a filter. For ultrafilters on a countable set regularity coincides with uniformity, but for uncountable sets we have a difficult set theoretic problem: Is every uniform ultrafilter regular? Even on ω_1 this is independent of ZFC, assuming the consistency of large cardinals.

It is useful to have also a parametrized version of regularity: A filter D is said to be (κ, λ)-regular, if there is a family $E \subseteq D$, which is (κ, λ)-regular, that is, $|E| = \lambda$ and $\bigcap F = \emptyset$ for all $F \subseteq E$ with $|F| = \kappa$. Thus ordinary regularity on a set of size λ means the same as (ω, λ)-regularity. The smaller the regular family is required to be, the easier it is for a filter to have one. A filter that has a regular family of size κ is called κ-regular.

5. The Compactness Theorem

The close relationship between regularity of filters and the Compactness Theorem is manifested by a simple but striking connection, reminiscent of the ultraproduct proof of the Compactness Theorem of first order logic. To describe the connection, let us call an abstract logic L $[\kappa,\lambda]$-compact if the following condition holds: If T and T_1 are sets of L-sentences such that $|T_1| = \lambda$ and whenever $T_2 \subseteq T_1$ is of cardinality $< \kappa$, the theory $T \cup T_2$ has a model, then $T \cup T_1$ has a model. A logic is compact if it satisfies $[\omega,\lambda]$-compactness for all λ. A condition usually called κ-compactness is in this terminology $[\kappa,\kappa]$-compactness. The extension $L(\exists^{\geq\kappa})$ of first order logic was introduced by Andrzej Mostowski in [14]. Here $\exists^{\geq\kappa}$ is the generalized quantifier

$$\mathfrak{M} \models \exists^{\geq\kappa} x\phi(x,\vec{a}) \iff |\{b \in M : \mathfrak{M} \models \phi(b,\vec{a})\}| \geq \kappa.$$

Mostowski asked whether $L(\exists^{\geq\kappa})$ is $\aleph_0$-compact and observed that $L(\exists^{\geq\aleph_0})$ is certainly not. In 1963 Gerhard Fuhrken [6] proved that $L(\exists^{\geq\kappa})$ is $\aleph_0$-compact if $\aleph_0$ is small for κ (i.e. if $\lambda_n < \kappa$ for $n < \omega$, then $\prod_{n<\omega}\lambda_n < \kappa$). The method Fuhrken used was ultraproducts. Let us say that a filter D on a set I has the Łos Property for a logic L if for all $\{\mathfrak{M}_i : i \in I\}$ there is an extension $\mathfrak{N}$ of $\prod \mathfrak{M}_i/D$ such that for all ϕ in L:

$$(5.1) \qquad \mathfrak{N} \models \phi([f_1],...,[f_n]) \iff \{i \in I : \mathfrak{M}_i \models \phi(f_1(i),...,f_n(i))\} \in D.$$

It is easy to see that a logic L is compact if and only if every ultrafilter has the Łos Property for L. If every ultrafilter has the stronger Łos Property that actually $\mathfrak{N} = \prod \mathfrak{M}_i/D$, then L is the first order logic. The Łos Property can be proved for $\phi \in L(\exists^{\geq\kappa})$ if $\aleph_0$ is small for κ, and for $L(\exists^{\geq\kappa_n})_{n<\omega}$, if $\aleph_0$ is small for each κ_n. The $\aleph_0$-compactness follows from this immediately.

Vaught [18] proved $\aleph_0$-compactness of $L(\exists^{\geq\aleph_1})$ and Chang [3] extended this to $L(\exists^{\geq\kappa^+})$ by proving $(\omega_1,\omega) \to (\kappa^+,\kappa)$, when $\kappa^{<\kappa} = \kappa$. Jensen [7] extended this to all κ under the assumption GCH+$\square_\kappa$, which he showed to follow from $V = L$. Keisler [8] proved with a different method $\aleph_0$-compactness of $L(\exists^{\geq\kappa})$ when κ is a singular strong limit cardinal.

It is curious that $L(\exists^{\geq|\mathbb{N}|^+})$ and $L(\exists^{\geq|\mathbb{R}|^+})$ are $\aleph_0$-compact, but $L(\exists^{\geq|\mathbb{N}|})$ is not; for $L(\exists^{\geq|\mathbb{R}|})$ the question is open.

Apart from the first order logic, the most famous fully compact logics are the extensions of first order logic by the cofinality quantifiers

$$\mathfrak{M} \models Q^{\mathrm{cf}}_{\kappa} xy\phi(x, y, \vec{a}) \iff \mathrm{cof}(\phi^{\mathfrak{M}}) = \kappa,$$

where κ is a regular cardinal.

The topological logic L^t, introduced in [13] (see also [4]) is like monadic second order logic except that the set-quantifiers range over a topology and have the limitation that one is only allowed to write $\exists X(y \in X \wedge \phi)$ if ϕ is a formula containing no positive occurrences of X. All ultrafilters have the Łos property relative to topological logic.

The tight connection between regularity of filters and compactness of logics is expressed by the following ([12]):

Proposition 5.1. *Suppose L is any abstract logic. L is $[\kappa, \lambda]$-compact if and only if there is a (κ, λ)-regular filter on $P_{<\kappa}(\lambda)$ which has the Łos property for L.*

Proof – We prove only the easy direction. Suppose a set T of L-sentences is given. Suppose T_1 is another set of L-sentences such that $|T_1| = \lambda$ and whenever $T_2 \subseteq T_1$ is of cardinality $< \kappa$, the theory $T \cup T_2$ has a model. We may assume the filter D is on $I = P_{<\kappa}(T_1)$. Let $\{E_\phi : \phi \in T_1\}$ be a (κ, λ)-regular family in D. For any $i \in I$, the set $\{\phi : i \in E_\phi\}$ has cardinality $< \kappa$, so there is a model $\mathfrak{M}_i$ such that

$$\mathfrak{M}_i \models T \cup \{\phi : i \in E_\phi\}.$$

Let $\mathfrak{N}$ be an extension of $\prod_i \mathfrak{M}_i/D$ such that (5.1) holds. If $\phi \in T_1$, then

$$E_\phi \subseteq \{i \in I : \mathfrak{M}_i \models \phi\},$$

for if $j \in E_\phi$, then $\mathfrak{M}_j \models \phi$, whence $j \in \{i \in I : \mathfrak{M}_i \models \phi\}$. Thus $\mathfrak{N} \models \phi$ by (5.1). □

6. The Ehrenfeucht-Fraïssé Game

If the Ehrenfeucht-Fraïssé game is played on two reduced powers $\mathfrak{M}^\lambda/D$ and $\mathfrak{N}^\lambda/D$, rather on arbitrary structures, the situation is remarkably balanced.

Both players play elements which are essentially sequences of elements from the models $\mathfrak{M}$ and $\mathfrak{N}$. A result of Benda [2] and its improvement by Shelah [16, Theorem VI.1.8] shows that for regular D we can glue together pieces of winning strategies of Eloise in finite Ehrenfeucht-Fraïssé games on $\mathfrak{M}$ and $\mathfrak{N}$ and create one strategy of length α ($\alpha < \lambda^+$) on $\mathfrak{M}^\lambda/D$ and $\mathfrak{N}^\lambda/D$. By using $\Box^{fin}_{\lambda,D}$ we can stretch this result to games of length λ^+. This is quite an important step up, for if the models are of size λ^+, then they are isomorphic, and we have got isomorphism of the regular reduced powers.

The basic result about filters and reduced powers has the following form: If the filter is (κ, λ)-regular, then Eloise has a winning strategy in the one-directional Ehrenfeucht-Fraïssé game EF_α ($\alpha < \lambda^+$) on $\mathfrak{M}$ and $\mathfrak{N}^\lambda/D$ with a payoff determined by κ. By using the square-like principle $\Box^{fin}_{\lambda,D}$ from [10] we can stretch this result to games of length λ^+. Then if $|N| \leq \lambda$ and $2^\lambda = \lambda^+$, the cardinality of $\mathfrak{N}^\lambda/D$ is λ^+, and then $\mathfrak{N}^\lambda/D$ is a universal structure in the model theoretic sense.

Apart from the above model theoretic applications of the principle $\Box^{fin}_{\lambda,D}$ and regular filters we discuss the relationship of $\mathfrak{M}^\lambda/D$ to other square-like principles and to the transfer principle $(\aleph_1, \aleph_0) \to (\lambda^+, \lambda)$.

We observed earlier that if $\mathfrak{M}$ and $\mathfrak{N}$ are elementarily equivalent κ-saturated structures then Eloise has a winning strategy in $\mathrm{EF}_\kappa(\mathfrak{M}, \mathfrak{N})$. If D is an ω-incomplete κ-good filter, then the reduced powers $\mathfrak{M}^\lambda/D$ and $\mathfrak{N}^\lambda/D$ are κ-saturated, so such a long winning strategy can be found for Eloise. We now observe that since we have a better control of the reduced powers $\mathfrak{M}^\lambda/D$ and $\mathfrak{N}^\lambda/D$ than over arbitrary κ-saturated structures, we obtain the same conclusion from the weaker assumption that D is merely regular. In fact, we prove a quite tight balance between the regularity properties of the filter and existence of winning strategies for Eloise in reduced products ([16]):

Theorem 6.1. *Suppose D is a λ-regular filter on I. Suppose $\mathfrak{M}$ and $\mathfrak{N}$ are structures of a vocabulary of size $\leq \lambda$. If Eloise has a winning strategy in $\mathrm{EF}_n(\mathfrak{M}_i, \mathfrak{N}_i)$ for each $n < \omega$ and each $i \in I$, then she has a winning strategy in $\mathrm{EF}_\alpha(\prod \mathfrak{M}_i/D, \prod \mathfrak{N}_i/D)$ for each $\alpha < \lambda^+$.*

Proof – Let $\mathfrak{M} = \prod \mathfrak{M}_i/D$ and $\mathfrak{N} = \prod \mathfrak{N}_i/D$. Let $E = \{E_\xi : \xi < \lambda\}$ be the λ-regular family in D. Let $w(i) = \{\xi : i \in E_\xi\}$ (finite). Let

$$\phi_\xi(x_{\beta(0,\xi)}, ..., x_{\beta(n(\xi),\xi)}), \xi < \lambda$$

be a complete list of all first order formulas with variables among x_ξ, $\xi < \alpha$. Each $w(i)$ contains only finitely many ordinals ξ. For each ξ the formula ϕ_ξ has the free variables $x_{\beta(0,\xi)}, ..., x_{\beta(n(\xi),\xi)}$. If we put these together, we get another finite list:

$$\bigcup_{\xi \in w(i)} \{x_{\beta(0,\xi)}, ..., x_{\beta(n(\xi),\xi)}\} = \{x_{\gamma(0,i)} < ... < x_{\gamma(m(i),i)}\}.$$

Let L_i be a common finite vocabulary of all $\phi_\xi(x_{\beta(0,\xi)}, ..., x_{\beta(n(\xi),\xi)})$, where $i \in E_\xi$. For each $i < \lambda$ let τ^i be a winning strategy of Eloise in $\mathrm{EF}_{n(i)}(\mathfrak{M}_i \restriction L_i, \mathfrak{N}_i \restriction L_i)$. We describe the winning strategy of Eloise in $\mathrm{EF}_\alpha(\mathfrak{M}, \mathfrak{N})$. Eloise makes sure that the following condition is maintained: Suppose the position in the game $\mathrm{EF}_\alpha(\mathfrak{M}, \mathfrak{N})$ is

(6.1)

$\mathfrak{M}$	$\mathfrak{N}$
a_0/D	b_0/D
...	...
a_ξ/D	b_ξ/D
...	...
a_β/D	

Here $a_\xi : I \to M$ and $b_\xi : I \to N$. In this position Abelard has just made a move a_β/D and waits for Eloise's move. Eloise has to define a function $b_\beta : I \to N$. Let us fix $i \in I$. Eloise pays attention only to the case that $\beta = \gamma(k,i)$. She has made sure that in such a case the position

(6.2)

$\mathfrak{M}$	$\mathfrak{N}$
$a_{\gamma(0,i)}(i)$	$b_{\gamma(0,i)}(i)$
...	...
$a_{\gamma(k-1,i)}(i)$	$b_{\gamma(k-1,i)}(i)$

is a position in the game $\mathrm{EF}_{n(i)}(\mathfrak{M}_i \restriction L_i, \mathfrak{N}_i \restriction L_i)$ while Eloise uses τ^i. Now Eloise pretends that Abelard has made one more move, namely $a_{\gamma(k,i)}(i)$, in this finite game. The winning strategy τ^i gives her a response $b_{\gamma(k,i)}(i)$ and this is her move $b_\beta(i)$. □

Why did we have to assume $\alpha < \lambda^+$? What goes wrong if $\alpha = \lambda^+$? The amazing thing is that nothing goes actually wrong, but we have to be more careful in the way we index variables and formulas, and to be able to do this we need something like a square-principle, and this may fail if λ is singular strong limit. We discuss this further below.

Variant 1

The Ehrenfeucht-Fraïssé game of $L(\exists^{\geq\lambda^+})$ has the following additional moves apart from the moves the ordinary EF_α: During any round Abelard can pick a subset X of one of the models instead of an element. The only requirement is that $|X| \geq \lambda^+$. Eloise responds by picking a subset Y of the other model. Again we require that $|Y| \geq \lambda^+$. Next Abelard picks an element $y \in Y$ and finally Eloise picks an element $x \in X$. Now the game continues as if the players had played (x, y) in the ordinary game. If $\kappa^\omega = \kappa$, then the above theorem holds for the Ehrenfeucht-Fraïssé game for $L(\exists^{\geq\kappa^+})$ and for countable index set I.

Variant 2

Let $\mathrm{EF}_\alpha^\lambda(\mathfrak{M}, \mathfrak{N})$ denote the Ehrenfeucht-Fraïssé game with α moves, where the players play sequences of length $< \lambda$. That is, Abelard starts by playing $\{a_\xi^0 : \xi < \zeta_0\}$, where $\zeta_0 < \lambda$, in one of the models, say $\mathfrak{M}$. Then Eloise plays $\{b_\xi^0 : \xi < \zeta_0\}$ in the other model, in this case in $\mathfrak{N}$. After the first round Eloise is winning if the mapping $a_\xi^0 \mapsto b_\xi^0$ is a partial isomorphism. The same is repeated for α rounds. In the end a mapping $\pi = \{(a_\xi^\beta, b_\xi^\beta) : \xi < \zeta_\beta, \beta < \alpha\}$, where each ζ_b is less than λ, is generated, and Eloise wins if this is a partial isomorphism $\mathfrak{M} \to \mathfrak{N}$. Benda [2] proved the following beautiful result:

> *Suppose D is a (κ, λ)-regular ω-incomplete filter on I. If Eloise has a winning strategy in $\mathrm{EF}_n^\kappa(\mathfrak{M}_i, \mathfrak{N}_i)$ for each $n < \omega$ and each $i \in I$, then she has a winning strategy in $\mathrm{EF}_\omega^{\lambda^+}(\prod \mathfrak{M}_i/D, \prod \mathfrak{N}_i/D)$.*

This is proved in principle just like Theorem 6.1. The main difference is that the game has only ω moves but each move is a massive sequence. Such a sequence of moves could be subsumed by α moves for some $\alpha < \lambda^+$, so this seems to reduce to Theorem 6.1; but note that the filter is not λ-regular, only (κ, λ)-regular. If we had a λ-regular filter, Benda's result would follow from Theorem 6.1.

Variant 3

There is a canonical way to define reduced products for topological models. The EF game for topological logic is described in detail in [4]. Theorem 6.1 holds in topological model theory.

7. Square-Principles

We now introduce the weak square principle, introduced in [10], useful for taking full advantage of regular filters. Suppose D is a filter on λ. The principle states:

$\square^{fin}_{\lambda,D}$: There exist finite sets C^ξ_α and integers n_ξ for each $\alpha < \lambda^+$ and $\xi < \lambda$ such that for each ξ, α

i. $C^\xi_\alpha \subseteq \alpha + 1$

ii. If $B \subset \lambda^+$ is a finite set of ordinals and $\alpha < \lambda^+$ is such that $B \subseteq \alpha+1$, then $\{\xi : B \subseteq C^\xi_\alpha\} \in D$

iii. $\beta \in C^\xi_\alpha$ implies $C^\xi_\beta = C^\xi_\alpha \cap (\beta + 1)$

iv. $|C^\xi_\alpha| < n_\xi$

This has a slightly simpler but still interesting formulation, avoiding the filter D, in the case $\lambda = \omega_1$:

$\square^{fin}_{\omega_1}$: There are sets C^n_α, where $\alpha < \omega_1$ and $n < \omega$, with the following properties

i. $n < m$ implies $C^n_\alpha \subseteq C^m_\alpha$

ii. $\alpha + 1 = \bigcup_{n<\omega} C^n_\alpha$

iii. $\beta \in C^n_\alpha$ implies $C^n_\beta = C^n_\alpha \cap (\beta + 1)$

iv. $|C^n_\alpha| < n + 1$

The nice thing about this principle is that it is provable in ZFC. To see this, we use induction on $\alpha < \omega_1$, and make sure on the way that

$$\text{v.}\quad \lim_{n\to\infty} \frac{|C^n_\alpha|}{n+1} = 0.$$

The sets C^n_α are defined in a straightforward way from the sets C^n_β, where $\beta < \alpha$. Let $C^n_{\alpha+1} = C^n_\alpha \cup \{\alpha+1\}$. For a limit $\alpha < \omega_1$, let $(\alpha_n)_{n<\omega}$ be a strictly increasing sequence converging to α. Choose $n_0 < n_1 < \ldots$ so that

$$\alpha_0 \in C^{n_0}_{\alpha_1}, \alpha_1 \in C^{n_1}_{\alpha_2}, \ldots$$

Note, that

$$C^{n_0}_{\alpha_0} \subseteq C^{n_1}_{\alpha_1} \subseteq C^{n_2}_{\alpha_2}, \ldots$$

We define

$$C^j_\alpha = \begin{cases} C^j_{\alpha_0} \cup \{\alpha\}, & \text{if } j < n_0, \\ C^j_{\alpha_1} \cup \{\alpha\}, & \text{if } n_0 \leq j < n_1, \\ C^j_{\alpha_2} \cup \{\alpha\}, & \text{if } n_1 \leq j < n_2, \\ \text{etc.} \end{cases}$$

This works perfectly, except for the condition iv. But here v comes in handy. The conditions iv and v can be easily satisfied by modifying the construction as follows: when we choose the numbers $n_0 < n_1 < \ldots$, we make sure that $n \geq n_{i+1}$ implies $n_i \cdot |C^n_{\alpha_{i+1}}| < n$. Let us see in detail why ii holds. For successor ordinals α the claim follows from induction hypothesis. Let us then assume α is limit, and $\delta < \alpha$, say $\delta < \alpha_j$. By induction hypothesis, $\delta \in C^m_{\alpha_j}$ for some m. Let i be large enough for $\max(m, j) \leq n_i$ to hold. Then $C^m_{\alpha_j} \subseteq C^{n_i}_{\alpha_j} \subseteq C^{n_i}_{\alpha_{i+1}}$, whence $\delta \in C^{n_i}_\alpha$.

For cardinals $\lambda > \omega_1$ we have to use a filter D and the above full version of $\square^{fin}_{\lambda,D}$. Since we now move away from provable principles, in order to see the genealogy of this principle, let us recall some related more familiar ones.

The ordinary $\square_\lambda$ states that each $\alpha < \lambda^+$ of cofinality $> \omega$ has a club C_α of order type $\leq \lambda$ such that if $\beta \in \lim(C_\alpha)$, then $C_\beta = C_\alpha \cap \beta$. Such coherent sets C_α can be used to construct various smaller sets C^ξ_α. For example, for singular strong limit λ, the principle $\square_\lambda$ implies the following principle, introduced in [15]:

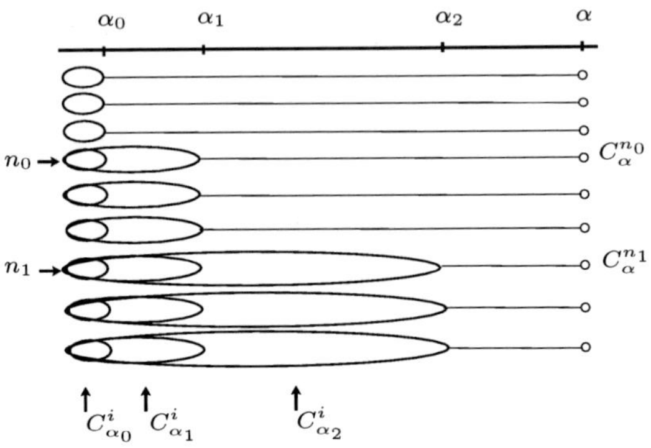

Figure 1 – Construction of the sets C^n_α.

$\square^a_\lambda$: There are sets C^ξ_α, where $\alpha < \lambda^+$ and $\xi < \mathrm{cof}(\lambda)$, with the following properties

i. $\xi < \zeta$ implies $C^\xi_\alpha \subseteq C^\zeta_\alpha$

ii. $\alpha + 1 = \bigcup_{\xi < \mathrm{cof}(\lambda)} C^\xi_\alpha$

iii. $\beta \in C^\xi_\alpha$ implies $C^\xi_\beta = C^\xi_\alpha \cap (\beta + 1)$

iv. For every $\xi < \mathrm{cof}(\lambda)$ there are $< \lambda$ non-isomorphic structures of the form $(C^\xi_\alpha, <, \{(\gamma, \beta) : \gamma \in C^\xi_\beta\})$, $\alpha < \lambda^+$, $\beta < \alpha$.

The inductive construction of the sets C^ξ_α from the $\square_\lambda$-sequence (C_α) is easy: Let (λ_ξ), $\xi < \mathrm{cof}(\lambda)$, be a cofinal increasing sequence of cardinals in λ. We make sure $|C^\xi_\alpha| \leq \lambda_\xi$. This implies condition iv above, since λ is a strong limit. Let $C^\xi_{\alpha+1} = C^\xi_\alpha \cup \{\alpha + 1\}$. For each limit $\alpha < \lambda^+$ with $|C_\alpha| \leq \lambda_\xi$ we let C^ξ_α be the set $\bigcup\{C^\xi_\beta : \beta \in \lim(C_\alpha), \beta < \alpha\}$, and otherwise $C^\xi_\alpha = \emptyset$. This needs a small correction when $\alpha > \sup \lim(C_\alpha)$, but gives anyway the right idea, and the same idea will recur in other constructions below. For strongly inaccessible λ one can prove $\square^a_\lambda$ even without assuming $\square_\lambda$.

If λ is a limit cardinal, then any sequence witnessing $\square^a_\lambda$ also witnesses the following principle from [15] and [1]:

$\mathcal{S}_\lambda$: There are sets $\langle C^\xi_\alpha : \alpha < \lambda^+, \xi < \mathrm{cof}(\lambda)\rangle$ such that

Items *i* – *iii* as above in $\square^a_\lambda$

iv. For every $\xi < \mathrm{cof}(\lambda)$, $\sup\{\mathrm{otp}(C^\xi_\alpha) : \alpha < \lambda^+\} < \lambda$.

Conversely, if λ is a singular strong limit, then any sequence witnessing $\mathcal{S}_\lambda$ also witnesses $\Box^a_\lambda$. Continuing with singular strong limit λ, by considering closures of increasing sequences in the sets C^ξ_α, one can prove that $\mathcal{S}_\lambda$ implies the weak square $\Box^*_\lambda$, [1].

A different kind of weakening of $\Box^a_\lambda$, following already from $\lambda^{<\lambda} = \lambda$, is:

$\Box^b_\lambda$: Items *i – iii* as above in $\Box^a_\lambda$

iv. For every $\xi < \mathrm{cof}(\lambda)$ there are $\leq \lambda$ non-isomorphic structures of the form $(C^\xi_\alpha, <, \{(\gamma, \beta) : \gamma \in C^\xi_\beta\})$, $\alpha < \lambda^+$.

We shall see below that even the weaker principle $\Box^b_\lambda$ implies $\mathcal{S}_\lambda$ for limit κ.

We now discuss principles, closely related to the above versions of the square, which are actually equivalent to model theoretic statements. In the model theory of $L(\exists^{\geq\lambda^+})$, and even more so in the model theory of first order logic, we have to deal all the time with non-standard, in particular non-well-ordered models. This is simply a consequence of the Compactness Theorem, and characteristic of model theory. Thus, in order for the existence of the sequences referred to in the principles $\Box^a_\lambda$ and $\Box^b_\lambda$ to have a model theoretic proof, we have to modify the principles. The modified "non-standard" versions are:

$\Box^{a^*}_\lambda$: There is a λ^+-like linear order L, sets $\langle C^\zeta_a : a \in L, \zeta < cf(\lambda)\rangle$, equivalence relations $\langle E^\zeta : \zeta < cf(\lambda)\rangle$, and functions $\langle f^\zeta_{a,b} : \zeta < \lambda, a \in L, b \in L\rangle$ such that the following conditions hold:

At first we have three conditions which are like the corresponding conditions in $\Box^a_\lambda$:

i. $\xi < \zeta$ implies $C^\xi_a \subseteq C^\zeta_a$

ii. $\{b : b \leq_L a\} = \bigcup_{\xi < \mathrm{cof}(\lambda)} C^\xi_a$

iii. $b \in C^\xi_b$ implies $C^\xi_b = \{c \in C^\xi_a : c \leq_L b\}$.

The next five conditions are an adaption of condition iv of $\Box^a_\lambda$ to the situation where the order L may be non-well-ordered and hence we cannot count on uniqueness of isomorphisms between isomorphic sets C^ξ_a:

iv. E^ζ is an equivalence relation on L with $< \lambda$ equivalence classes.

v. If $\zeta < \xi < \mathrm{cof}(\lambda)$, then E^ξ refines E^ζ.

vi. If $aE^\zeta b$, then $f^\zeta_{a,b}$ is an order-preserving one to one mapping from C^ζ_a onto C^ζ_b such that for $d \in C^\zeta_a, dE^\zeta f^\zeta_{a,b}(d)$.

vii. If $\zeta < \xi < cf(\lambda)$ and $aE^\xi b$, then $f^\zeta_{a,b} \subseteq f^\xi_{a,b}$.

viii. If $f^\zeta_{a,b}(a_1) = b_1$, then $f^\zeta_{a_1,b_1} \subseteq f^\zeta_{a,b}$.

ix. If $a \in C^\zeta_b$ then $\neg E^\zeta(a,b)$.

$\square^{b^*}_\lambda$: Items *i – iii*, v – *ix* as above, and

iv. E^ζ is an equivalence relation on L with $\leq \lambda$ equivalence classes.

Note again that $\square^{a^*}_\lambda$ implies trivially $\square^{b^*}_\lambda$. Moreover, $\square^a_\lambda$ implies $\square^{a^*}_\lambda$, and $\square^b_\lambda$ implies $\square^{b^*}_\lambda$. Thus $\square^{b^*}_\lambda$ is true if $\lambda^{<\lambda} = \lambda$, and also if λ is a singular strong limit cardinal and $\square_\lambda$ holds.

We are now ready to establish the position of $\square^{fin}_{\lambda,D}$ among the various square-like principles:

Proposition 7.1. *Suppose λ is a cardinal satisfying $\square^{b*}_\lambda$. Then every regular filter D on λ generated by λ sets satisfies $\square^{fin}_{\lambda,D}$.*

Proof – We give an outline and refer to [10] for more details. Let $T_\zeta = \{a/E^\zeta : a \in L\}$. Thus $|T_\zeta| \leq \lambda$. The relation $t_1 <_\zeta t_2 \iff (\exists a_1 \in t_1)(\exists a_2 \in t_2)(a_1 \in C^\zeta_{a_2})$ defines a tree order on T_ζ. If $a <_L b$, let $\xi(a,b)$ be the least ξ such that $a \in C^\xi_b$. Let $typ_0(a,b) = \langle a/E^{\xi(a,b)}, b/E^{\xi(a,b)}\rangle$. If $a_1 <_L ... <_L a_n$, let

$$typ(a_1, ..., a_n) = \{\langle l, m, typ_0(a_l, a_m)\rangle | 1 \leq l < m \leq n\}$$

and $\Gamma = \{typ(\vec{a}) : \vec{a} \in L^{<\omega}\}$. Thus $|\Gamma| \leq \lambda$. For $t = typ(\vec{a}), \vec{a} \in L^n$, we use $len(t)$ to denote the length of $\vec{a}$, and $\xi(\vec{a})$ to denote $\max\{\xi(a_l, a_n) : 0 \leq l < n\}$. We define on Γ a directed partial order $\leq_\Gamma$ as follows: $typ(\vec{a}) \leq_\Gamma typ(\vec{b})$ if $\vec{a} = \vec{a'}$ and $\vec{b} = \vec{b'}$ for some $\vec{a'}$ and $\vec{b'}$ such that $\vec{a'}$ is a subsequence of $\vec{b'}$. Suppose $t \in \Gamma$, $t = typ(b_0, \ldots b_{n-1})$, and $a \in L$ such that there exists $k < n$ with $b_k E^{\xi(b_0,\ldots,b_{n-1})} a$. Such a k is necessarily unique. Then let

$$C^t_a = \{f^{\xi(b_0,\ldots,b_{n-1})}_{b_k,a}(b_l) : l < k\}.$$

Otherwise, let $C^t_a = \emptyset$. Note that the definition of C^t_a is independent of the choice of $b_0, \ldots b_{n-1}$, thanks to the conditions i–ix of $\Box^{b*}_\lambda$, in particular the condition ix. Let F be the filter on Γ generated by the λ sets $\{t \in \Gamma : u \leq_L t\}$, where $u \in \Gamma$. We can now see that the sets C^t_a, the numbers $len(t)$ and the filter F satisfy conditions i–iv of $\Box^{fin}_{\lambda,F}$ with L instead of λ^+.: For ii suppose $B \subseteq L$ is finite. Let $a \in L$ be above all elements of B. Let $\vec{a}$ enumerate $B \cup \{a\}$ in increasing order and let $u = typ(\vec{a})$. Clearly $t \geq_\Gamma u \Rightarrow B \subseteq C^t_a$. Thus $\{t : B \subseteq C^t_a\} \in F$.

To get $\Box^{fin}_{\lambda,F}$ on λ^+ we observe that since L is λ^+-like, we can assume that $\langle \lambda^+, < \rangle$ is a submodel of $\langle L, <_L \rangle$. Then we define $\bar{C}^t_\alpha = C^t_\alpha \cap \{\beta : \beta < \alpha\}$. The new sets $\bar{C}^t_a$ satisfy $\Box^{fin}_{\lambda,F}$. Having D a filter on Γ instead of λ is immaterial as $|\Gamma| = \lambda$.

Finally we introduce a variation which gives $\Box^{fin}_{\lambda,D}$ instead of $\Box^{fin}_{\lambda,F}$. Let $\{A_\alpha : \alpha < \lambda\}$ be a regular family in D, and let $\{Z_\alpha : \alpha < \lambda\}$ be a family generating D. We define a function $h : \lambda \to \lambda$ as follows. Suppose $\xi < \lambda$. Then let $h(\xi)$ be an element of the intersection of the finitely many Z_α for which $\xi \in A_\alpha$. Now define $\tilde{C}^\xi_\alpha = C^{h(\xi)}_\alpha$. Define also $\tilde{n}_\xi = n_{h(\xi)}$. Now the sets $\tilde{C}^\xi_\alpha$ and the numbers $\tilde{n}_\xi$ satisfy $\Box^{fin}_{\lambda,D}$. □

We noted above that if λ is a limit cardinal, then $\Box^a_\lambda$ implies $\mathcal{S}_\lambda$. We now derive $\mathcal{S}_\lambda$ from $\Box^{fin}_{\lambda,D}$.

Proposition 7.2. *Suppose λ is a limit cardinal and D is a regular filter on λ generated by λ sets. Then $\Box^{fin}_{\lambda,D}$ implies $\mathcal{S}_\lambda$.*

Proof – Suppose $\mathcal{A} = \{A_\alpha : \alpha < \lambda\}$ is a family of sets generating D, closed under finite intersections. Let $\langle \lambda_\alpha : \alpha < \mathrm{cof}(\lambda) \rangle$ be an increasing cofinal sequence in λ such that $\lambda_0 \geq \omega$. Let us represent λ as the continuously increasing union of set L_α, $\alpha < \mathrm{cof}(\lambda)$, such that $|L_\alpha| \leq \lambda_\alpha$ and the set $\{A_\gamma : \gamma \in L_\alpha\}$ is closed under finite intersections. For $\zeta < \mathrm{cof}(\lambda)$ and $\alpha < \lambda^+$, let

$$\bar{C}^\zeta_\alpha = \{\beta \leq \alpha : (\exists \gamma \in L_\zeta)(A_\gamma \subseteq \{\xi : \beta \in C^\xi_\alpha\})\}.$$

Now $\langle \bar{C}^\zeta_\alpha : \alpha < \mathrm{cof}(\lambda) \rangle$ is a continuously increasing sequence of subsets of ζ of cardinality $\leq \lambda_\alpha$, covering $\alpha + 1$. If $\xi \in \bar{C}^\zeta_\alpha$, then $\bar{C}^\zeta_\xi = \bar{C}^\zeta_\alpha \cap \xi$. $\sup\{otp(\bar{C}^\zeta_\alpha) : \zeta < \lambda^+\} \leq \lambda^+_\alpha$. □

Thus $\square^{fin}_{\lambda,D}$ is true, if $\lambda^{<\lambda} = \lambda$, and also if λ is a singular strong limit cardinal and $\square_\lambda$ holds. Since $\mathcal{S}_\lambda$ implies $\square^*_\lambda$, the principle $\square^{fin}_{\lambda,D}$ fails for singular strong limit λ above a strongly compact cardinal.

8. The Completeness Theorem

In this section we establish the relationship between the combinatorial principles of the previous section and various model theoretic principles. The simplest and oldest model theoretic principle in this context is the two cardinal transfer principle $(\kappa^+, \kappa) \to (\lambda^+, \lambda)$ which says the following: If a first order sentence has a model of cardinality κ^+ in which the interpretation of the unary predicate P has cardinality κ, then the sentence has a model of cardinality λ^+ in which the interpretation of P has cardinality λ. By Vaught's Two Cardinal Theorem, $(\lambda^+, \lambda) \to (\aleph_1, \aleph_0)$ holds for all λ. The transfer principle $(\kappa^+, \kappa) \to (\lambda^+, \lambda)$ permits a transfer of results from a context based on κ to a context based on λ. A succinct way to formulate such results in the adoption of the generalized quantifier $\exists^{\geq\kappa^+}$. Keisler [9] introduced the axioms

A1. $\neg\exists^{\geq\lambda^+} x(x = y \vee x = z)$

A2. $\forall x(\phi \to \psi) \to (\exists^{\geq\lambda^+} x\phi \to \exists^{\geq\lambda^+} x\psi)$

A3. $\exists^{\geq\lambda^+} x\phi(x) \to \exists^{\geq\lambda^+} y\phi(y)$

A4. $\exists^{\geq\lambda^+} x\exists y\phi \to (\exists x\exists^{\geq\lambda^+} y\phi \vee \exists^{\geq\lambda^+} x\exists y\phi)$

and proved the Keisler Completeness Theorem for the logic $L(\exists^{\geq\aleph_1})$ with these axioms. The **Completeness Theorem** of $L(\exists^{\geq\kappa^+})$ is the statement that any sentence of $L(\exists^{\geq\kappa^+})$ which is syntactically consistent with the above axioms, has a model. If we have $(\aleph_1, \aleph_0) \to (\lambda^+, \lambda)$, we can argue as follows: Suppose ϕ is a sentence of $L(\exists^{\geq\lambda})$ (syntactically) consistent with the above axioms. Then by the Completeness Theorem of $L(\exists^{\geq\aleph_1})$, ϕ has a model as a sentence of $L(\exists^{\geq\aleph_1})$. By the transfer principle (and some coding, called the Fuhrken Reduction, see [5]), ϕ has a model as a sentence of $L(\exists^{\geq\lambda^+})$. Thus $(\aleph_1, \aleph_0) \to (\lambda^+, \lambda)$ allows us to transfer the completeness of the above axiomatization for $L(\exists^{\geq\aleph_1})$ to the completeness for $L(\exists^{\geq\lambda^+})$. Conversely, let us assume the Completeness Theorem for $L(\exists^{\geq\lambda^+})$. Suppose a first order sentence ϕ has a model of cardinality $\aleph_1$ in which the interpretation of the unary predicate P has cardinality $\aleph_1$. Then

$$\phi \wedge \exists^{\geq\lambda^+} x(x = x) \wedge \neg\exists^{\geq\lambda^+} xP(x)$$

is certainly (syntactically) consistent with the Keisler axioms A1 – A4. By the Completeness Theorem for $L(\exists^{\geq\lambda^+})$, this sentence has a model. Thus ϕ has a model of cardinality $\geq \lambda^+$ in which the interpretation of P has cardinality $\leq \lambda$. This is the main idea of the equivalence of $(\aleph_1, \aleph_0) \to (\lambda^+, \lambda)$ and the Completeness Theorem of $L(\exists^{\geq\lambda^+})$.

Shelah [15] proved that $(\aleph_1, \aleph_0) \to (\lambda^+, \lambda)$, and hence the Completeness Theorem of $L(\exists^{\geq\lambda^+})$, is equivalent to $\square_\lambda^{b^*}$. To obtain $\square_\lambda^{b^*}$, he formulates the existence of the λ^+-like linear order L, the sequence of sets (C_a^ξ), and the sequence of equivalence relations E^ξ with functions $f_{a,b}^\xi$, as the existence of a model for a sentence ϕ of $L(\exists^{\geq\lambda^+})$. By $(\aleph_1, \aleph_0) \to (\lambda^+, \lambda)$ it suffices to construct a model for ϕ when ϕ is thought of as a sentence of $L(\exists^{\geq\aleph_1})$. By the absoluteness of consistency in $L(\exists^{\geq\aleph_1})$ it suffices to force a model for ϕ. This is the basic idea of deriving $\square_\lambda^{b^*}$ from $(\aleph_1, \aleph_0) \to (\lambda^+, \lambda)$. Conversely, if $\square_\lambda^{b^*}$ is assumed, the trees T_ζ defined in the proof of Proposition 7.1 are used to build a model needed for the Completeness Theorem of $L(\exists^{\geq\lambda^+})$. Thus the model is not constructed as the union of a linear chain of models, as in Keisler's proof of the Completeness Theorem of $L(\exists^{\geq\aleph_1})$, but as a direct limit of a tree of models.

The Omitting Types Theorem for $L(\exists^{\geq\lambda^+})$ is defined a little differently than for first order logic. This modification was introduced by Keisler in [9]. Suppose T is a theory in $L(\exists^{\geq\lambda^+})$ and $p(x)$ a type. We say that the theory strongly λ-omits $p(x)$ if whenever a formula of the form

$$(Q_i\vec{y_i})_{i<\alpha}\exists x \bigwedge_{j<\delta} \psi_j(x, \vec{y}),$$

where $\delta < \lambda$ and each Q_i is either $\exists$ or $\exists^{\geq\lambda^+}$, is consistent with T, then there is some $\phi(x) \in p(x)$ such that

$$(Q_i\vec{y_i})_{i<\alpha}\exists x(\neg\phi(x) \wedge \bigwedge_{j<\delta} \psi_j(x, \vec{y}))$$

is consistent with T. The **Omitting Types Theorem** for $L(\exists^{\geq\lambda^+})$ says: If T is a consistent $L(\exists^{\geq\lambda^+})$-theory of cardinality at most λ and p_i, $i < \lambda$, are types strongly λ-omitted by T, then T has a model which omits each p_i. Shelah [15] proved that the Omitting Types Theorem for $L(\exists^{\geq\lambda^+})$ is equivalent to $\Box_\lambda^{a^*}$.

9. Isomorphism and Embedding Theorems

Assume L is of cardinality $\leq \lambda$ and for each $i < \lambda$ let $\mathfrak{M}_i$ and $\mathfrak{N}_i$ be elementarily equivalent L-structures. We already pointed out that if D is a regular filter on λ, then Eloise has a winning strategy in the game EFγ on $\prod_i \mathfrak{M}_i/D$ and $\prod_i \mathfrak{N}_i/D$ for each $\gamma < \lambda^+$. We show now that under the stronger assumption $\Box_{\lambda,D}^{fin}$, Eloise has a winning strategy even in the game EF_{λ^+}. This makes a big difference because, if we assume that the models $\mathfrak{M}_i$ and $\mathfrak{N}_i$ are of size $\leq \lambda^+$, and $2^\lambda = \lambda^+$, then the models $\prod_\xi \mathfrak{M}_\xi/D$ and $\prod_\xi \mathfrak{N}_\xi/D$ are of size $\leq \lambda^+$. Thus by the remark made before, if Eloise has a winning strategy in EF_{λ^+}, the reduced powers are actually isomorphic. The results of this section are from [10] and [11].

Theorem 9.1. *Suppose λ is a cardinal and D is a regular filter on λ. The following conditions are equivalent:*

1. $\Box_{\lambda,D}^{fin}$.
2. *Let L be a vocabulary of cardinality $\leq \lambda$ and for each $i < \lambda$ let $\mathfrak{M}_i$ and $\mathfrak{N}_i$ be elementarily equivalent L-structures. Then Eloise has a winning strategy in the game EF_{λ^+} on $\prod_\xi \mathfrak{M}_\xi/D$ and $\prod_\xi \mathfrak{N}_\xi/D$.*

Proof – We prove only the direction 1. → 2. and refer to [11] for the other direction. Let $\mathfrak{M} = \prod_\xi \mathfrak{M}_\xi/D$ and $\mathfrak{N} = \prod_\xi \mathfrak{N}_\xi/D$. Let the sets C_α^ξ and the integers n_ξ be as in $\Box_{\lambda,D}^{fin}$. The crucial idea of the proof is the following: When the Ehrenfeucht-Fraïssé game $\mathrm{EF}_{\lambda^+}(\mathfrak{M}, \mathfrak{N})$ is played, the players are actually playing λ Ehrenfeucht-Fraïssé games simultaneously, namely the games $\mathrm{EF}_{\lambda^+}(\mathfrak{M}_\xi, \mathfrak{N}_\xi)$. However, note that Eloise is not assumed to have a winning strategy in such long games on $\mathfrak{M}_\xi$ and $\mathfrak{N}_\xi$. So how is she able to win them all simultaneously if she cannot even win a single game? The coherent finite sets C^ξ_α are used. She plays simultaneously, not the long games $\mathrm{EF}_{\lambda^+}(\mathfrak{M}_\xi, \mathfrak{N}_\xi)$, but the finite games $\mathrm{EF}_{n_\xi}(\mathfrak{M}_\xi, \mathfrak{N}_\xi)$. The little pieces of finite winning plays for Eloise can be glued together, thanks to the strong coherence qualities of the sets C^ξ_α, into one long winning play for Eloise.

Let $\langle \phi_\alpha : \alpha < \lambda\}$ a complete listing of first order formulas of language L. Let $\langle X_\alpha : \alpha < \lambda\}$ a regular family in D. Let E be the following equivalence relation on λ:

$$\zeta E \xi \iff (\forall \alpha < \lambda)(\zeta \in X_\alpha \leftrightarrow \xi \in X_\alpha).$$

Let $\langle Y_\alpha : \alpha < \lambda\rangle$ be a list of the equivalence classes of E. If $\epsilon < \lambda$, let $\zeta(\epsilon)$ be the unique $\zeta < \lambda$ such that $\epsilon \in Y_\zeta$. Let $\Delta_\xi = \{\phi_\alpha : Y_\xi \subseteq X_\alpha\}$. Note that Δ_ξ is finite and $X_\alpha \subseteq \{\epsilon : \phi_\alpha \in \Delta_{\zeta(\epsilon)}\}$, so for D-many ϵ the formula ϕ_α is in $\Delta_{\zeta(\epsilon)}$. Let $L_\xi \subseteq L$ be the finite common vocabulary of the formulas in Δ_ξ. For each $\epsilon < \lambda$ let σ_ϵ be a winning strategy for Eloise in the game $\mathrm{EF}_{n_{\zeta(\epsilon)}}$ on the models $\mathfrak{M}_{\zeta(\epsilon)} \restriction L_{\zeta(\epsilon)}$ and $\mathfrak{N}_{\zeta(\epsilon)} \restriction L_{\zeta(\epsilon)}$.

A good position is a sequence $\langle (f_\beta, g_\beta) : \beta < \alpha\rangle$, where $\alpha < \lambda^+$, and for all $\beta < \alpha$ we have $f_\beta \in \prod_\epsilon M_{\zeta(\epsilon)}$, $g_\beta \in \prod_\epsilon N_{\zeta(\epsilon)}$, and if $\xi < \lambda$, then $\langle (f_\eta(\epsilon), g_\eta(\epsilon)) : \eta \in C^\epsilon_\beta\rangle$ is a play according to σ_ϵ on the models $M_{\zeta(\epsilon)}$ and $N_{\zeta(\epsilon)}$. Note that in a good position the equivalence classes of the functions f_β and g_β determine a partial isomorphism of the reduced products. To see this we argue as follows: Suppose α rounds have been played and we are in a good position. Let $\phi_\gamma([f_{\beta_1}], \ldots, [f_{\beta_k}])$ be an atomic formula holding in $\prod_\epsilon M_{\zeta(\epsilon)}/D$, where $\beta_1 < \ldots < \beta_k < \alpha$, and let $A = \{\epsilon < \lambda : \{\beta_1, \ldots, \beta_k\} \subseteq C^\epsilon_a\} \in D$. Then since $B = \{\epsilon < \lambda : \mathfrak{M}_{\zeta(\epsilon)} \models \phi_\gamma(f_{\beta_1}(\epsilon), \ldots, f_{\beta_k}(\epsilon))\} \in D$, we must have $A \cap B \cap X_\gamma \in D$. On the other hand, for $\epsilon \in A \cap B \cap X_\gamma$ by coherence we must have that $\beta_1, \ldots, \beta_k \in C^\epsilon_{\beta_k}$ and therefore since we are in a good position $\langle (f_\eta(\epsilon), g_\eta(\epsilon)) : \eta \in C^\epsilon_{\beta_k}\rangle$ is a play according to σ_ϵ. Then $\langle (f_\epsilon(\xi), g_\epsilon(\xi)) : \epsilon \in C^\xi_{\zeta_k}\rangle$ determines a partial isomorphism of the structures $\mathfrak{M}_{\zeta(\epsilon)} \restriction L_{\zeta(\epsilon)}$ and $\mathfrak{N}_{\zeta(\epsilon)} \restriction L_{\zeta(\epsilon)}$. Since this was the case for all $\xi \in A \cap B \cap X_\gamma \in D$, we get $\prod_\epsilon \mathfrak{N}_{\zeta(\epsilon)}/D \models \phi_\gamma([g_{\beta_1}], \ldots, [g_{\beta_k}])$.

The strategy of Eloise is to keep the position of the game good and thereby win the game. So suppose β rounds have been played and Eloise has been able to keep the position "good". Then for all $\gamma < \beta$ and $\epsilon < \lambda$, $\langle (f_\eta(\epsilon), g_\eta(\epsilon)) : \eta \in C^\epsilon_\gamma\rangle$ is a play according to σ_ϵ. Now suppose Abelard plays f_β. We show that Eloise can play g_β so that $\langle (f_\gamma, g_\gamma) : \gamma \leq \beta \rangle$ remains good. Let $\epsilon < \lambda$. Let us look at $\langle (f_\eta(\epsilon), g_\eta(\epsilon)) : \eta \in C^\epsilon_\beta \setminus \{\beta\} \rangle$. This is a play according to the strategy σ_ϵ. Since $|C^\epsilon_\beta| < n_\epsilon$, Eloise can play one more move in EF_{n_ϵ} on $\mathfrak{M}_{\zeta(\epsilon)}$ and $\mathfrak{N}_{\zeta(\epsilon)}$ with Abelard playing $f_\beta(\epsilon)$. Let $g_\beta(\epsilon)$ be the answer of Eloise in this game according to σ_ϵ. The values $g_\beta(\epsilon)$, $\epsilon < \lambda$, constitute the function g_ν. We have shown that Eloise can maintain a good position. □

The proof of the above theorem is quite versatile, like that of Theorem 6.1. It is basically a strategy for organizing a large number of finite sequences of moves into a long sequence of moves.

Corollary 9.1. *Assume GCH and λ regular. Let L be a vocabulary of cardinality $\leq \lambda$ and for each $\xi < \lambda$ let M_ξ and N_ξ be two elementarily equivalent L-structures. If D is a regular filter on λ, then $\prod_i \mathfrak{M}_i/D \cong \prod_i \mathfrak{N}_i/D$.*

Thus if GCH is assumed, we can associate with every complete first order theory T of a regular cardinality λ and with every regular filter D on λ an invariant, namely the isomorphism type of the structure $\mathfrak{M}^\lambda/D$. Here $\mathfrak{M}$ is an arbitrary model of T of cardinality λ. The invariant is independent of the choice of $\mathfrak{M}$. GCH is essential here: Shelah [17] has proved the consistency of the existence two elementarily equivalent countable graphs $\mathfrak{M}$ and $\mathfrak{N}$ and a (regular) ultrafilter D such that $\mathfrak{M}^\omega/D \not\cong \mathfrak{N}^\omega/D$.

Since we know that $\square^{fin}_{\lambda,D}$ can fail, we get the result from Theorem 9.1 that there can be elementarily equivalent structures $\mathfrak{M}$ and $\mathfrak{N}$ of a vocabulary of size $\leq \lambda$ such that Eloise has winning strategy in the game $\mathrm{EF}_\alpha(\mathfrak{M}^\lambda/D, \mathfrak{N}^\lambda/D)$ for each $\alpha < \lambda^+$ but not in the game $\mathrm{EF}_{\lambda^+}(\mathfrak{M}^\lambda/D, \mathfrak{N}^\lambda/D)$. If GCH and $(\aleph_1, \aleph_0) \not\twoheadrightarrow (\aleph_{\omega+1}, \aleph_\omega)$, which is consistent relative to the consistency of a supercompact cardinal, then $\aleph_\omega$ is a singular strong limit, so $\square^{fin}_{\aleph_\omega,D}$ fails for some (any) regular filter D on $\aleph_\omega$ generated by $\aleph_\omega$ sets, and hence there are elementarily equivalent models $\mathfrak{M}$ and $\mathfrak{N}$ of cardinality $\aleph_\omega$ such that

$$\mathfrak{M}^{\aleph_\omega}/D \not\cong \mathfrak{N}^{\aleph_\omega}/D.$$

This observation shows the utility of establishing exact relationships between set theoretical principles and model theoretic principles.

The proof of Theorem 9.1 is interesting even in the case of the one-directional game $\mathrm{EF}^{\rightarrow}_{\kappa}(\mathfrak{M}, \mathfrak{N})$, that is, if Abelard is restricted to play only in one "side", for example only in the models $\mathfrak{M}_\xi$, leaving the models $\mathfrak{N}_\xi$ completely untouched. Naturally we then need to assume less about the models $\mathfrak{M}_\xi$ and $\mathfrak{N}_\xi$. They need not be elementarily equivalent, as long as existential statements true in $\mathfrak{M}_\xi$ are true in $\mathfrak{N}_\xi$. This idea leads to various embedding theorems all equivalent to $\square^{fin}_{\lambda,D}$.

Since there are a variety of different embedding results, we introduce a new concept that fits them all: Suppose Δ is a set of formulas of any logic of vocabulary L. The set of Δ-existential formulas is the set of formulas of the form

$$\exists x_1 ... \exists x_n (\phi_1 \wedge ... \wedge \phi_n),$$

where each ϕ_i is in Δ. The set of weakly Δ-existential formulas is the set of formulas of the above form, where each ϕ_i is in Δ or is the negation of a formula in Δ. If $\mathfrak{M}$ and $\mathfrak{N}$ are L-structures and $h : M \to N$, we say that h is a Δ-homomorphism $\mathfrak{M} \to \mathfrak{N}$ if h preserves the truth of Δ-formulas. If h preserves also the truth of negations of Δ-formulas, it is called a Δ-embedding.

Theorem 9.2. *Suppose λ is a cardinal and D is a regular filter on λ. Then the following conditions are equivalent*

1. $\square^{fin}_{\lambda,D}$.

2. *Let $\mathfrak{M}$ and $\mathfrak{N}$ be a models in a vocabulary of cardinality $\leq \lambda$. We assume the cardinality of $\mathfrak{M}$ is at most λ^+. Suppose Δ is a set of first order formulas in L and D a regular filter on λ such that either Δ consists of Horn formulas only, or D is an ultrafilter. We assume that every weakly Δ-existential sentence true in $\mathfrak{M}$ is true also in $\mathfrak{N}$. Then there is a Δ-embedding of $\mathfrak{M}$ into $\mathfrak{N}^\lambda/D$. (We can replace "weakly Δ-existential" by "Δ-existential" in the Theorem, if we only want a Δ-homomorphism.)*

Corollary 9.2. *Assume $\square^{fin}_{\lambda,D}$. If $\mathfrak{M}$ is a model with vocabulary $\leq \lambda$, and D is a regular ultrafilter over λ, then $\mathfrak{M}^\lambda/D$ is λ^{++}-universal, i.e. if $\mathfrak{M}'$ is of cardinality $\leq \lambda^+$, and $\mathfrak{M}' \equiv \mathfrak{M}$, then $\mathfrak{M}'$ is elementarily embeddable into the ultrapower $\mathfrak{M}^\lambda/D$.*

The proof of Theorem 9.2 is an adaption of the proof of Theorem 9.1 from the context of EF_{λ^+} into the context of $\mathrm{EF}^{\rightarrow}_{\lambda^+}$.

We end with a diagram of the relationships of the various combinatorial principles and model theoretic principles:

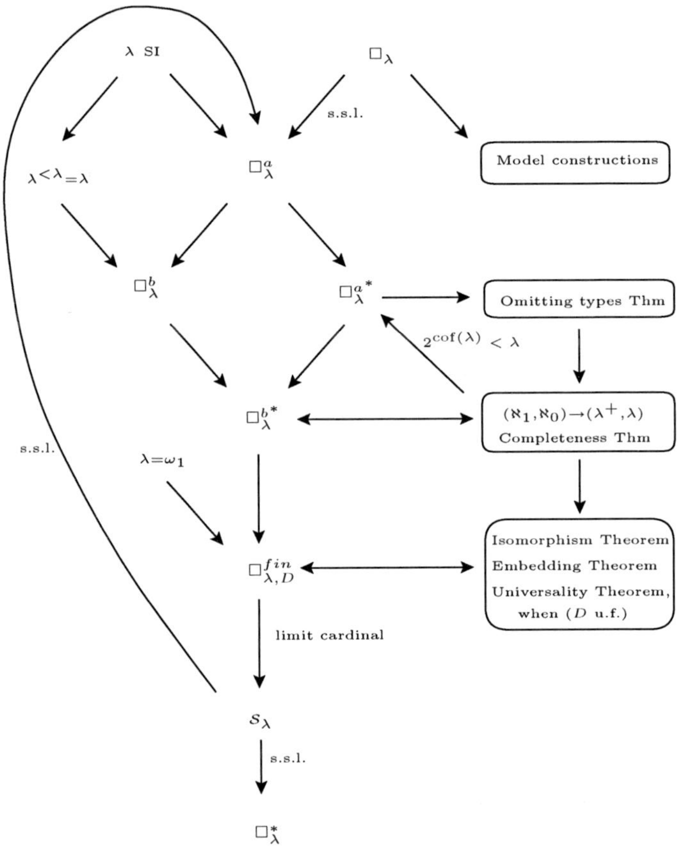

Figure 2 – Square principles and model theory. (SI = strongly inaccessible, s.s.l. = singular strong limit)

References

[1] BEN-DAVID, S. and SHELAH, S.: The two-cardinals transfer property and resurrection of supercompactness, *Proc. Amer. Math. Soc.* **124** (9) (1996), 2827–2837.

[2] BENDA, M.: Reduced products and nonstandard logics, *J. Symbolic Logic* **34** (1969), 424–436.

[3] CHANG, C.C.: A note on the two cardinal problem, *Proc. Amer. Math. Soc.* **16** (1965), 1148–1155.

[4] FLUM, J. and ZIEGLER, M.: *Topological model theory*, Lecture Notes in Mathematics **769**, Springer, Berlin (1980).

[5] FUHRKEN, G.: Skolem-type normal forms for first-order languages with a generalized quantifier, *Fund. Math.* **54** (1964), 291–302.

[6] FUHRKEN, G.: Languages with added quantifier "there exist at least $\aleph_\alpha$", in *Theory of Models* (Proc. 1963 Internat. Sympos. Berkeley), North-Holland, Amsterdam (1965), 121–131.

[7] JENSEN, R.B.: The fine structure of the constructible hierarchy, *Ann. Math. Logic* **4** (1972), 229–308; erratum, ibid. 4 (1972), 443. With a section by Jack Silver.

[8] KEISLER, H.J.: Models with orderings, in *Logic, Methodology and Philos. Sci.* **III** (Proc. Third Internat. Congr., Amsterdam, 1967), North-Holland, Amsterdam (1968), 35–62.

[9] KEISLER, H.J.: Logic with the quantifier "there exist uncountably many", *Ann. Math. Logic* **1** (1970), 1–93.

[10] KENNEDY, J. and SHELAH, S.: On regular reduced products, *J. Symbolic Logic* **67** (3) (2002), 1169–1177.

[11] KENNEDY, J. and SHELAH, S.: More on regular reduced products, *J. Symbolic Logic* **69** (4) (2004), 1261–1266.

[12] MAKOWSKY, J.A. and SHELAH, S.: Positive results in abstract model theory: a theory of compact logics, *Ann. Pure Appl. Logic* **25** (3) (1983), 263–299.

[13] McKee, T.A.: Infinitary logic and topological homeomorphisms, *Z. Math. Logik Grundlagen Math.* **21** (5) (1975), 405–408.

[14] Mostowski, A.: On a generalization of quantifiers, *Fund. Math.* **44** (1957), 12–36.

[15] Shelah, S.: "Gap 1" two-cardinal principles and the omitting types theorem for $L(Q)$, *Israel J. Math.* **65** (2) (1989), 133–152.

[16] Shelah, S.: *Classification theory and the number of nonisomorphic models*, Studies in Logic and the Foundations of Mathematics **92**, North-Holland Publishing Co., Amsterdam (1990), second edition.

[17] Shelah, S.: Vive la différence. I. Nonisomorphism of ultrapowers of countable models, in *Set theory of the continuum* (Berkeley, CA, 1989), Math. Sci. Res. Inst. Publ. **26**, Springer, New York (1992), 357–405.

[18] Vaught, R.L.: The completeness of logic with the added quantifier "there are uncountably many", *Fund. Math.* **54** (1964), 303–304.

Set Theory of Infinite Imperfect Information Games

Benedikt Löwe

Contents

1. Introduction

For most mathematicians, the term "game theory" evokes associations of applications in Economics and Computer Science: it is often associated with the Prisoner's Dilemma or other applications in the social sciences. Game theory is not perceived as an area of mathematics but rather as an area in which mathematics is applicable.

In addition to this famous area of game theory that has been repeatedly honoured by the Nobel Prize for Economics, games also play an important rôle in mathematics and computer science, often connected with logic. We use the name Logic & Games for the research field that connects techniques from logic in order to investigate games, or game theoretic methods in order to prove theorems of logic. This field itself is broad and has many subfields that are as diverse as logic itself. One of its subfields is the set theoretic study of infinite games.

It was the Polish school of topologists and measure theorists that connected game theory to set theory. According to Steinhaus [48, p. 464], Banach and Mazur knew in the 1930s that there is a non-determined infinite game (constructed by a use of the axiom of choice) and that there is a connection between games and the Baire property.

Gale and Stewart in their seminal paper [11] presented the general theory of infinite games, and Mycielski and Steinhaus proposed a set theoretic analysis of game-related axioms in [41].

In the 1960s, the early set theoretic investigation of the game theoretic axioms proposed by Mycielski and Steinhaus was mainly done by Mycielski and a growing group of Californian logicians, among them John Addison, Tony Martin, Yiannis Moschovakis and Bob Solovay at the University of California at Berkeley and Los Angeles. In the 1970s, the Los Angeles area set theory seminar with researchers from UCLA and the California Institute of Technology (including prominently the researchers mentioned earlier and Alekos Kechris, John Steel, and later Hugh Woodin) became known as "the Cabal" and its regular meetings together with a conference series called the *Very Informal Gathering* produced a theory that is now known as Cabal-style descriptive set theory and was published in the four proceedings volumes of the Cabal seminar [25, 22, 23, 24] and codified in Moschovakis' text book [40].

The Cabal has investigated the consequences of game theoretic axioms in set theory, and they have unveiled deep connections between these axioms and the foundations of mathematics and metamathematics. On their way, they have also developed a lot of set theoretic techniques for dealing with infinite games. This theory has been developed in the setting of two-player perfect information games. Although the Cabal has been looking at variants of determinacy with different sets of possible moves (*e.g.*, $\mathsf{AD}_{\mathbb{R}}$) and variants of determinacy for games of different transfinite lengths [47, 43], they didn't give up the general perfect information structure.

In this survey paper, we shall extend results from Cabal-style descriptive set theory to a broader class of games, namely to a type of imperfect information games used in statistics and called Blackwell games. The imperfection of the information in Blackwell games is very minor: instead of playing infinitely many times in turn, the players play infinitely many times simultaneously and reveal their moves after each round. Many game-theoretic phenomena cannot be modelled with Blackwell games. However, Blackwell games are so far the only type of imperfect information games that have been investigated from the point of view of set theory. It will turn out in Section 5 (Theorem 5.2) that if we restrict our attention to set-theoretic strength, even this puny amount of imperfect information can be avoided, and we end up doing an analysis of perfect information games played with probabilistic strategies.

In Sections 2 and 3, we introduce the basic notions of the theory of determinacy of perfect and imperfect information games, including the protagonist of this paper, the axiom of Blackwell determinacy Bl-AD. We relate the development of the 1990s on Blackwell determinacy in Sections 4 and 5, in particular, we discuss Martin's equivalence conjecture (Conjecture 4.1) and the Martin-Vervoort Zero-One Laws (Theorem 5.1 and Corollary 5.1). These developments led to a considerable simplification of the formulation of axioms of Blackwell determinacy that opened up the possibility of generalization in several directions. In Sections 6 and 7, we discuss results pointing towards the truth of Martin's conjecture: the computation of the consistency strength of Bl-AD due to Martin, Neeman and Vervoort (Theorem 6.1) and some partial results concerning infinitary combinatorics and hierarchies of sets and functions due to the present author.[1] In Sections 8 and 9, we finally discuss weaker and stronger versions of BI-AD. The paper closes with a list of open questions in the appendix.

2. Games in Set Theory

Our paper is written for readers with some background in set theory. We are using standard notation from set theory throughout the paper. As usual in set theory, we shall be working on Baire space ω^ω with the standard topology generated by the basic open sets $[s] := \{x \in \omega^\omega \ ; \ s \subseteq x\}$ for finite sequences $s \in \omega^{<\omega}$. It is well-known that this topological space is homeomorphic to the irrational numbers, so it is safe to call its elements **real numbers**.

Throughout we shall work in the theory $\mathsf{ZF} + \mathsf{AC}_\omega(\mathbb{R})$. This small fragment of the axiom of choice is necessary for the definition of axioms of Blackwell determinacy. In Sections 7.2 and 7.3, we shall extend our basic theory to $\mathsf{ZF} + \mathsf{DC}$ and $\mathsf{ZF} + \mathsf{DC}_\mathbb{R}$, respectively, but not without explicitly mentioning it in the results.

We look at infinite games of the following type: we have two players, called player I and player II, and a fixed payoff set $A \subseteq \omega^\omega$. The players play infinitely often natural numbers, player I begins with x_0, player II answers with x_1, then player I plays x_2 and so on. After infinitely many rounds of the game, they have produced a sequence $x := \langle x_i \ ; \ i \in \omega \rangle \in \omega^\omega$ which they compare to the payoff set A. If $x \in A$, then player I wins; otherwise, player II wins. We shall be using the standard notation for infinite games: If $x \in \omega^\omega$ is the sequence of moves for player I and $y \in \omega^\omega$ is the sequence of moves for player II, we let $x * y$ be the sequence constructed by playing x against y, *i.e.*,

$$(x * y)(n) := \begin{cases} x(k) & \text{if } n = 2k, \\ y(k) & \text{if } n = 2k+1. \end{cases}$$

[1]A few years ago, the present author has published another survey on Blackwell games entitled "*Consequences of the Axiom of Blackwell Determinacy*" [29]. That survey focusses on set-theoretic properties that provably hold under the assumption of BI-AD; the material corresponds roughly to the content of Section 7 of this paper.

Conversely, if $x \in \omega^\omega$ is a run of a game, then we let x_{I} be the part played by player I and x_{II} be the part played by player II, *i.e.*, $x_{\mathrm{I}}(n) = x(2n)$ and $x_{\mathrm{II}}(n) = x(2n+1)$. The game just described will be denoted by $\mathrm{G}(A)$.

Let us denote by ω^{Even} the set of finite sequences of natural numbers of even length, by ω^{Odd} the set of such sequences of odd length; then we call a function $\sigma : \omega^{\mathrm{Even}} \to \omega$ a **(pure) strategy for player I** and a function $\sigma : \omega^{\mathrm{Odd}} \to \omega$ a **(pure) strategy for player II**. Clearly, you can let strategies play against each other, recursively producing an infinite sequence of natural numbers. If σ is a strategy for player I and τ is a strategy for player II, then we denote the recursively defined outcome of playing σ against τ by $\sigma * \tau$.

A strategy σ for player I (τ for player II) is called **winning in** $\mathrm{G}(A)$ if for every strategy τ for player II (σ for player I), we have $\sigma * \tau \in A$ ($\sigma * \tau \notin A$). We call a set A **determined** if one of the two players has a winning strategy.

As an aside, let us mention that for pure strategies it is not even necessary to win against all strategic opponents; it is enough to win against opponents that do not react to our moves ("blindfolded opponents"): A real $x \in \omega^\omega$ defines **blindfolded strategies** σ_x and τ_x by

$$\sigma_x(s) := \tau_x(s) := x\left(\left\lceil \frac{\mathrm{lh}(s)-1}{2} \right\rceil\right).$$

We write $\sigma * x := \sigma * \tau_x$ and $x * \tau = \sigma_x * \tau$.

Proposition 2.1. *If σ is a pure strategy for player I and A is an arbitrary payoff set, then the following are equivalent for the game on A:*

i. For all $x \in \omega^\omega$, we have $\sigma * x \in A$, and

ii. σ is a winning strategy.

The analogous statement holds for strategies for player II.

PROOF – We only have to show *i* ⇒ *ii*. Without loss of generality, assume that σ is a strategy for player I. Let τ be a pure counterstrategy such that $\sigma * \tau \notin A$. Let $x := (\sigma * \tau)_{\mathrm{II}}$. Then $\sigma * x = \sigma * \tau_x = \sigma * \tau \notin A$. □

Games of the type $\mathrm{G}(A)$ have been investigated by Zermelo in [53] as a formalization of the game of chess. The notion of determinacy was introduced by Gale and Stewart in [11] where they also proved two fundamental theorems about this notion: they proved that every open set $A \subseteq \omega^\omega$ is determined, and they constructed (using the axiom of choice) a non-determined set. Generalizing the open determinacy theorem, Wolfe [52] proved the determinacy of all $\mathbf{\Sigma}^0_2$ sets, then Morton Davis [7] extended this to the determinacy of all $\mathbf{\Sigma}^0_3$ sets. At the very next level lurked the first metamathematical surprise: while the determinacy of all $\mathbf{\Sigma}^0_4$ sets is provable in ZFC [45], any proof of it must essentially use set theoretic techniques (due to a theorem of Harvey Friedman and Tony Martin).[2] This development culminated in Tony Martin's 1975 proof of Borel determinacy [36].

In fact, Borel determinacy is the best possible answer in ZFC: in Gödel's constructible universe, there is a non-determined Π^1_1 set. More precisely, the determinacy of all Π^1_1 sets is equivalent to the existence of $0^\#$ by a theorem of Harrington [12]. Again, this equivalence theorem was the start of a new development in which many researchers connected the theory of determinacy to large cardinal axioms.

In a very natural way, this investigation converged with other developments in seemingly unrelated areas of set theory to give a tremendously deep understanding of the structure of the set-theoretic continuum combining descriptive set theory, the theory of large cardinals and core model theory.

In light of the fact that Gale and Stewart had constructed a non-determined set (using the axiom of choice), it is surprising that one of the most central axioms of this theory was the full **axiom of determinacy** AD stating that all sets are determined. Despite the fact that it contradicts AC, the axiom AD allows a very detailed structure theory in its models (in particular, AD implies that all sets of reals have so-called regularity properties like Lebesgue measurability, the perfect set property and the Baire property; *cf.* Section 7.1) and models of AD show up in prominent places as soon as the set theoretic

[2]The exact result is: Let ZC^- be Zermelo set theory without power set, let $\mathsf{ZFC}^- := \mathsf{ZC}^- + \mathit{Ersetzung}$, let $\mathsf{Ers}(\Sigma_1)$ be *Ersetzung* restricted to Σ_1 formulas, and let β_0 be the least ordinal such that $\mathbf{L}_{\beta_0} \models \mathsf{ZFC}^-$ (if it exists). Then:

$$\mathsf{ZC}^- + \mathsf{Ers}(\Sigma_1) + \mathsf{Det}(\Sigma^0_4) \vdash \text{“}\beta_0 \text{ exists”}.$$

universe is rich enough.[3]

A discussion of the rich theory of determinacy and its interesting history is far beyond the scope of this survey, and we refer the reader to [18, Sections 27-32]. Some of the consequences of AD are discussed in Sections 7.1, 7.2 and 7.3.

3. When is an imperfect information game determined?

If you show the definitions and results of Section 2 to a typical game theorist, he or she will note that all we have said so far concerns a very special class of games. Our games discussed so far have the following properties:

- They are games of perfect information: At each stage, both players know the exact situation of the game. This excludes that moves are made simultaneously ("scissors-stone-paper"), that cards are hidden or private knowledge of one player, and similar phenomena.
- They are games of perfect recall: Both players have unlimited storage for the moves that have been made.
- In our games we have absolute noncooperation; not only are our games zero-sum, but we always have winner and loser and no other relevant parameter. This excludes phenomena of compromise and trade-off.
- We consider only two-player games. Coalitions only play a role for games with more than two players.

It would be very interesting to develop set theoretic infinite versions of games of full imperfect information, of imperfect recall, of cooperation and with more than two players. However, the only area that has been investigated so far is a rather small class of imperfect information games. It is those games that we shall now look into.

[3]The precise statement of this is the following theorem of Woodin: "Assume that there are ω Woodin cardinals and a measurable cardinal above them. Then $\mathbf{L}(\mathbb{R}) \models \mathsf{ZF} + \mathsf{AD}$."

Games with imperfect information were investigated by Johann von Neumann whose famous minimax theorem (1928) is the imperfect information analogue of Zermelo's theorem of the determinacy of finite perfect information games. Infinite versions of von Neumann's games were introduced by David Blackwell in [3] who proved that if the payoff set of a certain game modelled with probability measures is a countable unions of closed sets, then one of the players can approximate an optimal strategy.[4]

Blackwell's games are very close to the games of Section 2: we still have the two players and a fixed payoff set $A \subseteq \omega^\omega$. The players play infinitely often natural numbers x_i, but this time they don't do this in turn, but always play their numbers x_{2i} and x_{2i+1} simultaneously. As before, after infinitely many rounds of the game, they have produced a sequence $x := \langle x_i\ ;\ i \in \omega\rangle \in \omega^\omega$ which they compare to the payoff set A. If $x \in A$, then player I wins, otherwise, player II wins. We denote this game by $\mathrm{B}(A)$.

For player I, there is almost no change: in $\mathrm{G}(A)$, being the one who has to move first, he was already at a disadvantage. The change for player II, however, is profound: in $\mathrm{G}(A)$, he was allowed to base his move x_{2i+1} on his knowledge of x_{2i}. As these two moves are now played simultaneously, this is not possible anymore. This also changes the notion of strategy: in the game $\mathrm{B}(A)$, both players I and II have to use functions from ω^{Even} to ω.

In such a situation, the notion of winning strategy (and hence the notion of determinacy) is not helpful anymore, as the following example shows.

We consider the Gale-Stewart and Blackwell variants of the game of "playing different reals" in which player I wins if for all i, players I and II have played different numbers in round i. More formally, let $M := \{x \in \omega^\omega\ ;\ \forall i(x(2i) \neq x(2i+1)\}$, a closed subset of ω^ω. This game will serve as an illustrative example throughout this paper. Clearly, player II has a winning strategy in the game $\mathrm{G}(M)$.

Remark 3.1 - None of the players has a winning strategy in $\mathrm{B}(M)$.

PROOF – A winning strategy of player I in $\mathrm{B}(M)$ would be one in $\mathrm{G}(M)$ contradicting the fact that player II has a winning strategy there.

---

[4] Blackwell calls these games "Games with Slightly Imperfect Information" in his [4].

Now let σ be a strategy for player II. In the Blackwell game, this means that $\sigma : \omega^{\text{Even}} \to \omega$. Recursively define a real x as follows: Assume that $x \restriction n$ is already defined, then let $x(n) := \sigma(x \restriction n * \sigma) + 1$. Then clearly, if player I plays x, he will never play the same as σ in the same round, and thus win. □

Despite the fact that the notion of winning strategy doesn't yield an analysis of the game $\mathrm{B}(M)$, there is a clear intuition that the game is easier for player I: after all, the players make infinitely many moves independently, and if only one of them produces different results, player I wins. So, just randomizing at each step will eventually pay off for player I.

We shall now mould this intuition into a precise definition:

Let $\mathrm{Prob}(\omega)$ be the set of probability measures on ω. We call a function $\sigma : \omega^{\text{Even}} \to \mathrm{Prob}(\omega)$ a **Blackwell strategy for player I** and a function $\tau : \omega^{\text{Odd}} \to \mathrm{Prob}(\omega)$ a **Blackwell strategy for player II** if for all $n, m \in \omega$ and $s \in \omega^{\text{Even}}$, we have $\tau(s^\frown\langle n\rangle) = \tau(s^\frown\langle m\rangle)$.

If σ and τ are Blackwell strategies for players I and II, respectively, then they allow to describe the (randomized) behaviour of the two players: Player I chooses his first move x_0 according to the probability measure $\sigma(\langle\ \rangle)$, then player II consults his strategy about the measure $\tau(\langle x_0\rangle)$ (which may not depend on x_0 if τ is a Blackwell strategy) and plays according to that probability measure, and so on.

Let

$$\nu(\sigma,\tau)(s) := \begin{cases} \sigma(s) & \text{if } \mathrm{lh}(s) \text{ is even, and} \\ \tau(s) & \text{if } \mathrm{lh}(s) \text{ is odd.} \end{cases}$$

Then for any $s \in \omega^{<\omega}$, we can define

$$\mu_{\sigma,\tau}([s]) := \prod_{i=0}^{\mathrm{lh}(s)-1} \nu(\sigma,\tau)(s\restriction i)(\{s(i)\}).$$

Using $\mathsf{AC}_\omega(\mathbb{R})$ to pick Borel codes, this generates a Borel probability measure on ω^ω which can be seen as a measure of how well the strategies σ and τ performs against each other. If B is a Borel set, $\mu_{\sigma,\tau}(B)$ is interpreted as the probability that the result of the game ends up in the set B when player I randomizes according to σ and player II according to τ.

For any Blackwell strategy σ for player I or τ for player II we can define a measure for its quality (the **value of the strategy**) by

$$\mathrm{val}_{\mathrm{I}}^{A}(\sigma) := \inf\{\mu^{-}_{\sigma,\tau}(A)\,;\ \tau \text{ is Blackwell strategy for player II}\}, \text{ and}$$

$$\mathrm{val}_{\mathrm{II}}^{A}(\tau) := \sup\{\mu^{+}_{\sigma,\tau}(A)\,;\ \sigma \text{ is Blackwell strategy for player I}\}.$$ [5]

Now we define the **value sets** for player I and player II by

$$V_{\mathrm{I}}(A) := \{\mathrm{val}_{\mathrm{I}}^{A}(\sigma)\,;\ \sigma \text{ is a Blackwell strategy for player I}\}, \text{ and}$$

$$V_{\mathrm{II}}(A) := \{\mathrm{val}_{\mathrm{II}}^{A}(\tau)\,;\ \tau \text{ is a Blackwell strategy for player II}\}.$$

Then $V_{\mathrm{II}}(A)$ lies entirely above $V_{\mathrm{I}}(A)$ in the sense that for all $v \in V_{\mathrm{II}}(A)$ and $v^* \in V_{\mathrm{I}}(A)$ we have $v \geq v^*$. If now these two sets $V_{\mathrm{I}}(A)$ and $V_{\mathrm{II}}(A)$ touch each other at a point p (depicted in Figure 1), then the outcome of the game is determined in the following sense: if both players play rationally, then the probability that player I wins is arbitrarily close to p. In this case, we say that the payoff set **has a value**.

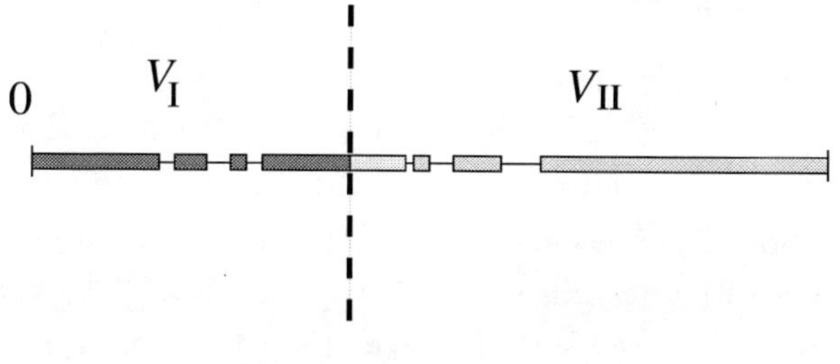

Figure 1 – Value sets V_{I} and V_{II} touch each other: the outcome is stochastically determined if both players approximate optimal play

In the other case, when the sets $V_{\mathrm{I}}(A)$ and $V_{\mathrm{II}}(A)$ don't touch each other, then the interval between the supremum v^- of $V_{\mathrm{I}}(A)$ (the **lower value**) and the infimum v^+ of $V_{\mathrm{II}}(A)$ (the **upper value**) is an area of indeterminacy: Player I can bound his chance of winning from below by v^- and player II can bound

[5] Here, μ^+ denotes outer measure and μ^- denotes inner measure with respect to μ in the usual sense of measure theory. If A is Borel, then $\mu^+(A) = \mu^-(A) = \mu(A)$ for Borel measures μ.

player I's chance of winning from above by v^+, but since these are not the same number, the real outcome can be somewhere in the interval (as depicted in Figure 2).

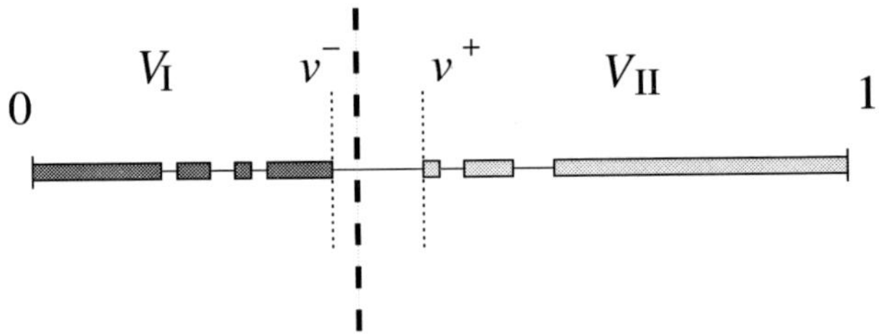

Figure 2 – Value sets V_I and V_{II} don't touch: the payoff set does not have a value

What sets have a value? A first restriction is given by the following very simple example: Let $H := \{x \in \omega^\omega\,;\, x(0) \geq x(1)\}$. This is a clopen payoff set that corresponds to the game "who knows the larger natural number?".

Remark 3.2 - The game $\mathrm{B}(H)$ does not have a value.

PROOF – We shall show that the lower value of this game is 0 and the upper value is 1: Let σ be any strategy for player I. We shall show that it has value $< \varepsilon$ for every $\varepsilon > 0$. Since all moves after the first two are irrelevant for the payoff, a strategy for player I is just a probability measure on ω. We know that $\sum_{i\in\omega} \sigma(\{i\}) = 1$, so find a natural number n such that $\sum_{i<n} \sigma(\{i\}) > 1 - \varepsilon$. Then take the trivial strategy τ that assigns the total measure to the number n. The probability that player I wins against τ using the strategy σ is $< \varepsilon$. The proof that the upper value is 1 is exactly the same. □

This simple game shows that we should restrict our games to finitely many possible moves, for instance, binary choices. Therefore, for a pointclass Γ, we shall write $\mathsf{Val}(\Gamma)$ for "Every set $A \subseteq 2^\omega$ such that $A \in \Gamma$ has a value." It is not hard to see [28, Corollary 4.4] that 2 can be replaced by any finite set with more than one element (if Γ is reasonably closed).

Proofs of Blackwell determinacy had a development not unsimilar to the development of determinacy proofs described in Section 2. As mentioned, von Neumann's Minimax theorem takes the role of Zermelo's theorem. In [3], Blackwell proved the Blackwell determinacy of all $\mathbf{\Sigma}^0_2$ sets. There were results extending Blackwell's result by Orkin [44] and Maitra and Sudderth [35], before Blackwell in an extended abstract [4] revived the interest in the foundational questions. In particular, he asked whether all Borel sets are Blackwell determined. Instigated by this, Marco Vervoort [49, 50] proved $\mathbf{\Sigma}^0_3$ Blackwell determinacy, before Martin solved Blackwell's question (*cf.* Section 4).

Before continuing with describing the developments of the 1990s (Section 4), let us stop for a moment and ask ourselves what we can do with these new games.

Theorem 3.1 (Vervoort, 1995). *Let* $\mathbf{\Gamma}$ *be a boldface pointclass. Assume* $\mathsf{Val}(\mathbf{\Gamma})$. *Then all sets in* $\mathbf{\Gamma}$ *are Lebesgue measurable.*

Proof – Let $A \subseteq 2^\omega$ be a set in $\mathbf{\Gamma}$. Let us define a continuous function as follows: for $s \in 2^{2n}$, we let

$$\widehat{s}(k) := |s(2k) - s(2k+1)|, \text{ and}$$

let $\pi : 2^\omega \to 2^\omega$ be the function induced by $s \mapsto \widehat{s}$. For given strategies σ and τ, let $\widehat{\mu_{\sigma,\tau}}$ be the pull-back measure with respect to π, *i.e.*,

$$\widehat{\mu_{\sigma,\tau}}(A) := \mu_{\sigma,\tau}(\pi^{-1}[A]).$$

If either of σ or τ is the "randomize" strategy assigning probability $\frac{1}{2}$ to both 0 and 1, then $\widehat{\mu_{\sigma,\tau}}$ is Lebesgue measure on 2^ω.

Now let $A^* := \pi^{-1}[A] \in \mathbf{\Gamma}$. We get that

$$\sup V_{\mathrm{I}}(A^*) \leq \widehat{\mu_{\sigma,\tau}}^-(A) \leq \widehat{\mu_{\sigma,\tau}}^+(A) \leq \inf V_{\mathrm{II}}(A^*),$$

and since $\sup V_{\mathrm{I}}(A^*)$ and $\inf V_{\mathrm{II}}(A^*)$ coincide by assumption, A is Lebesgue measurable. □

Vervoort's proof of Theorem 3.1 gave rise to a new proof of Lebesgue measurability from determinacy due to Tony Martin which was published in [38].

The proof of Theorem 3.1 is specific for Lebesgue measurability and is not easily adapted to other regularity properties like the Baire property or the perfect set property. We shall discuss this in more detail in Section 7.1.

4. Martin's Theorem and Conjecture

A priori, there is no immediate connection between the axioms $\mathsf{Det}(\mathbf{\Gamma})$ and $\mathsf{Val}(\mathbf{\Gamma})$. If player II has a winning strategy in $\mathrm{G}(A)$, then he may use the extra information that player II has not at his disposal in the game $\mathrm{B}(A)$. If the game $\mathrm{B}(A)$ has a value, especially if the value is neither 0 or 1, then this doesn't give a hint on how to construct a winning strategy in $\mathrm{G}(A)$.

This lack of connection is illustrated by our game "playing different reals" from Section 3:

Proposition 4.1. *Let* $M := \{x \in 2^{\mathbb{N}}\,;\, \forall n(x(2n) \neq x(2n+1)\}$*. Player II has a winning strategy in* $\mathrm{G}(M)$ *while the value of* $\mathrm{B}(M)$ *is* 1.

Proof – We already indicated in the discussion after Remark 3.1 that $\mathrm{B}(M)$ is easier for player I than for player II. Using the definitions from Section 3, we can this make more precise now:

We shall show that $\sup V_{\mathrm{I}} = 1$. For any strategy τ for player II, the "randomize" strategy $\sigma(s)(\{0\}) = \sigma(s)(\{1\}) = \frac{1}{2}$ (which we already used in the proof of Theorem 3.1) will give $\mu_{\sigma,\tau}(M) = 1$. □

As a consequence, we cannot expect a simple translation of good strategies in $\mathrm{G}(A)$ into good strategies in $\mathrm{B}(A)$. However, Martin showed that you can still mimick $\mathrm{B}(A)$ with a set of perfect information games. In particular, fixing a set $A \subseteq 2^\omega$, Martin defined for each $v \in (0,1]$ a set A_v that intuitively expresses bounds for the lower and upper values:

Lemma 4.1 (Martin, 1998). *If player I has a winning strategy for* $\mathrm{G}(A_v)$*, then* $\sup V_{\mathrm{I}}(A) \geq v$*. If player II has a winning strategy for* $\mathrm{G}(A_v)$*, then* $\inf V_{\mathrm{II}} \leq v$.

Proof – For the definition of A_v, *cf.* [37, p. 1569]. The two statements of the lemma are [37, Lemma 1.3] and [37, Lemma 1.6]. □

Martin used Lemma 4.1 to derive the following result:

Theorem 4.1 (Martin, 1998). *If* $\mathbf{\Gamma}$ *is a boldface pointclass, then* $\mathsf{Det}(\mathbf{\Gamma})$ *implies* $\mathsf{Val}(\mathbf{\Gamma})$.

Of the two possible implications between $\mathsf{Det}(\mathbf{\Gamma})$ and $\mathsf{Val}(\mathbf{\Gamma})$, this was probably the one less expected to hold. Since having a value talks about a much more general class of strategies, it was natural to assume that $\mathsf{Val}(\mathbf{\Gamma})$ should be stronger than the axiom that only talks about pure strategies, a rather restricted class of strategies.

The uniform nature of Martin's result and the fact that having a value was seen as the intuitively stronger of the two properties led to the natural conjecture:

Conjecture 4.1 (Martin). *Let* $\mathbf{\Gamma}$ *be a boldface pointclass. Then* $\mathsf{Det}(\mathbf{\Gamma})$ *and* $\mathsf{Val}(\mathbf{\Gamma})$ *are equivalent.*

Martin's Conjecture 4.1 has proved to be more difficult than people thought at first and it is still open in its general form. The most intriguing open instance of Conjecture 4.1 is the question whether "Every set $A \subseteq 2^\omega$ has a value" implies AD.

5. Zero-One Laws and Optimal Strategies

In a surprising development of the years 1999 and 2000, it became clear that the issue of imperfect information does not play any rôle for the question raised by Martin as Conjecture 4.1.

For this, let us look at a class of games that lie between $\mathrm{G}(A)$ and $\mathrm{B}(A)$: they are played like $\mathrm{G}(A)$, *i.e.*, the players move in turn and there is no information hidden for either of the players, but they are played with probabilistic strategies. Formally, we call functions $\sigma : \omega^{\mathrm{Even}} \to \mathrm{Prob}(\omega)$ **mixed strategies for player I** and functions $\tau : \omega^{\mathrm{Odd}} \to \mathrm{Prob}(\omega)$ **mixed strategies for player II**. Note that every mixed strategy for player I is also a Blackwell strategy for player I, but not so for player II. Mixed strategies σ and τ give rise to a Borel probability measure on ω^ω as did Blackwell strategies, and we define **mixed values** for mixed strategies in analogy to the values of Section 3. The game $\mathrm{M}(A)$ is the game with payoff A played with mixed strategies. We say that A is **Blackwell determined** if the upper and lower mixed value of $\mathrm{M}(A)$ coincide. We also say that A **has a mixed value**. For a pointclass Γ, we write $\mathsf{Bl\text{-}Det}(\Gamma)$ for "all sets in Γ are Blackwell determined".

Proposition 5.1. *For every boldface pointclass* $\mathbf{\Gamma}$, *we have that* $\mathsf{Val}(\mathbf{\Gamma})$ *implies* $\mathsf{Bl\text{-}Det}(\mathbf{\Gamma})$.

PROOF – You can easily see $\mathrm{M}(A)$ as a Blackwell game by defining $\pi : x \mapsto x^*$ via

$$\begin{aligned} x^*(2n) &:= x(4n), \text{ and} \\ x^*(2n+1) &:= x(4n+3). \end{aligned}$$

Then $\mathrm{M}(A)$ and $\mathrm{B}(\pi^{-1}[A])$ are the same game. □

The mixed game $\mathrm{M}(A)$ is essentially a perfect information game, and this has important consequences for the possible mixed values. As a first step, Martin and Vervoort could prove a first **Zero-One Law** for mixed games in the year 2000:

Theorem 5.1 (Martin-Vervoort Zero-One Law, 2000). *Suppose A has a mixed value. Then the mixed value is either* 0 *or* 1.

With the Martin Vervoort Zero-One Law in mind, we can define a very strong property for mixed strategies:

A mixed strategy σ for player I is now called **optimal for** A if for every mixed strategy τ for player II, there is a Borel subset $B \subseteq A$ such that $\mu_{\sigma,\tau}(B) = 1$. Similarly, a mixed strategy τ for player II is called **optimal for** A if for every mixed strategy σ for player I, there is a Borel superset $B \supseteq A$ such that $\mu_{\sigma,\tau}(B) = 0$.

Vervoort was able to use the Martin-Vervoort Zero-One-Law to prove the existence of optimal strategies:

Corollary 5.1 (Vervoort Strong Zero-One Law, 2000). *Let* $\mathbf{\Gamma}$ *be a boldface pointclass. Suppose that* $\mathsf{Bl\text{-}Det}(\mathbf{\Gamma})$ *holds and* $A \in \mathbf{\Gamma}$. *Then there is either an optimal strategy for player I or an optimal strategy for player II in* $\mathrm{M}(A)$.

PROOF – Proofs of both Theorem 5.1 and Corollary 5.1 can be found as [51, Theorems 5.3.3 and 5.3.4] or [39, Lemmas 3.7 and 3.10]. □

Using the Zero-One Laws, we can now prove the converse of Proposition 5.1 and rephrase our definition of Blackwell determinacy:

Theorem 5.2 (Martin). *For boldface pointclasses* $\mathbf{\Gamma}$, *the statements* $\mathsf{Val}(\mathbf{\Gamma})$ *and* $\mathsf{Bl\text{-}Det}(\mathbf{\Gamma})$ *are equivalent.*

Proof – This result has been mentioned in passing in [39, p. 619]. One just has to check that the proof of Lemma 4.1 still goes through with optimal strategies instead of winning strategies. □

Synthetizing what we have said about Blackwell determinacy so far, we can give a new but equivalent definition of the notion as follows: We call a set A **Blackwell determined** if either player I or player II has an optimal strategy in the game $\mathrm{M}(A)$. By the Strong Zero-One Law, this will give the same notion restricted to boldface pointclasses. In Sections 8 and 9, we shall use this definition as the basis for generalizations.

6. Consistency Strength of Blackwell Determinacy

Martin's proof of Theorem 4.1 immediately yields upper bounds for the consistency strength of axioms of Blackwell determinacy: for instance, if there is a measurable cardinal, then $\mathsf{Bl\text{-}Det}(\mathbf{\Pi}^1_1)$ must hold.

If Martin's Conjecture 4.1 is true, then all of these upper bounds should be sharp and the corresponding lower bound should be provable as well. As a first result, Martin proved the existence of sharps from $\mathsf{Bl\text{-}Det}(\mathbf{\Pi}^1_1)$ thereby establishing the equivalence of determinacy and Blackwell determinacy at the $\mathbf{\Pi}^1_1$ level. (This result is published with a full proof in [30, Theorem 3.8].)

It was open in the years 1998 and 1999 whether you could get any stronger large cardinal assumptions out of Blackwell determinacy. Then, the Zero-One Law (Theorem 5.1) allowed Martin, Neeman and Vervoort to simulate the proof of the Third Periodicity Theorem [40, Section 6E] with mixed strategies. In particular, they showed

Lemma 6.1 (Martin-Neeman-Vervoort). *Call a pointclass* Γ **weakly scaled** *if every set in* Γ *admits a scale* $\{\varphi_n\,;\,n\in\omega\}$ *such that every* φ_n *is a* Γ*-norm. If* Γ *is a weakly scaled pointclass and* $\Delta := \Gamma\cap\breve{\Gamma}$, *then* $\mathsf{Bl\text{-}Det}(\Delta)$ *implies* $\mathsf{Det}(\Delta)$.

From this, you can get the equivalence between Blackwell determinacy and determinacy for many pointclasses (thus proving many instances of Martin's Conjecture 4.1):

Theorem 6.1 (Martin-Neeman-Vervoort). *Let* $\mathbf{\Gamma}$ *be either* $\mathbf{\Delta}^1_{2n}$, $\mathbf{\Sigma}^1_{2n}$, $\mathbf{\Delta}^1_{2n+1}$, $\Game^n(<\omega^2\text{-}\mathbf{\Pi}^1_1)$, *or* $\mathscr{P}(\omega^\omega) \cap \mathbf{L}(\mathbb{R})$. *Then* $\mathsf{Bl\text{-}Det}(\mathbf{\Gamma})$ *implies* $\mathsf{Det}(\mathbf{\Gamma})$.

PROOF – This is Theorem 5.1, Corollary 5.3, Theorem 5.4, Theorem 5.6, and Theorem 5.7 in [39]. □

Theorem 6.1 settles the question of the consistency strength of the axiom of Blackwell determinacy Bl-AD completely: If $\mathbf{V} \models$ Bl-AD, then this is still true in $\mathbf{L}(\mathbb{R})$, but there we get $\mathbf{L}(\mathbb{R}) \models$ AD by Theorem 6.1. Consequently, the consistency strength of Bl-AD is the same as that of AD.

7. Approaching Equivalence

While Theorem 6.1 settles the question about the consistency strength of Blackwell determinacy, it leaves open the most vexing question that we mentioned before: does Bl-AD imply AD?

Looking at the type of AD-arguments that still work under the assumption of Bl-AD, we see that the most serious problem with Bl-AD is that pure strategies generate continuous choice-like functions whereas even optimal mixed strategies don't generate a function at all. In other words: We are missing a principle that given an optimal strategy for player I in $\mathrm{M}(A)$ allows a parametrised choice of an element of the payoff set A (and similarly for player II and the complement of A). Such a principle called the **parametrised choice principle** (PCP) has been defined in [33] where we show that (in the base theory $\mathsf{ZF} + \mathsf{AC}_\omega(\mathbb{R}) + \mathsf{Bl\text{-}AD}$), AD and PCP are equivalent.

Thus, PCP can be seen as the difference between AD and Bl-AD. If we can show PCP from Bl-AD, Martin's Conjecture 4.1 would be proved.

The direct approaches of proving determinacy from Blackwell determinacy have not been successful so far. A different approach is to take some of the very characteristic consequences of the axiom of determinacy and prove that they hold under BI-AD.

This has been successful in some cases. We focus here on three very characteristic features of AD-set theory: regularity properties of all sets of reals (Section 7.1), the very concrete combinatorial theory of small cardinals (up to $\aleph_{\varepsilon_0}$ and slightly higher; Section 7.2), and the existence of global hierarchies of sets of reals (Section 7.3).

7.1. Regularity Properties.

One of the most basic consequences of AD is the fact that sets of reals have all desirable properties. These properties are typically called **regularity properties**, and examples are Lebesgue measurability, the Baire property, and the perfect set property.[6]

Working in ZF + AD, Mycielski and Swierczkowski showed in [42] that every set is Lebesgue measurable. A well-known topological game argument due to Banach and Mazur [20, 8H] establishes that every set has the Baire property. Finally, Davis proved in [7] that every set has the perfect set property. Most of these proofs use a modest amount of coding, but typically link the regularity property of a set A directly to the determinacy of some set A^* simply definable from A.

Let us describe this in more detail for the perfect set property: Fix some bijection $\pi : \omega \to 2^{<\omega}$. We now recursively define a function $\pi : \omega^\omega \to \omega^\omega$: Given some $x \in \omega^\omega$, let $x^{(0)} := \pi(x(0))$, $x^{(2i+1)} := x^{(2i)\frown}\langle x(2i+1)\rangle$, and $x^{(2i+2)} := x^{(2i+1)\frown}\pi(x(2i+2))$. We write $N_{x,i}$ for the length of the sequence $x^{(i)}$. Then the function π defined by $\pi(x) := \bigcup_{i\in\omega} x^{(i)}$ is continuous.

For $A \subseteq 2^\omega$, let $A^* := \{x\,;\,\pi(x) \in A\}$ be the π-preimage of A. There is a well-known connection between A^* and the perfect set property of A:

Fact 7.1. *For every set A, A has the perfect set property if and only if A^* is determined.*

PROOF – *Cf.* [20, Theorem (21.1)]. □

[6] A set A has the **Baire property** if there is a Borel set B such that $B\triangle A$ is a countable union of nowhere dense sets. It has the **perfect set property** if it is either countable or contains a nonempty perfect subset.

Thus, AD implies that every set of reals has the perfect set property. What if we replace AD by Bl-AD? From Theorem 3.1 we know that every set is Lebesgue measurable, but as we mentioned earlier, its proof was very heavily using the affinity of the definition of Blackwell determinacy to measures and doesn't seem to have analogues for other regularity properties.

One could ask whether analogues of Fact 7.1 could hold for Blackwell determinacy. Unfortunately, Hjorth showed that the answer is no. This argument was sketched in in [37, p. 1580]. We give a more detailed proof here: A measure μ is called **atomless** if every singleton has μ-measure 0. A set is called **universally zero** if it has μ-measure 0 for every atomless Borel measure μ.

Theorem 7.1 (Hjorth). *Assume* AC. *Then there is a set A with the following properties:*

1. *A^* is Blackwell determined, but*

2. *A does not have the perfect set property.*

Proof – Using the axiom of choice, pick one element from each WO_α and call the resulting set A. This is clearly an uncountable set which (by the boundedness lemma) cannot have a perfect subset. Another boundedness argument shows easily that A is universally null [27, Proposition 3.4].

We shall now show that player II has an optimal strategy in $\mathrm{M}(A^*)$. This strategy is again the "randomize strategy" τ assigning $\frac{1}{2}$ to both 0 and 1. We shall show that for any strategy σ for player I, $\mu_{\sigma,\tau}(A^*) = 0$. Look at the pullback measure ν by the function π, *i.e.*, the measure defined by $\nu(X) = \mu_{\sigma,\tau}(\pi^{-1}[X])$.

For any fixed $x \in 2^\omega$, we have

$$\pi^{-1}[\{x\}] \subseteq \{\, y\,;\, \forall n(\, y(2n+1) = x(N_{y,2n})\,)\,\}\,,$$

and consequently, $\mu_{\sigma,\tau}(\pi^{-1}[\{x\}]) = 0$. Therefore, ν is an atomless measure, and thus $\nu(A) = \mu_{\sigma,\tau}(A^*) = 0$. □

As a consequence of Theorem 7.1, any proof of the perfect set property from Blackwell determinacy cannot follow the simple path outlined by Fact

7.1. In fact, it is still open whether the perfect set property can be proved from Blackwell determinacy alone.

Question 7.1. *Does* Bl-AD *imply that all sets of reals have the perfect set property?*

Similarly, nothing is known about other regularity properties. For the regularity property sometimes called "Martin measurability"[7], there is a construction similar to Theorem 7.1 in [27, Theorem 6.3]: Under the assumption of AC + CH, there is a set which is closed under Turing equivalence and Blackwell determined, but not Martin measurable.

Question 7.2. *Does* Bl-AD *imply that all sets of reals have the Baire property?*

Question 7.3. *Does* Bl-AD *imply that all Turing closed sets of reals are Martin measurable ("Turing determinacy")?*

7.2. Infinitary Combinatorics.

While the consequences of AD for regularity properties of sets of real are the most well-known properties of the playful universe, the combinatorial structure theory for small cardinals like $\aleph_1$, $\aleph_2$, $\aleph_{\omega+1}$ and many others is definitely the most striking property. It starkly contrasts with the fact that ZFC cannot even determine the continuum function for these cardinals: under AD many if not all cardinals have some strong combinatorial properties, and there is a very clear pattern which cardinals have what properties.

If we let

$$\Theta := \sup\{\xi\ ;\ \xi \text{ is the length of a prewellordering of } \omega^\omega\},$$

then this ordinal measures the range of the effect that the reals can have on the combinatorics of ordinals. It is known that AD implies that Θ is a reasonably large cardinal: for instance, $\aleph_\Theta = \Theta$ by a result of Solovay's [18, Exercise 28.17],

[7] Let X be a set closed under Turing equivalence, *i.e.*, if $x \equiv_T y$ and $x \in X$, then $y \in X$. Then we say that X is **Martin measurable** if it contains a Turing cone or is disjoint from a Turing cone.

and all of the cardinals that we called "small cardinals" in the last paragraph are well below Θ.

Moreover, all of the small cardinals can be explicitly identified as ultrapowers with concretely given measures. The identification of cardinals as concrete ultrapowers together with the determination of their combinatorial properties is called a **measure analysis.** It is rather peculiar that measure analyses have little to do with actual games: typically, we need the strong partition property (see below) of one cardinal that allows us to define measures on it. The proof of the strong partition property uses games, but the subsequent analysis doesn't. For an historical account and introduction of the basics, we refer the reader to [18, Chapter 28]. If the reader is interested in the full structure theory of cardinals, we refer him or her to [15] or [17]. In this section, we shall use DC in order to guarantee that ultrapowers by measures are well-founded. We can therefore identify an ultrapower of an ordinal by a measure with the ordinal isomorphic to it via the Mostowski collapse.

Let κ be a cardinal. We say that κ has the **strong partition property** if $\kappa \to (\kappa)^\kappa$ holds, and that κ has the **weak partition property**, if for all $\lambda < \kappa$, $\kappa \to (\kappa)^\lambda$ holds. Both the weak and the strong partition property of any uncountable cardinal severely violate the axiom of choice, and they are extremely strong combinatorial properties.

Theorem 7.2 (Martin (1971), Kleinberg (1977)). *Assume* AD. *Then* $\aleph_1$ *has the strong partition property, and the club filter* $\mathcal{C}$ *on* $\aleph_1$ *is a normal measure (i.e., a* σ*-complete normal ultrafilter).*[8] *Assuming* DC, *the recursively defined ultrapowers*

$$\begin{aligned}\kappa_1 &:= \aleph_1,\\ \kappa_{n+1} &:= {\kappa_n}^{\aleph_1}/\mathcal{C}\end{aligned}$$

can be computed as $\kappa_n = \aleph_n$ *and all of them are Jónsson cardinals.*

Proof – *Cf.* [18, Theorem 28.12]. A full proof of the Kleinberg part of this theorem can be found in [26]. □

[8] That $\mathcal{C}$ is a normal measure on $\aleph_1$ follows from the strong partition property, but had been proved directly by Solovay before Martin's 1971 result.

Theorem 7.2 gives an indication of how the strong partition property generates measures that are used to concretely describe small cardinals. We define the projective ordinals by

$$\boldsymbol{\delta}^1_n := \sup\{\xi\ ;\ \xi \text{ is the length of a prewellordering of } \omega^\omega \text{ in } \boldsymbol{\Delta}^1_n\}.$$

The ordinal $\boldsymbol{\delta}^1_n$ can be seen as the limit of influence that the $\boldsymbol{\Delta}^1_n$ sets and functions have on combinatorics on the ordinals. Kunen and Martin proved in 1971 [19, Theorem 3.12] that under AD, we have $\boldsymbol{\delta}^1_{2n+2} = (\boldsymbol{\delta}^1_{2n+1})^+$, so it is enough to compute the odd projective ordinals. This computation was done by Steve Jackson:

Theorem 7.3 (Jackson). *Define* $\mathbf{e}_1 := 0$ *and* $\mathbf{e}_{n+1} = \omega^{\mathbf{e}_n}$ *and assume* AD *and* DC . *Then for every* $n \in \omega$,

$$\boldsymbol{\delta}^1_{2n+1} = \aleph_{\mathbf{e}_{2n+1}+1},$$

and all odd projective ordinals have the strong partition property.

Proof – *Cf.* [13] and [14]. □

Using the combinatorial properties for the projective ordinals, Jackson was able to give a measure analysis for the cardinals below the supremum of the projective ordinals, $\aleph_{\varepsilon_0}$. This analysis uses Jackson's notion of a **description**; there is a very simple and concrete measure analysis for cardinals below $\boldsymbol{\delta}^1_5$ due to Jackson and his student Khafizov [16]. The underlying methodology of **canonical measure assignments** is described in detail in [17].

Jackson's analysis can be pushed even further and works up to $\aleph_{\omega_1}$ [15]. Even beyond that, we find many strong partition cardinals below Θ by work of Kechris, Kleinberg, Moschovakis and Woodin [21]:

Theorem 7.4. *Assume* AD *and* DC. *Then there are cofinally many strong partition cardinals below* Θ.

All in all, we get a picture as in the left column of Figure 3: we have the canonical measure assignment analysis below $\boldsymbol{\delta}^1_5$, then Jackson's description analysis up to $\boldsymbol{\delta}^1_\omega = \aleph_{\varepsilon_0+1}$, and even beyond to $\aleph_{\omega_1}$, and finally much less information (but still cofinally many strong partition cardinals) between $\aleph_{\omega_1}$ and Θ.

It is this picture that we should like to recreate under the assumption of BI-AD. However, progress has been modest. One of the very first steps on the way to computing the projective ordinals is to ascertain that they are actually cardinals. The proof of this under AD uses the Moschovakis Coding Lemma. We write CL for the following statement:

> Let $\leq$ be a $\mathbf{\Delta}^1_n$ prewellordering of $X \subseteq \omega^\omega$ with length ξ and associated norm φ. Then for every function $f : \xi \to \mathscr{P}(\omega^\omega)$ there is a $g : \xi \to \mathscr{P}(\omega^\omega)$ with
>
> 1. For all $\eta < \xi$, $g(\eta) \subseteq f(\eta)$,
> 2. for all $\eta < \xi$, if $f(\eta)$ is nonempty, then so is $g(\eta)$, and
> 3. $\{\langle x, y\rangle \ ; \ x \in X \ \& \ y \in g(\varphi(x))\}$ is a $\mathbf{\Sigma}^1_n$ set.

Theorem 7.5 (Moschovakis). *Assume* AD. *Then* CL *holds.*

It is unknown whether CL follows from BI-AD, and it is in general unknown whether the $\boldsymbol{\delta}^1_n$ are cardinals.

Question 7.4. *Does* BI-AD *imply* CL*?*

We can however, still show some results for cardinals at the bottom of the hierarchy depicted in Figure 3 by following the ideas of the proof of Theorem 7.2:

Theorem 7.6. *Assume* BI-AD. *Then $\aleph_1$ has the strong partition property, and for all natural numbers n, the odd projective ordinals $\boldsymbol{\delta}^1_{2n+1}$ have the countable partition property, i.e., for all $\alpha < \omega_1$, the partition relation $\boldsymbol{\delta}^1_{2n+1} \to (\boldsymbol{\delta}^1_{2n+1})^\alpha$ holds.*[9]

[9] As in Footnote 8, we get as a consequence that the club filter on $\aleph_1$ is a normal measure. That this holds under BI-AD had been proved before Theorem 7.6 by Vervoort (unpublished).

We get a canonical measure assignment for all cardinals $\aleph_n$ which also determines their cofinalities as

$$\mathrm{cf}(\aleph_n) := \begin{cases} \aleph_0 & \text{if } n = 0, \\ \aleph_1 & \text{if } n = 1, \\ \aleph_2 & \text{otherwise} \end{cases}$$

and allows us to compute $\boldsymbol{\delta}^1_3$:

Corollary 7.1. *Assume* BI-AD. *Then* $\boldsymbol{\delta}^1_3 = \aleph_{\omega+1}$.

PROOF – By Theorem 7.6, we know that $\boldsymbol{\delta}^1_3$ is a regular cardinal. By Theorem 6.1, we get PD and that every real has a sharp. This allows us to apply results of Martin that say that $\boldsymbol{\delta}^1_2 = \aleph_2 < \boldsymbol{\delta}^1_3 \leq \aleph_{\omega+1}$. But with the cofinality function as given above, and since $\boldsymbol{\delta}^1_3$ is regular, $\aleph_{\omega+1}$ is the only possible choice left for it. □

Since we don't even know that $\boldsymbol{\delta}^1_4$ is a cardinal, we cannot compute it under the assumption of BI-AD.

Question 7.5. *Does* BI-AD *imply that* $\boldsymbol{\delta}^1_4 = \aleph_{\omega+2}$*?*

As mentioned, the measure analysis due to Jackson does not really use games to a large extent. A lot of Jackson's analysis can be derived abstractly from the strong partition properties and some additional combinatorial properties of cardinals. This raises the question whether we can prove these strong partition results for higher projective ordinals:

Question 7.6. *Does* BI-AD *imply that* $\boldsymbol{\delta}^1_3$ *(or in general,* $\boldsymbol{\delta}^1_{2n+1}$*) has the strong partition property?*

7.3. Hierarchies of sets of reals.

Almost as striking as the detailed combinatorial analysis of cardinals is the fact that AD gives us a wellordered hierarchy of sets of reals. From a topological point of view, there is no reducibility relation on sets of reals more natural than **Wadge reducibility** and **Lipschitz reducibility** defined by

$A \leq_{\mathrm{W}} B :\iff$ there is a continuous function f with $A = f^{-1}[B]$, and

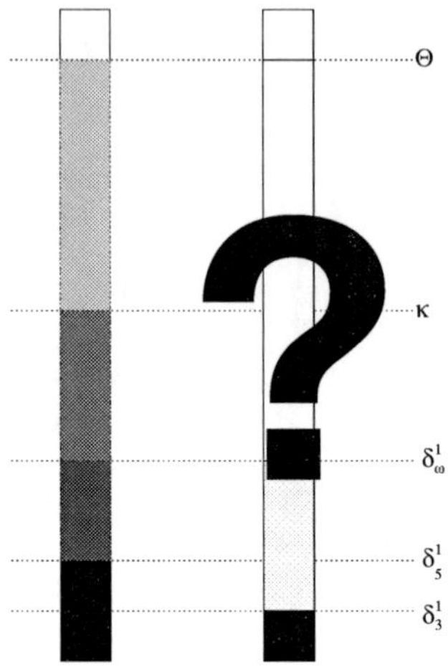

Figure 3 – Left. The extent of measure analyses under AD. The darkest area corresponds to the analysis via canonical measure assignments up to $\boldsymbol{\delta}^1_5$, the dark area to the analysis via description theory up to $\boldsymbol{\delta}^1_\omega = \aleph_{\varepsilon_0+1}$ and $\kappa := \aleph_{\omega_1}$, the light area corresponds to a cofinal sequence of strong partition cardinals. *Right.* The extent of measure analyses under Bl-AD. Almost nothing is known beyond $\boldsymbol{\delta}^1_3$.

$$A \leq_{\mathrm{L}} B :\Longleftrightarrow \text{there is a Lipschitz function } f \text{ with } A = f^{-1}[B].$$

These relations are partial preorders inducing equivalence relations $\equiv_{\mathrm{W}}$ and $\equiv_{\mathrm{L}}$ which in turn give rise to a degree structure (with a partial ordering). Under the assumption of the axiom of choice, these partial preorders can be quite wild:

Proposition 7.1. *If A is an uncountable set of reals without a nonempty perfect subset, then there is some Π^0_2 set P such that A and P are incomparable in the $\leq_{\mathrm{W}}$ relation.*

Proof – Consider $P := \{x\,;\, \forall n \exists k > n(x(k) = 0)\}$. It is easy to see that if $P \leq_{\mathrm{W}} X$, then X contains a nonempty perfect set. So, by our assumption, we have that $P \not\leq_{\mathrm{W}} A$. But we cannot have $A \leq_{\mathrm{W}} P$, as then A would be $\boldsymbol{\Pi}^0_2$ as a continuous preimage of a Π^0_2 set. This would contradict the fact that all Borel sets have the perfect set property. Therefore A and P are $\leq_{\mathrm{W}}$-incomparable. □

Both $\leq_{\mathrm{W}}$ and $\leq_{\mathrm{L}}$ have an equivalent definition in terms of games: For an element $x \in \omega^\omega$ we say that it is **finite** if $D_x := \{n\,;\, x(n) > 0\}$ is finite, and for a nonfinite x, we let $d_x : \omega \to \omega$ be the increasing enumeration of D_x. Then we can define $\widehat{x}(n) := x(d_x(n)) - 1$.[10] For given A and B, we can define $X^{\mathrm{L}}_{A,B} := \{x\,;\, x_{\mathrm{I}} \in A \not\leftrightarrow x_{\mathrm{II}} \in B\}$ and $X^{\mathrm{W}}_{A,B} := \{x\,;\, x_{\mathrm{II}}$ is finite or $x_{\mathrm{I}} \in A \not\leftrightarrow \widehat{x_{\mathrm{II}}} \in B\}$.

Proposition 7.2. *For any two sets of reals A and B, we have:*

1. *$A \leq_{\mathrm{L}} B$ if and only if player II has a winning strategy in the game $\mathrm{G}(X^{\mathrm{L}}_{A,B})$.*
2. *$A \leq_{\mathrm{W}} B$ if and only if player II has a winning strategy in the game $\mathrm{G}(X^{\mathrm{W}}_{A,B})$.*

Now that we reduced the existence of continuous and Lipschitz functions to game-theoretic terms, we get a structure result under AD.

Lemma 7.1 (Wadge's Lemma). *For any two Borel sets A and B, either $A \leq_{\mathrm{W}} B$ or $\omega^\omega \backslash B \leq_{\mathrm{W}} A$. Moreover, assuming AD, this is true for any two sets of reals.*

Much more is true: Using a technique called the **Martin-Monk technique**, Martin was able to prove (in $\mathsf{ZF} + \mathsf{AD} + \mathsf{DC}_{\mathbb{R}}$) the wellfoundedness of the relations $\leq_{\mathrm{W}}$ and $\leq_{\mathrm{L}}$ and Steel and van Wesep completely determined the structure of the Wadge hierarchy: Both hierarchies are semi-linear (*i.e.*, antichains have length at most 2) and wellfounded. The antichains of length 2 are called **non-selfdual pairs**, and these pairs occur in the Lipschitz hierarchy exactly at level α where $\mathrm{cf}(\alpha) > \omega$ and $\alpha = 0$. In the Wadge hierarchy, the blocks of selfdual classes of length ω_1 collapse to a single class, and hence the Wadge hierarchy has non-selfdual pairs at every second successor levels and limit levels of uncountable cofinality.

This pattern is depicted in Figure 4 and is highly characteristic of AD. For more details, we refer the reader to the survey paper [2].

[10] This operation corresponds to considering the move 0 as a passing move. We think of x as a sequence of real moves (where $x(n) > 0$ represents the number $x(n) - 1$) and passes (coded by $x(n) = 0$). The real $\widehat{x}$ corresponds to deleting the passes from the real x.

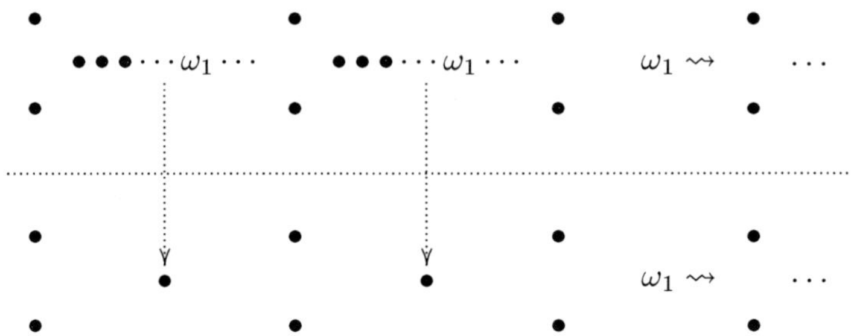

Figure 4 – The Lipschitz hierarchy (above the dotted line) with blocks of ω_1 classes between non-selfdual pairs, and the Wadge hierarchy (below the dotted line) where these blocks collapse to a single class. At levels of uncountable cofinality, both hierarchies have a non-selfdual pair.

Of course, the game-theoretic description in Proposition 7.2 immediately suggests a Blackwell version of these hierarchies: we define

$$A \leq_{\mathrm{BIL}} B :\Longleftrightarrow \text{ player II has an optimal strategy in } \mathrm{G}(X^{\mathrm{L}}_{A,B}), \text{ and}$$

$$A \leq_{\mathrm{BIW}} B :\Longleftrightarrow \text{ player II has an optimal strategy in } \mathrm{G}(X^{\mathrm{W}}_{A,B}).$$

Can we get a similar structure theory for the Blackwell Wadge and Lipschitz hierarchy under the assumption of Bl-AD? The basic theory of Blackwell Lipschitz degrees works smoothly:

Theorem 7.7. *Assume* Bl-AD. *Say that a Blackwell Lipschitz degree* $\mathbf{d}$ *is called a* ***successor degree*** *if there is a degree* $\mathbf{p} <_{\mathrm{BIL}} \mathbf{d}$ *such that there is no* $\mathbf{e}$ *with* $\mathbf{p} <_{\mathrm{BIL}} \mathbf{e} <_{\mathrm{BIL}} \mathbf{d}$. *We say that* $\mathbf{d}$ *is* ***of countable cofinality*** *if there is a sequence* $\langle [A_n]_{\mathrm{BIL}}\,;\, n \in \omega\rangle$ *without a greatest element such that*

$$A \equiv_{\mathrm{BIL}} \bigoplus_{n\in\omega} A_n.$$

Then the Blackwell Lipschitz degrees are semi-linearly ordered, every successor degree is a selfdual degree, and the non-selfdual degrees are exactly the first two degrees and the non-successor degrees which are not of countable cofinality.

However, the Martin-Monk technique does not seem to work with BI-AD. As a consequence, we do not know whether our new hierarchy is wellfounded.

Question 7.7. *Does* BI-AD+$\mathsf{DC}_{\mathbb{R}}$ *imply that the Blackwell Lipschitz hierarchy is wellfounded?*

Since our analysis of $\leq_{\mathrm{BIW}}$ and $\leq_{\mathrm{BIL}}$ did not yield the results we hoped for, we could look for other global hierarchies that we know to exist under the assumption of AD. One of them is the hierarchy generated by the First Periodicity Theorem, investigated as the "Steel hierarchy" in [6] and [10].

Instead of sets of reals, we look at surjective functions $\varphi : \omega^\omega \to \alpha$ (for some ordinal α) called **norms**. Each norm φ defines an order

$$\leq_\varphi := \{\langle x, y\rangle\,;\, \varphi(x) \leq \varphi(y)\}.$$

For two norms φ and ψ, we say that φ is **FPT-reducible to** ψ (for "**F**irst **P**eriodicity **T**heorem"; in symbols: $\varphi \leq_{\mathrm{FPT}} \psi$) if there is a continuous function $F : \omega^\omega \to \omega^\omega$ such that for all $x \in \omega^\omega$, we have

$$\varphi(F(x)) \leq \psi(x).$$

As with $\leq_{\mathrm{W}}$ and $\leq_{\mathrm{L}}$, FPT-reducibility can be expressed in game terms. Given φ and ψ, we look at the set $X_{\varphi,\psi} := \{x\,;\, x_{\mathrm{II}} \text{ is finite or } \varphi(x_{\mathrm{I}}) > \psi(\widehat{x_{\mathrm{II}}})\}$.

Proposition 7.3. *For any two norms φ and ψ, we have $\varphi \leq_{\mathrm{FPT}} \psi$ if and only if player II has a winning strategy in* $\mathrm{G}(X_{\varphi,\psi})$.

Theorem 7.8 (Moschovakis). *Assume* AD *and* $\mathsf{DC}_{\mathbb{R}}$. *Then $\leq_{\mathrm{FPT}}$ is a prewellordering.*

PROOF – This is essentially the proof of the First Periodicity Theorem of Moschovakis [1]. For more details, *cf.* [40, 6B]. □

As with the Martin-Monk technique, the proof of Theorem 7.8 involves filling an infinite diagram of games using strategies. However, this proof is much simpler than the Martin-Monk proof and can be adapted to the Blackwell determinacy situation. In analogy with $\leq_{\mathrm{FPT}}$ and Proposition 7.3, we define

$$\varphi \leq_{\mathrm{BIFPT}} \psi :\iff \text{player II has an optimal strategy in } \mathrm{G}(X_{\varphi,\psi}).$$

Theorem 7.9. *Assume* Bl-AD *and* $\mathsf{DC}_{\mathbb{R}}$. *Then* $\leq_{\mathrm{BlFPT}}$ *is a prewellordering.*

PROOF – *Cf.* [31, Theorem 3.2]. □

Theorem 7.9, along with Theorem 3.1 is one of the few consequences of Bl-AD that is known to hold for all sets of reals and not just for a definable initial segment of the Wadge hierarchy.

8. Weaker Axioms of Blackwell Determinacy

So far, we have looked at three types of games on a given payoff set A: the perfect information game with pure strategies $\mathrm{G}(A)$ ("Gale-Stewart game"), the game with slightly imperfect information with Blackwell strategies $\mathrm{B}(A)$ ("Blackwell game"), and the perfect information game with mixed strategies $\mathrm{M}(A)$. All of these games were symmetric in the sense that both players were using strategies of the same type.

Proposition 2.1 told us that Gale-Stewart games allow some flexibility concerning the status of the two players: if player I has a strategy winning against all blindfolded strategies, then he has a strategy winning against all pure strategies. This is not true for optimal strategies as was indicated by Proposition 4.1:

Remark 8.1 - Consider once again the game of playing different reals with payoff set $M := \{x \in 2^{\mathbb{N}}\,;\, \forall n(x(2n) \neq x(2n+1)\}$ as in Section 3. We have already seen (Proposition 4.1) that player II has a winning strategy in $\mathrm{G}(M)$ and player I has an optimal strategy in $\mathrm{B}(M)$. Clearly, in $\mathrm{M}(M)$, player II has an optimal strategy (use the winning strategy from $\mathrm{G}(M)$).

But at the same time, in $\mathrm{M}(M)$, player I has a strategy with the following property: for every blindfolded strategy τ for player II, $\mu_{\sigma,\tau}(M) = 1$.

So, player I has a strategy which is optimal against blindfolded opponents, but cannot be optimal against all opponents.

PROOF – Essentially, this is the proof of Proposition 4.1: we showed that the "randomize" strategy $\sigma(s)(\{0\}) = \sigma(s)(\{1\}) = \frac{1}{2}$ is optimal in $\mathrm{B}(M)$, but every blindfolded strategy is a Blackwell strategy. □

Remark 8.1 suggests the definition of weaker optimality properties. If we denote the classes of mixed, pure and blindfolded strategies with $\mathcal{S}_{\text{mixed}}$, $\mathcal{S}_{\text{pure}}$, and $\mathcal{S}_{\text{blindfolded}}$, respectively, and let $\mathcal{S}$ be any of these classes, then we can say that a mixed strategy σ is **$\mathcal{S}$-optimal** for the payoff set $A \subseteq \omega^\omega$ if for all $\tau_* \in \mathcal{S}$ for player II, $\mu^-_{\sigma,\tau_*}(A) = 1$ (and similarly for mixed strategies of player II). Clearly, $\mathcal{S}_{\text{mixed}}$-optimality is the same as optimality in the sense of Section 5. Proposition 2.1 can now be rephrased as "for a pure strategy, being a winning strategy and being $\mathcal{S}_{\text{blindfolded}}$-optimal are equivalent".

We can now call a set $A \subseteq \omega^\omega$ **purely** or **blindfoldedly Blackwell determined** if either player I or player II has an $\mathcal{S}_{\text{pure}}$-, or $\mathcal{S}_{\text{blindfolded}}$-optimal strategy, respectively, and we write $\mathsf{pBl\text{-}AD}$, and $\mathsf{blBl\text{-}AD}$ for the full axioms claiming pure or blindfolded Blackwell determinacy for all sets.

Note that Remark 8.1 raises the question whether $\mathsf{blBl\text{-}AD}$ behaves like a determinacy statement at all: blindfolded Blackwell determinacy lacks the basic dichotomy property of determinacy theory, as in $\mathrm{M}(M)$, both players have an $\mathcal{S}_{\text{blindfolded}}$-optimal strategy.

It turns out that despite of this behaviour, the axiom $\mathsf{blBl\text{-}AD}$ and its definable fragments still give us logical strength:

Theorem 8.1 (Martin). *The statements* $\mathsf{blBl\text{-}Det}(\boldsymbol{\Pi}^1_1)$, $\mathsf{pBl\text{-}Det}(\boldsymbol{\Pi}^1_1)$, *and* $\mathsf{Det}(\boldsymbol{\Pi}^1_1)$ *are equivalent.*

Proof – *Cf.* [30, Corollary 3.9] □

Theorem 8.2. *The axioms* $\mathsf{blBl\text{-}AD}$ *and* $\mathsf{pBl\text{-}AD}$ *imply the existence of an inner model with a strong cardinal.*

Proof – *Cf.* [30, Corollary 4.9]. □

There is no general method to deal with $\mathsf{blBl\text{-}AD}$: it is still unknown whether $\mathsf{blBl\text{-}AD}$ implies that all sets are Lebesgue measurable, and since this is used in the proof of Theorem 7.6, we cannot get the results on infinitary combinatorics from Section 7.2. We were able to derive a consequence violating the axiom of choice in [30, Lemma 3.12], though:

Corollary 8.1. *The axiom* $\mathsf{blBl\text{-}AD}$ *violates the axiom of choice.*

Moving from blindfolded Blackwell determinacy to pure Blackwell determinacy, we shall be able to show a conditional equivalence theorem. We identify the set of pure strategies with the set $\mathbb{N}^{(\omega^{<\omega})}$. For any mixed strategy τ, we shall define a probability measure V_τ on $\mathbb{N}^{(\omega^{<\omega})}$ which we shall call the **Vervoort code of** τ. If p_0, ..., p_n are elements of $\omega^{<\omega}$ and N_0, ..., N_n are natural numbers, define

$$\mathrm{V}_\tau(\{\tau^* \,;\, \tau^*(p_0) = N_0 \,\&\, ... \,\&\, \tau^*(p_n) := N_n\}) = \prod_{i=0}^{n} \tau(p_i)(\{N_i\}),$$

and let V_τ be the unique extension of this function to the Borel σ-algebra on $\mathbb{N}^{(\omega^{<\omega})}$.

Theorem 8.3 (Vervoort). *If σ and τ are mixed strategy and B is a Borel subset of ω^ω, then*

$$\begin{aligned} \mu_{\sigma,\tau}(B) &= \int \mu_{\sigma,x}(B)\, d\mathrm{V}_\tau(x) \\ &= \int \mu_{x,\tau}(B)\, d\mathrm{V}_\sigma(x) \end{aligned}$$

PROOF – This is essentially [49, Theorem 6.6]. □

Theorem 8.4. *Let $\mathbf{\Gamma}$ be a boldface pointclass such that all sets in $\mathbf{\Gamma}$ are Lebesgue measurable. Let $A \in \mathbf{\Gamma}$ and assume that A is purely Blackwell determined. Then A is Blackwell determined. Moreover, every $\mathcal{S}_{\mathrm{pure}}$-optimal strategy is actually $\mathcal{S}_{\mathrm{mixed}}$-optimal.*

PROOF – Note that for any σ and τ, the $\mu_{\sigma,\tau}$-measurability of a set A is equivalent to the Lebesgue measurability of some continuous preimage of A, so in our situation, all sets in $\mathbf{\Gamma}$ are $\mu_{\sigma,\tau}$-measurable.

Now since A is purely Blackwell determined, let σ be (without loss of generality) an $\mathcal{S}_{\mathrm{pure}}$-optimal strategy for player I. Let τ be an arbitrary mixed strategy for player II. We shall show that for every Borel superset $B \supseteq A$, we have $\mu_{\sigma,\tau}(B) = 1$. Because A is $\mu_{\sigma,\tau}$-measurable, this proves the claim.

By our assumption, we know that for every pure strategy τ^* and every Borel $B \supseteq A$, we have $\mu_{\sigma,\tau^*}(B) = 1$. But this means that the function

$$f : \mathbb{N}^{(\omega^{<\omega})} \to \mathbb{R} : x \mapsto \mu_{\sigma,x}(B)$$

is the constant function with value 1.

Using Theorem 8.3, we get

$$\mu_{\sigma,\tau}(B) = \int \mu_{\sigma,x}(B)\, d\mathrm{V}_\tau(x) = \int f(x)\, d\mathrm{V}_\tau(x) = 1.$$

□

Corollary 8.2. *Let* $\mathbf{\Gamma}$ *be a boldface pointclass such that all sets in* $\mathbf{\Gamma}$ *are Lebesgue measurable. Then* $\mathsf{pBl\text{-}Det}(\mathbf{\Gamma})$ *implies* $\mathsf{Bl\text{-}Det}(\mathbf{\Gamma})$.

Using the Martin-Neeman-Vervoort Equivalence Theorem 6.1, we can bootstrap some equivalence for pure Blackwell determinacy from Corollary 8.2:

Corollary 8.3. *Let* $\mathbf{\Gamma} \in \{\mathbf{\Delta}^1_2, \mathbf{\Sigma}^1_2, \mathbf{\Delta}^1_3\}$. *Then* $\mathsf{pBl\text{-}Det}(\mathbf{\Gamma})$ *implies* $\mathsf{Det}(\mathbf{\Gamma})$.

Proof – "$\mathbf{\Delta}^1_2$ & $\mathbf{\Sigma}^1_2$": By Theorem 8.1, we have $\mathsf{Det}(\mathbf{\Pi}^1_1)$ and thus by the usual Solovay unfolding argument, every $\mathbf{\Sigma}^1_2$ set is Lebesgue measurable. Now Corollary 8.2 and Theorem 6.1 finish the proof.

"$\mathbf{\Delta}^1_3$": By the first argument, we get $\mathsf{Det}(\mathbf{\Sigma}^1_2)$. Again, we use Solovay unfolding to get the Lebesgue measurability of all $\mathbf{\Sigma}^1_3$ sets and finish the proof with Corollary 8.2 and Theorem 6.1. □

The methods of Theorem 6.1 do not extend to prove the equivalence of $\mathsf{Bl\text{-}Det}(\mathbf{\Sigma}^1_3)$ and $\mathsf{Det}(\mathbf{\Sigma}^1_3)$ (*cf.* [39, p. 633]). This halts our progress on the third level of the projective hierarchy. Of course, $\mathsf{pBl\text{-}Det}(\mathbf{\Sigma}^1_3)$ gives $\mathsf{Det}(\mathbf{\Delta}^1_3)$ by Corollary 8.3, hence the Lebesgue measurability of all $\mathbf{\Sigma}^1_3$ sets and thus $\mathsf{Bl\text{-}Det}(\mathbf{\Sigma}^1_3)$ by Corollary 8.2.

Both for blindfolded Blackwell determinacy and pure Blackwell determinacy, it is open whether they are equivalent to Blackwell determinacy (and then, if Martin's Conjecture 4.1 is true) to determinacy.

Question 8.1. *Are* pBl-AD *or* blBl-AD *equivalent to* AD*?*

Building on the results from [30], Greg Hjorth (2002, personal communication) found a proof of "blBl-AD implies the existence of an inner model with a Woodin cardinal" using Cabal-style descriptive set theory and inner model theory.

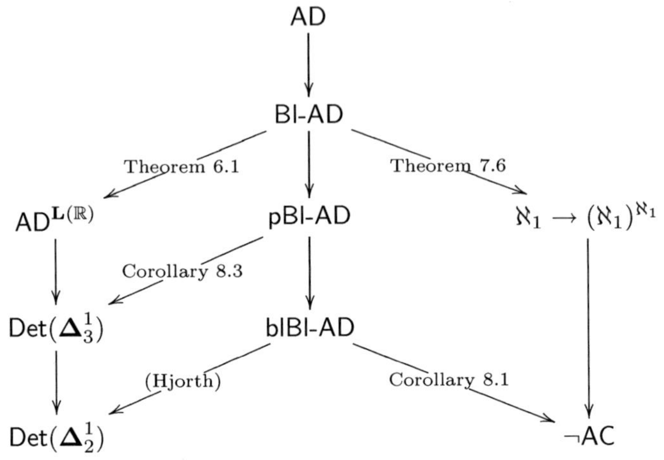

Figure 5 – Diagram of axioms of determinacy and consequences

9. Stronger Axioms of Blackwell Determinacy

Axioms of determinacy have been extended in several directions (*cf.* [47, p. 96]: researchers have investigated long games (*i.e.*, transfinite games of length $\alpha > \omega$) and games with more than countably many moves (*e.g.*, real number moves). We write AD^α for the axiom of determinacy for games of length α and AD_X for the axiom of determinacy for games in which the players play elements of X in each move.

For perfect information games, there is a limit to both α and X:

Theorem 9.1. *The axioms* AD^{ω_1}, AD_{ω_1} *and* $\mathsf{AD}_{\mathscr{P}(\mathbb{R})}$ *are inconsistent.*

Proof – For AD^{ω_1} take the following game: Player I plays 0s for some countable time α. After he plays the first 1, player II has to react by using the next ω rounds to produce an element of WO_α. If the game is determined, player II must have a winning strategy, but this would be a choice function for the family $\{\mathrm{WO}_\alpha\,;\,\alpha < \omega_1\}$ and thus an uncountable wellordered set of reals which would contradict AD.

The games for AD_{ω_1} and $\mathsf{AD}_{\mathscr{P}(\mathbb{R})}$ are similar: let player I play α or WO_α itself instead of α many 0s. □

Also, Blass has proved that there is a link between long games and games with uncountably many moves (*cf.* Theorem 9.4). In this section, we shall investigate what is known about the Blackwell analogues of these extensions.

9.1. Uncountable sets of moves.

When we want to extend the notion of Blackwell determinacy to subsets of X^ω (in place of ω^ω), almost all of the definitions are extended in the obvious way: mixed strategies become functions from $X^{<\omega}$ to Prob, and the rest of the definitions stay the same using this new extended definition of mixed strategy.

We have to be careful, however, as our definition of Blackwell determinacy involves extending the function $\mu_{\sigma\tau}$ defined on the basic open sets to the entire σ-algebra of the Borel sets. In the case of ω^ω, we needed $\mathsf{AC}_\omega(\mathbb{R})$ in order to show that the set of interpretations of Borel codes is the Borel σ-algebra.

When moving to X^ω, there are two potential problems:

1. Our definition of the measure $\mu_{\sigma,\tau}$ depends on the values of σ and τ on singleton sets. For uncountable sets X, this might not be enough to properly define a measure. We can easily deal with this problem by requiring the measures to have countable support (*cf.* [39, Remark 3.1]): Let $\text{Prob}(X)$ the set of probability measures on X with countable support, *i.e.*, those measures μ on X such that there is a countable subset $Y \subseteq X$ such that $\mu(Y) = 1$.

2. If X is large, then the extension of $\mu_{\sigma,\tau}$ from a function defined on the basic open sets to a function defined on the σ-algebra of Borel sets may require more choice than we have at our disposal in our base theory $\mathsf{ZF} + \mathsf{AC}_\omega(\mathbb{R})$. This is not the case if there is a surjection from $\mathbb{R}$ onto $\mathbb{R} \times X^\omega$ [32, Proposition 2].

Proposition 9.1. *For $X \in \{\mathbb{R}, \omega_1, \omega_1 \cup \mathbb{R}, {\omega_1}^\omega\}$, any function $\mu_{\sigma,\tau}$ defined on the basic open sets is extendible to the Borel σ-algebra.*

Using Proposition 9.1, we can now use the obvious analogues of the definitions in Section 3 to define the notion of **Blackwell determinacy** for sets in

X^ω if $X \in \{\mathbb{R}, \omega_1, \omega_1 \cup \mathbb{R}, {\omega_1}^\omega\}$, and define $\mathsf{Bl\text{-}AD}_X$ to mean "Every $A \subseteq X^\omega$ is Blackwell determined".

These new axioms of Blackwell determinacy obviously give the following implicational structure:

$$\begin{array}{ccccc} & & \mathsf{Bl\text{-}AD}_{{\omega_1}^\omega} & & \\ & & \downarrow & & \\ & & \mathsf{Bl\text{-}AD}_{\omega_1 \cup \mathbb{R}} & & \\ & \swarrow & & \searrow & \\ \mathsf{Bl\text{-}AD}_{\mathbb{R}} & & & & \mathsf{Bl\text{-}AD}_{\omega_1} \\ & \searrow & & \swarrow & \\ & & \mathsf{Bl\text{-}AD} & & \end{array}$$

In the following, we shall show that the top two of the strong axioms of Blackwell determinacy are inconsistent. Note that since there is a surjection from $\mathscr{P}(\mathbb{R})$ onto $\omega_1 \cup \mathbb{R}$ this means that there cannot be a reasonable consistent version of $\mathsf{Bl\text{-}AD}_{\mathscr{P}(\mathbb{R})}$.

Lemma 9.1. *If μ is a probability measure on X with countable support then there is a unique $\mathbf{m}_\mu > 0$ such that*

$$\{x\,;\,\mu(\{x\}) = \mathbf{m}_\mu\} \text{ is nonempty and finite, and}$$

$$\{x\,;\,\mu(\{x\}) > \mathbf{m}_\mu\} = \varnothing.$$

Proof – For $\varepsilon > 0$, let $X_\varepsilon := \{x\,;\,\mu(\{x\}) \geq \varepsilon\}$. By σ-additivity, it is impossible that any set X_ε is infinite. Now pick $z \in \mathrm{supp}(\mu)$ and let $\varepsilon := \mu(\{z\})$, so that X_ε is finite and nonempty. Define

$$\mathbf{m}_\mu := \max\{\mu(\{x\})\,;\, x \in X_\varepsilon\}.$$

□

Theorem 9.2. *Let X be linearly ordered by $\preceq$, $n \in \mathbb{N}$, $A \subseteq X^n$ and τ be a optimal strategy for player II in the game on $X^n \backslash A$. Then there is a definable function $\hat{\tau}$ defined on the mixed strategies such that $\mathrm{ran}(\hat{\tau}) \subseteq A$.*

Proof – Given a mixed strategy σ, we define recursively

$$x_{2i} := \min_{\preceq}\{x\,;\,\sigma(\langle x_0, \ldots, x_{2i-1}\rangle)(\{x\}) = \mathbf{m}_{\sigma(\langle x_0,\ldots,x_{2i-1}\rangle)}\}, \text{ and}$$

$$x_{2i+1} := \min_{\preceq}\{x\,;\, \tau(\langle x_0, \ldots, x_{2i}\rangle)(\{x\}) = \mathbf{m}_{\tau(\langle x_0,\ldots,x_{2i}\rangle)}\}.$$

By Lemma 9.1, the minimum is taken over a nonempty finite linearly ordered set, so this sequence is definable in ZF. We let $\hat{\tau}(\sigma)$ be the sequence $\langle x_0, \ldots, x_{n-1}\rangle$ and claim that for all σ, $\hat{\tau}(\sigma) \in A$:

Clearly, $\mathbf{m} := \mu_{\sigma,\tau}(\{\langle x_0, \ldots, x_{n-1}\rangle\})$ is the finite product of the positive numbers $\mathbf{m}_{\sigma(\langle x_0,\ldots,x_{2i-1}\rangle)}$ and $\mathbf{m}_{\tau(\langle x_0,\ldots,x_{2i}\rangle)}$, and therefore strictly positive. If $\langle x_0, \ldots, x_{n-1}\rangle \notin A$, then $\mu_{\sigma,\tau}(X^n \backslash A) \geq \mathbf{m}$, contradicting the optimality of τ. □

For sets Y and Z, we define the class of Y-Z-choice games $\mathrm{CG}_{Y,Z}(A)$ as follows: If $A : Y \to \mathscr{P}(Z)$ is a family of nonempty subsets of Z indexed by elements of Y, then the game $\mathrm{CG}_{Y,Z}(A)$ is the two-round game in which player I plays an element $y \in Y$, player II follows up with playing an element $z \in Z$, and player II wins if $z \in A(y)$.

Theorem 9.3. *If $X := Y \cup Z$ is linearly ordered and* $\mathsf{Bl\text{-}AD}_X$ *(is defined and) holds, then* $\mathsf{AC}_Y(Z)$ *holds.*[11]

Proof – Let $A : Y \to \mathscr{P}(Z)$ be a family of nonempty sets. Because the sets are nonempty, player I cannot have an optimal strategy in $\mathrm{GC}_{Y,Z}(A)$. Let τ be an optimal strategy for player II. Let σ_y be defined by $\sigma_y(\varnothing)(\{y\}) := 1$. Then

$$f : y \mapsto \hat{\tau}(\sigma_y)$$

is a definable choice function by Theorem 9.2. □

Corollary 9.1. *The following implications between axioms of Blackwell determinacy and choice principles hold:*

1. $\mathsf{Bl\text{-}AD}_{\mathbb{R}}$ *implies* $\mathsf{AC}_{\mathbb{R}}(\mathbb{R})$,

2. $\mathsf{Bl\text{-}AD}_{\omega_1 \cup \mathbb{R}}$ *implies* $\mathsf{AC}_{\omega_1}(\mathbb{R})$.

Consequently, $\mathsf{Bl\text{-}AD}_{\mathbb{R}}$ *is strictly stronger than* $\mathsf{Bl\text{-}AD}$, *and* $\mathsf{Bl\text{-}AD}_{\omega_1 \cup \mathbb{R}}$ *and* $\mathsf{Bl\text{-}AD}_{\omega_1^\omega}$ *are inconsistent.*

---

[11]The requirement that $\mathsf{Bl\text{-}AD}_X$ be defined is just a reminder that we need to use Proposition 9.1 or a similar result.

Proof – It is well-known that if $\mathbb{R}$ is not wellordered, then $\mathsf{AC}_{\mathbb{R}}(\mathbb{R})$ is false in $\mathbf{L}(\mathbb{R})$. But if $\mathsf{Bl\text{-}AD}_{\mathbb{R}}$ is true in $\mathbf{V}$, then $\mathbf{L}(\mathbb{R}) \models \mathsf{Bl\text{-}AD}$. As discussed in the proof of Theorem 9.1, the choice principle $\mathsf{AC}_{\omega_1}(\mathbb{R})$ contradicts AD, as it allows to construct an uncountable sequence of reals. Using Theorem 7.6, we can show that it contradicts $\mathsf{Bl\text{-}AD}$ as well. □

Of course, the gap left by Theorem 9.1 and Corollary 9.1 suggests the obvious question:

Question 9.1. *Is* $\mathsf{Bl\text{-}AD}_{\omega_1}$ *inconsistent?*

Another question that is raised by Corollary 9.1 is the one about the strength of $\mathsf{Bl\text{-}AD}_{\mathbb{R}}$: we know that if $\mathsf{Bl\text{-}AD}$ is consistent, then so is $\mathsf{Bl\text{-}AD} + \neg\mathsf{Bl\text{-}AD}_{\mathbb{R}}$. In the case of $\mathsf{AD}_{\mathbb{R}}$, much more is known: we know that the consistency strength of $\mathsf{AD}_{\mathbb{R}}$ is considerably higher than that of AD. Work of de Kloet, Kieftenbeld and the present author [9] suggests that we can follow the lines of Solovay's analysis of $\mathsf{AD}_{\mathbb{R}}$ from [46] and prove analogous theorems for $\mathsf{Bl\text{-}AD}_{\mathbb{R}}$. But:

Question 9.2. *Does* $\mathsf{Bl\text{-}AD}_{\mathbb{R}}$ *imply* $\mathsf{AD}_{\mathbb{R}}$*?*

9.2. Long Blackwell games.

The underlying idea used here to define long Blackwell games goes back to David de Kloet's undergraduate research thesis [8]. De Kloet introduces two different versions of the axiom of real Blackwell determinacy in this paper, the one that we discussed in Section 9.1 and another one that he calls the Euclidean variant. De Kloet's Euclidean variant is closely connected to what we shall call $\mathsf{Bl\text{-}AD}^{\omega^2}$.

We shall be defining Blackwell determinacy for games of length $\omega \cdot n$ and ω^2. We say that a **transfinite mixed strategy** is a function from sequences of natural numbers of length $< \omega^2$ into Prob. Given two transfinite mixed strategies σ and τ, we shall now define a Borel measure $\mu^n_{\sigma,\tau}$ on $\omega^{\omega \cdot n}$. As before, we shall be defining the measure on the basic open subsets and then (using $\mathsf{AC}_\omega(\mathbb{R})$) extend this to all Borel sets. The basic open sets of the topological space $\omega^{\omega \cdot n}$ are of the type

$$[s_0, ..., s_{n-1}] := \{x\,;\, \forall i < n\,(\, s_i \subseteq (x)_i\,)\},$$

where $(x)_i(k) := x(\omega \cdot i + k)$ for $i < n$.

The definition will be by recursion on n. If $n = 1$, we can just use the ordinary definition of $\mu_{\sigma,\tau}$ for ordinary mixed strategies. Let us assume that we have defined a measure $\mu^n_{\sigma,\tau}$ on $\omega^{\omega \cdot n}$ which we now want to extend to a measure $\mu^{n+1}_{\sigma,\tau}$ on $\omega^{\omega\cdot(n+1)}$.

Fix $x \in \omega^{\omega \cdot n}$ and $s \in \omega^{<\omega}$ and follow the general idea of the definition of $\mu_{\sigma,\tau}$ in the length ω games. Almost as in Section 3, we let

$$\nu_x(\sigma, \tau)(t) := \begin{cases} \sigma(x^\frown t) & \text{if } \mathrm{lh}(t) \text{ is even, and} \\ \tau(x^\frown t) & \text{if } \mathrm{lh}(s) \text{ is odd.} \end{cases}$$

Then for any $s \in \omega^{<\omega}$, we can define

$$f_{\sigma,\tau}(x, s) := \prod_{i=0}^{\mathrm{lh}(s)-1} \nu_x(\sigma, \tau)(s{\restriction}i)(\{s(i)\}).$$

Finally, we set

$$\mu^{n+1}_{\sigma,\tau}([s_0, ..., s_n]) := \int_{[s_0,...,s_{n-1}]} f_{\sigma,\tau}(x, s)\, d\mu^n_{\sigma,\tau}(x).$$

Clearly, the measures $\mu^{n+1}_{\sigma,\tau}$ and $\mu^n_{\sigma,\tau}$ **cohere**, *i.e.*,

$$\mu^{n+1}_{\sigma,\tau}([s_0, ..., s_{n-1}, \varnothing]) = \mu^n_{\sigma,\tau}([s_0, ..., s_{n-1}]),$$

and thus generate a Borel measure on ω^{ω^2} by the Kolmogorov consistency theorem.

We have now defined a Borel measure on ω^{ω^2} for each pair of mixed strategies σ and τ. The notion of **optimality** can now be literally copied from the context of length ω games in Section 3, and for any $\alpha \leq \omega^2$, the axiom $\mathsf{BI\text{-}AD}^\alpha$ can be defined as "For every set $A \subseteq \omega^\alpha$, one of the two players has an optimal strategy in $\mathrm{M}(A)$".

We now have a definition of $\mathsf{BI\text{-}AD}_\mathbb{R}$ from Section 9.1 and a definition of $\mathsf{BI\text{-}AD}^{\omega^2}$. Blass [5] proved that the perfect information analogues of these two axioms are equivalent:[12]

[12] *Cf.* [34] for an improvement to $\mathsf{AD}^{\omega \cdot 2}$ on the basis of a strong result due to Woodin.

Theorem 9.4 (Blass). *The following are equivalent:*

1. $\mathsf{AD}_{\mathbb{R}}$, *and*
2. AD^{ω^2}.

Can we do the same for the Blackwell axioms?

Question 9.3. *Are* $\mathsf{Bl\text{-}AD}_{\mathbb{R}}$ *and* $\mathsf{Bl\text{-}AD}^{\omega^2}$ *equivalent?*

Appendix: List of Open Questions

The central open question in this area is of course Martin's Conjecture 4.1. Its positive solution solves most of the open questions listed below (an asterisk marks those that would be solved by a proof of Martin's Conjecture). Some of the open questions have definable versions: for instance, "Does Bl-AD imply that all sets of reals have the perfect set property?" has the definable versions "If $\mathbf{\Gamma}$ is a boldface pointclass, does $\mathsf{Bl\text{-}Det}(\mathbf{\Gamma})$ imply that all sets in $\mathbf{\Gamma}$ (or $\exists^{\mathbb{R}}\mathbf{\Gamma}$) have the perfect set property?". Those questions with definable versions are marked with a dagger.

Conjecture 4.1	Bl-AD implies AD.
Question 7.1	Does Bl-AD imply that all sets of reals have the perfect set property?*†
Question 7.2	Does Bl-AD imply that all sets of reals have the Baire property?*†
Question 7.3	Does Bl-AD imply that all Turing closed sets of reals are Martin measurable?*†
Question 7.4	Does Bl-AD imply the Moschovakis Covering Lemma?*
Question 7.5	Does Bl-AD imply $\boldsymbol{\delta}^1_4 = \aleph_{\omega+2}$?*
Question 7.6	Does Bl-AD imply that $\boldsymbol{\delta}^1_3$ has the strong partition property?*
Question 7.7	Does Bl-AD + $\mathsf{DC}_{\mathbb{R}}$ imply that $\leq_{\mathrm{BlL}}$ is well founded?*

Question 8.1 Are pBI-AD or blBI-AD equivalent to AD?

Question 9.1 Is $\mathsf{BI\text{-}AD}_{\omega_1}$ inconsistent?

Question 9.2 Does $\mathsf{BI\text{-}AD}_{\mathbb{R}}$ imply $\mathsf{AD}_{\mathbb{R}}$?

Question 9.3 Are $\mathsf{BI\text{-}AD}_{\mathbb{R}}$ and $\mathsf{BI\text{-}AD}^{\omega^2}$ equivalent?

References

[1] ADDISON, J.W. and MOSCHOVAKIS, Y.N.: Some consequences of the axiom of definable determinateness, *Proceedings of the National Academy of Sciences USA* **59** (1968), 708–712.

[2] ANDRETTA, A.: The SLO principle and the Wadge hierarchy, in S. Bold, B. Löwe, T. Räsch, and J. van Benthem editors, *Foundations of the Formal Sciences V. Infinite Games* (Papers of the conference "Foundations of the Formal Sciences V" held in Bonn, November 26–29, 2004), to appear.

[3] BLACKWELL, D.: Infinite G_δ games with imperfect information, *Polska Akademia Nauk. Instytut Matematyczny. Zastosowania Matematyki* **10** (1969), 99–101.

[4] BLACKWELL, D.: Games with Infinitely Many Moves and Slightly Imperfect Information, in R.J. Nowakowski editor, *Games of no chance* (Combinatorial games at MSRI, Workshop, July 11–21, 1994 in Berkeley, CA, USA), Mathematical Sciences Research Institute Publications **29**, Cambridge University Press, Cambridge (1996), 407–408.

[5] BLASS, A.: Equivalence of two strong forms of determinacy, *Proceedings of the American Mathematical Society* **52** (1975), 373–376.

[6] CHALONS, C.: A unique integer associated to each map from E^ω to ω, *Comptes Rendus de l'Académie des Sciences, Série I Mathématiques* **331** (2000), 501–506.

[7] DAVIS, M.: Infinite games of perfect information, in Melvin Dresher, Lloyd S. Shapley, Alan W. Tucker editors, *Advances in Game Theory*, Annals of Mathematics Studies **52**, Princeton Univ. Press, Princeton (1964), 85–101.

[8] DE KLOET, D.: Real Blackwell Determinacy, *ILLC Publications* X-2005-05.

[9] DE KLOET, D., KIEFTENBELD, V., and LÖWE, B.: Solovay objects from Blackwell determinacy, in preparation.

[10] DUPARC, J.: The Steel Hierarchy of Ordinal Valued Borel Mappings, *Journal of Symbolic Logic* **68** (2003), 187–234.

[11] GALE, D. and STEWART, F.M.: Infinite Games with Perfect Information, in Harold W. Kuhn and Albert W. Tucker editors, *Contributions to the Theory of Games II*, Annals of Mathematics Studies **28**, Princeton Univ. Press, Princeton (1953), 245–266.

[12] HARRINGTON, L.A.: Analytic determinacy and $0^{\#}$, *Journal of Symbolic Logic* **43** (1978), 685–693.

[13] JACKSON, S.: AD and projective ordinals, in A.S. Kechris, D.A. Martin, and J.R. Steel editors, *Cabal Seminar 81–85* (Proc. Caltech-UCLA Logic Sem. 1981–85), Lecture Notes in Math. **1333**, Springer-Verlag, Berlin (1988), 117–220.

[14] JACKSON, S.: A computation of $\boldsymbol{\delta}^1_5$, *Memoirs of the American Mathematical Society* **140** (1999).

[15] JACKSON, S.: Structural Consequences of AD, in M. Foreman, A. Kanamori, and M. Magidor editors, *Handbook of Set Theory*, to appear.

[16] JACKSON, S. and KHAFIZOV, F.T.: Descriptions and Cardinals below $\boldsymbol{\delta}^1_5$, accepted for publication in *Journal of Symbolic Logic* (*ILLC Publications* PP-2006-09).

[17] JACKSON, S. and LÖWE, B.: Canonical Measure Assignments, submitted (*ILLC Publications* PP-2006-05).

[18] KANAMORI, A.: *The Higher Infinite, Large Cardinals in Set Theory from Their Beginnings*, Perspectives in Mathematical Logic, Springer-Verlag, Berlin (1994).

[19] KECHRIS, A.S.: AD and Projective Ordinals, in A.S. Kechris and Y.N. Moschovakis editors, *Cabal Seminar 76–77* (Proc. Caltech-UCLA

Logic Sem. 1976–77), Lecture Notes in Math. **689**, Springer, Berlin (1978), 91–132.

[20] KECHRIS, A.S.: *Classical Descriptive Set Theory*, Graduate Texts in Mathematics **156**, Springer-Verlag, New York (1995).

[21] KECHRIS, A.S., KLEINBERG, E.M., MOSCHOVAKIS, Y.N., and WOODIN, W.H.: The Axiom of Determinacy, strong partition properties and non-singular measures, in A.S. Kechris, D.A. Martin, and Y.N. Moschovakis editors, *Cabal Seminar 77–79* (Proc. Caltech-UCLA Logic Sem. 1977–79), Lecture Notes in Math. **839**, Springer, Berlin-New York (1981), 75–99.

[22] KECHRIS, A.S., MARTIN, D.A., and MOSCHOVAKIS, Y.N. editors: *Cabal Seminar 77–79* (Proc. Caltech-UCLA Logic Sem. 1977–79), Lecture Notes in Math. **839**, Springer, Berlin-New York (1981).

[23] KECHRIS, A.S., MARTIN, D.A., and MOSCHOVAKIS, Y.N. editors: *Cabal seminar 79–81* (Proc. Caltech-UCLA Logic Sem. 1979–81), Lecture Notes in Math. **1019**, Springer-Verlag, Berlin (1983).

[24] KECHRIS, A.S., MARTIN, D.A., and STEEL, J.R. editors: *Cabal Seminar 81–85* (Proc. Caltech-UCLA Logic Sem. 1981–85), Lecture Notes in Math. **1333**, Springer-Verlag, Berlin (1988).

[25] KECHRIS, A.S. and MOSCHOVAKIS, Y.N. editors: *Cabal Seminar 76–77* (Proc. Caltech-UCLA Logic Sem. 1976–77), Lecture Notes in Math. **689**, Springer, Berlin (1978).

[26] KLEINBERG, E.M.: *Infinitary Combinatorics and the Axiom of Determinateness*, Lecture Notes in Mathematics **612**, Springer-Verlag, Berlin-New York (1977).

[27] LÖWE, B.: Turing Cones and Set Theory of the Reals, *Archive for Mathematical Logic* **40** (2001), 651–664.

[28] LÖWE, B.: Playing with mixed strategies on infinite sets, *International Journal of Game Theory* **31** (2002), 137–150.

[29] LÖWE, B.: Consequences of Blackwell Determinacy, *Bulletin of the Irish Mathematical Society* **49** (2002), 43–69.

[30] LÖWE, B.: The simulation technique and its consequences for infinitary combinatorics under the axiom of Blackwell determinacy, *Pacific Journal of Mathematics* **214** (2004), 335–358.

[31] LÖWE, B.: A global wellordering of norms defined via Blackwell games, *Order* **22** (2) (2005), 85–92.

[32] LÖWE, B.: Extensions of the Axiom of Blackwell Determinacy, *ILLC Publications* PP-2005-24.

[33] LÖWE, B.: A parametrised choice principle and Martin's conjecture on Blackwell determinacy, *Mathematical Logic Quarterly* **52** (2006), 187–189.

[34] LÖWE, B. and ROHDE, P.: Games of length $\omega \cdot 2$, *Proceedings of the American Mathematical Society* **130** (2002), 1247–1248.

[35] MAITRA, A.P. and SUDDERTH, W.D.: An operator solution of stochastic games, *Israel Journal of Mathematics* **78** (1992), 33–49.

[36] MARTIN, D.A.: Borel Determinacy, *Annals of Mathematics* **102** (1975), 363–371.

[37] MARTIN, D.A.: The Determinacy of Blackwell Games, *Journal of Symbolic Logic* **63** (1998), 1565–1581.

[38] MARTIN, D.A.: A simple proof that Determinacy implies Lebesgue Measurability, *Rendiconti del Seminario Matematico dell'Università e del Politecnico di Torino* **61** (2003), 393–397.

[39] MARTIN, D.A., NEEMAN, I., and VERVOORT, M.: The Strength of Blackwell Determinacy, *Journal of Symbolic Logic* **68** (2003), 615–636.

[40] MOSCHOVAKIS, Y.N.: *Descriptive Set Theory*, Studies in Logic and the Foundations of Mathematics **100**, North-Holland Publishing Co., Amsterdam-New York (1980).

[41] MYCIELSKI, J. and STEINHAUS, H.: A mathematical axiom contradicting the axiom of choice, *Bulletin de l'Académie Polonaise des Sciences, Série des Sciences Mathématiques, Astronomiques et Physiques* **10** (1962), 1–3.

[42] MYCIELSKI, J. and ŚWIERCZKOWSKI, S.: On the Lebesgue measurability and the axiom of determinateness, *Fundamenta Mathematicae* **54** (1964), 67–71.

[43] NEEMAN, I.: *The determinacy of long games*, de Gruyter Series in Logic and its Applications **7**, Walter de Gruyter GmbH & Co. KG, Berlin (2004).

[44] ORKIN, M.: Infinite games with imperfect information, *Transactions of the American Mathematical Society* **171** (1972), 501–507.

[45] PARIS, J.B.: $\mathsf{ZF} \vdash \Sigma^0_4$ determinateness, *Journal of Symbolic Logic* **37** (1972), 661–667.

[46] SOLOVAY, R.M.: The independence of DC from AD, in A.S. Kechris and Y.N. Moschovakis editors, *Cabal Seminar 76–77* (Proc. Caltech-UCLA Logic Sem. 1976–77), Lecture Notes in Math. **689**, Springer, Berlin (1978), 171–183.

[47] STEEL, J.R.: Long Games, in A.S. Kechris, D.A. Martin, and J.R. Steel editors, *Cabal Seminar 81–85* (Proc. Caltech-UCLA Logic Sem. 1981–85), Lecture Notes in Math. **1333**, Springer-Verlag, Berlin (1988), 56–97.

[48] STEINHAUS, H.: Games, an informal talk, *American Mathematical Monthly* **72** (1965), 457–468.

[49] VERVOORT, M.R.: *Blackwell games*, Doctoraalscriptie, Universiteit van Amsterdam, (1995).

[50] VERVOORT, M.: Blackwell games, in Thomas S. Ferguson, Lloyd S. Shapley, and James B. MacQueen editors, *Statistics, probability and game theory* (papers in honor of David Blackwell), IMS Lecture Notes Monogr. Ser. **30**, Inst. Math. Statist., Hayward, CA (1996), 369–390.

[51] VERVOORT, M.: *Games, Walks and Grammars, Problems I've worked on*, Academisch Proefschrift ter verkrijging van de graad van doctor aan de Universiteit van Amsterdam (2000).

[52] WOLFE, P.: The strict determinateness of certain infinite games, *Pacific Journal of Mathematics* **5** (1955), 841–847.

[53] ZERMELO, E.: Über eine Anwendung der Mengenlehre auf die Theorie des Schachspiels, in E.W. Hobson, A.E.H. Love editors, *Proceedings of the Fifth International Congress of Mathematicians* (Cambridge 1912) **2**, Cambridge (1913), 501–504.

Forcing with F_σ- and with Summable Filters

Heike Mildenberger

Contents

1. Introduction

Often forcings look very similar, and their generic filters share some distinguished properties, e.g., generating a P-point with no rapid ultrafilters Rudin-Keisler below it. We look at three notions of forcing, two of them adding such a special object:

The first is the forcing $\mathbf{P}_1 = (c_0 \setminus \ell^1, \leq^*)$, where c_0 is the set of real sequences that converge to 0, and ℓ^1 is the set of absolutely summable sequences. The binary relation $\leq^*$ is the partial order of eventual domination. The stronger condition is the $\leq^*$-smaller sequence which is still not summable.

The second forcing, $\mathbf{P}_2$, has as its domain all F_σ-filters on ω, and a stronger condition is a superset of a weaker one.

We take the forcing $\mathbf{P}_3 = ([\omega]^\omega, \subseteq^*)$ as a third object of study, because its comparison with the summable filter forcing brought up techniques that we shall be using, among others, also for the investigation of the first two forcings. Our main results are:

Theorem 1.1. *If* $\mathrm{cf}(2^\omega) = \mathfrak{p}$, *then the three forcings are equivalent.*

This extends a result of Vojtáš [10], which says that under the same premise the first an the third forcing are equivalent.

It is open whether this is a result in ZFC. Our trials to build isomorphisms failed right at the beginning:

Theorem 1.2. *The map* $f \mapsto \left\{\omega \setminus X \mid \sum_{i \in X} f(i) < \infty\right\}$ *is not a complete embedding of* $\mathbf{P}_1$ *into* $\mathbf{P}_2$.

This should be compared to the much greater knowledge about $\mathbf{P}_1$ and $\mathbf{P}_3$ from our joint work with Fuchino, Shelah, and Vojtáš [3]:

Theorem 1.3. *It is consistent that there is no complete embedding from* $\mathbf{P}_1$ *into* $\mathbf{P}_3$.

Laflamme [7, 6.3] showed that the forcing P_2 is not as large as it looks:

Theorem 1.4. *The forcing* $\mathbf{P}_2$ *is equivalent to forcing with filters generated by closed (in the Cantor space topology) sets.*

In [5] Just and Krawczyk investigate isomorphy types of Boolean algebras of the form $\mathscr{P}(\omega)/I$ for three concrete ideals I, one of them F_σ, the other two $F_{\sigma\delta}$. We will also look at quotients of the form $\mathscr{S}/J$ where $\mathscr{S}$ is a subset of $\mathscr{P}(\omega)$ or a small subset of $\mathscr{P}(\mathscr{P}(\omega))$ (that can be coded in $\mathscr{P}(\omega)$), and our ideals J on ω or on $\mathscr{P}(\omega)$ are given by the fact, that the forcings as written above are not separative and that we need to look at the separative quotient for determining forcing equivalence. The denominator J_3 for the forcing $\mathbf{P}_3$ is the ideal Fin of finite subsets of ω, and the other J are described via the relations $\approx_i$ given at the end of this section. In our scenario the F_σ-filters (or their dual ideals) are the conditions in the numerator of the quotient, in contrast to the mentioned work by Just and Krawczyk.

The three forcings are very similar in their generics: From any generic of filter G of $\mathbf{P}_1$ we can define

$$U_G = \Big\{X \subseteq \omega \,|\, (\exists f \in G) \sum_{n \in \omega \setminus X} f(n) < \infty\Big\}.$$

Laflamme [7, Section 6] showed that forcing with $\mathbf{P}_2$ adds a P-point with no rapid ultrafilter below it in the Rudin-Keisler ordering. Laflamme's proof works analogously for the forcing $\mathbf{P}_1$ and gives the same result for the ultrafilter U_G for generics G of $\mathbf{P}_1$. The forcing $\mathbf{P}_3$ adds a Ramsey ultrafilter by [8]. Although Ramsey ultrafilters are rapid, the difference in the generic objects does not necessarily imply that the notions of forcing are different. Our second result is a partial negative answer to the following question:

Question 1.1. *Is there a "definable" complete embedding of the forcing* $(c_0 \setminus \ell^1, \leq^*)$ *in the forcing* (F_σ*-filters*, $\supseteq$)*?*

Why do we speak about summable ultrafilters? Vojtáš [10] showed that the forcing $(c_0 \setminus \ell^1, \leq^*)$ is equivalent to forcing with summable filters. A filter F on ω is called summable, if there is a sequence $\langle a_i \,|\, i < \omega\rangle$ of positive reals such that $\sum_i a_i = \infty$ and $F = F_{\langle a_i \,|\, i<\omega\rangle} := \{\omega \setminus X \,|\, \sum_{i \in X} a_i < \infty\}$. Since the filters $F_{\langle a_i \,|\, i<\omega\rangle}$ are F_σ, the forcing with F_σ-filters is at least as a partial order a superstructure of the forcing with summable filters.

In the remainder of the section we review the relevant definitions and some facts. We write $A \subseteq^* X$ iff $A \setminus X$ is finite. A set $A \in [\omega]^\omega$ is called a pseudointersection of $\mathscr{X}$ if $(\forall X \in \mathscr{X})(A \subseteq^* X)$. The pseudointersection number $\mathfrak{p}$ is defined as $\min\{|\mathscr{X}| \mid \mathscr{X} \subseteq [\omega]^\omega$ is closed under finite intersections and there is no pseudointersection for $\mathscr{X}\}$.

We equip 2^ω with the Cantor space topology, i.e., a base for the open sets is $\{[s] \mid s \in 2^{<\omega}\}$ where $[s] = \{f \in 2^\omega \mid s \subseteq f\}$. A subset of $\mathscr{P}(\omega)$ is F_σ, if the set of the characteristic functions of its members is an F_σ-subset in 2^ω.

A filter on ω is a subset of $[\omega]^\omega$ that is closed under supersets and finite intersections. An F_σ-filter is a filter on ω such that the set of the characteristic functions of the sets in the filter is an F_σ-set.

An ultrafilter U on ω is a P-point if every countable subset $\{A_i \mid i < \omega\} \subseteq U$ has a pseudointersection $X \in U$. An ultrafilter U on ω is rapid, if for every $f\colon \omega \to \omega$ there is some $X \in U$ such that $(\forall n)|X \cap f(n)| \leq n$. For two ultrafilters U, V on ω we say U is below V is the Rudin-Keisler ordering, and we write $U \leq_{RK} V$, if there is some $f\colon \omega \to \omega$ such that $U = f(V) := \{X \subseteq \omega \mid f^{-1}X \in V\}$.

We recall what "equivalence of notions of forcing" means, and we recall the notion of embeddability. For our positive result we shall use the regular Boolean algebras of the forcings and criteria when complete Boolean algebras are isomorphic. For the negative result, it is more efficient to work with the partial orders directly.

A partial order $\mathbf{P} = (P, \leq)$ is a set P equipped with a reflexive and transitive relation $\leq$. When considering partial orders as notions of forcing, we take the convention that the smaller condition is the stronger one. We write $r \perp q$ iff $\neg\exists s\,(s \leq r \wedge s \leq q)$. The following is from Kunen's book [6, 7.1,7.7]:

Definition 1.1. *Let $\mathbf{P} = (P, \leq_P)$ and $\mathbf{Q} = (Q, \leq_Q)$ be partial orders and $i\colon P \to Q$. The map i is a complete embedding iff*

1. $(\forall p, p' \in P)(p' \leq_P p \to i(p') \leq_Q i(p))$, *and*
2. $(\forall p, p' \in P)(p' \perp_P p \leftrightarrow i(p') \perp_Q i(p))$, *and*
3. $(\forall q \in Q)(\exists p \in P)(\forall p' \in P)(p' \leq_P p \to (i(p') \not\perp_Q q)$. *Such a p is called a reduction of q.*

Definition 1.2. *A dense embedding $i\colon P \to Q$ is a complete embedding such that in addition $i''P$ is dense in Q, i.e., $(\forall q \in Q)(\exists r \in i''P)(r \leq_Q q)$.*

Every generic extension $V[G]$ by a $\mathbf{Q}$-generic filter G over V contains (as a subset) a generic extension $V[i^{-1}G]$ by $\mathbf{P}$ iff there is a complete embedding $i\colon P \to Q$. For details, and proofs see [6, Chapter 7]. In terms of Boolean algebras, this is equivalent to having an embedding of the regular open algebras that preserves arbitrary meets and unions.

Two partial orders generate the same generic extensions (where G has to range over all generic filters) iff there is a dense embedding from one into the other, or equivalently, if their regular open algebras are isomorphic [6, Chapter 7].

A partial order $\mathbf{P} = (P, \leq)$ is separative, iff

$$\forall p, q \in P\,(p \not\leq q \longrightarrow \exists r \in P\ (r \leq p\ \wedge\ r \perp q))\,,$$

or, in topological terms, for $p \neq q \in P$ we have that $\mathrm{int}(\mathrm{cl}(\{p' \mid p' \leq p\})) \neq \mathrm{int}(\mathrm{cl}(\{q' \mid q' \leq q\}))$, where the interiors and closures are taken in the so-called cut topology on $(P, \leq)$, which is generated by the basic open sets $\{\{p' \mid p' \leq p\} \mid p \in P\}$. Hence the map $e_{\mathbf{P}} : p \mapsto \mathrm{int}(\mathrm{cl}(\{p' \mid p' \leqslant p\}))$ is a dense embedding into the Boolean algebra (minus its zero element) of regular open subsets of P, called $\mathrm{RO}(\mathbf{P})$.

In general, for a partial order $(P, \leq)$, $A \subseteq P$ is called regular open iff $\mathrm{int}(\mathrm{cl}(A)) = A$. As shown in [4, page 152], for any separative $(P, \leqslant)$ there is a unique complete Boolean algebra $\mathrm{RO}(\mathbf{P})$ into which — leaving out the Boolean algebra's zero element, of course —$\mathbf{P}$ is densely embedded. The same holds for not necessarily separative partial orders, only then the embedding is possibly not injective. Note that Definition 1.1 does not require that i be injective.

If a partial order $(P, \leq)$ is not separative, we may take the separative quotient (see [4, page 154]): We set $p \approx q$ iff $\forall r(r \perp p \leftrightarrow r \perp q)$. We denote the $\approx$-equivalence class of p by $p/{\approx}$ and the set of all equivalence classes by $P/{\approx}$. There is a partial order $\leq /{\approx}$ on the equivalence classes, defined by

$$p/{\approx}\ \leq/{\approx}\ \ q/{\approx}$$

iff

$$\forall r \in P(r \not\perp p \rightarrow r \not\perp q).$$

This is well defined, and $(P/{\approx}, \leq/{\approx})$ is separative. Note that neither of the $\mathbf{P}_i$ is separative. We write $\approx_i$ for the $\approx$-relation on $\mathbf{P}_i$. It is well-known that $\mathbf{P}_3/{\approx_3} = ([\omega]^\omega/\mathrm{Fin}, \subseteq^*)$, and Vojtáš [10, 11] investigated the incomparability relation in $\mathbf{P}_1/{\approx_1}$. Let us call $\mathbf{B}_i = \mathrm{RO}(\mathbf{P}_i/{\approx_i})$.

Now we can state the technical versions of the questions we are interested in: Are there complete (or even dense) embeddings from $\mathbf{P}_i/{\approx_i}$ to $\mathbf{P}_j/{\approx_j}$?

2. All three can be the same

In this section we proof Theorem 1.1. The techniques Vojtáš developed and used in his work on $\mathbf{P}_1$ and on $\mathbf{P}_3$ yield a proof, once it is shown that the sufficient conditions are fulfilled by $\mathbf{P}_2$ as they are fulfilled by $\mathbf{P}_1$.

We use Balcar and Simon's chapter on "Disjoint refinement" in the Handbook of Boolean Algebras [1], in order to show that if $\mathfrak{p} = \mathrm{cf}(2^\omega)$ then $\mathbf{B}_1$, $\mathbf{B}_2$ and $\mathbf{B}_3$ are isomorphic. The proof follows the outline given in [10], therefore we only sketch it here. We shall use the following for $\kappa = 2$, $\lambda = \mathfrak{p}$, $\tau = \mathrm{cf}(2^\omega)$, $\rho < \tau$.

Definition 2.1.

1. *A Boolean algebra $\mathbf{B}$ is (ρ, κ)-distributive iff for each set $\{P_\alpha \mid \alpha < \rho\}$ of maximal antichains P_α there is a maximal antichain Q such that for each $\alpha < \rho$ and each $q \in Q$ we have that*

   $$|\{p \in P_\alpha \mid p \cdot_{\mathbf{B}} q \neq 0\}| < \kappa.$$

   For $\kappa = 2$ this amounts to ordinary ρ-distributivity.

   A Boolean algebra $\mathbf{B}$ is (ρ, κ)-nowhere distributive iff for each $x \in B \setminus \{0\}$ the algebra $\mathbf{B} \restriction x$ is not (ρ, κ)-distributive.

2. *A subset C of the domain of a Boolean algebra $\mathbf{B}$ is λ-closed iff for all $\leq_{\mathbf{B}}$-descending sequences $\langle c_i \mid i < \gamma\rangle$ from C, $\gamma < \lambda$, there is some $c \in C$ such that for all $i < \gamma$, $c \leq_{\mathbf{B}} c_i$.*

3. *A subset C of the domain B of a Boolean algebra $\mathbf{B}$ is a π-base iff for every $b \in B$ there is some $c \in C$ such that $c \leq_{\mathbf{B}} b$.*

We recall Balcar's and Simon's result [1, Theorem 1.13]:

Theorem 2.1 (Balcar, Simon). *If $\tau, \lambda \geq \aleph_0$, $\kappa \geq 2$ and B is a (τ, κ)-nowhere distributive Boolean algebra having a λ-closed dense subset C and being $(\rho, 2)$-distributive for every $\rho < \tau$ and having a π-base of cardinality $\kappa^{<\lambda}$, then there is a dense subset $T \subseteq C$ of B such that $(T, \geq)$ is a tree of height τ such that each $t \in T$ has $\kappa^{<\lambda}$ successors.*

Now we apply Theorem 2.1 with $\lambda = \mathfrak{p}$ ($\mathbf{B}_i$ are $\mathfrak{p}$-closed), $\kappa = 2$ and $\tau = \mathrm{cf}(2^\omega)$ (the $\mathbf{B}_i$ are not $\mathrm{cf}(2^\omega)$-distributive, as shown in [10]). However, the $\mathbf{B}_i$ are ρ-distributive for every $\rho < \tau = \mathfrak{p}$, since they are $\mathfrak{p}$-closed. Now, all the premises of Theorem 2.1 are fulfilled, and it gives the same tree for all three $\mathbf{B}_i$'s. □

Remark 2.1 - Take $i \neq j$. Then an isomorphism from $\mathbf{B}_i$ to $\mathbf{B}_j$, call it e, is possibly inconstructive in the following sense: We are aiming at a lifting ψ_e such that the outer parts of the Diagram 1 would be commutative. The first step is: There are densely many $p \in P$ such that $e(\{p' \mid p' \leq p\})$ has the form $\{q' \mid q' \leq q\}$. For these p we set $\varphi_e(p) = q$. Then φ_e is defined on a dense subset, say $D/\approx_i$ of $\mathbf{P}_i/\approx_i$, and it is a complete embedding from $(D, \leq_{P_i})/\approx_i$ to $\mathbf{P}_j/\approx_j$. It is open whether there is a Borel-lifting ψ_e, such that the diagram is commutative:

$$\begin{array}{ccc}
(D, \leq_{\mathbf{P}_i} \restriction (D \times D)) & \xrightarrow[\psi_e]{} & \mathbf{P}_j \\
\downarrow \pi & & \downarrow \pi \\
(D, \leq_{\mathbf{P}_i} \restriction (D \times D))/\approx_i & \xrightarrow[\varphi_e]{} & \mathbf{P}_j/\approx_j \\
\downarrow e_{\mathbf{P}_i/\approx_i} & & \downarrow e_{\mathbf{P}_j/\approx_j} \\
\mathbf{B}_i & \xrightarrow[e]{} & \mathbf{B}_j
\end{array}$$

Diagram 1

On liftings of Boolean algebras (regarded as structures with finitary Boolean operations, not as complete algebras in the forcing sense) and in particular on liftings of maps from $\mathscr{P}(\omega)/I$ and $\mathscr{P}(\omega)/J$ for ideals I, J, the reader may consult the survey article [2] and the references given there.

In the forcing case, the question on the kind of lifting is analogous and multifaceted: Suppose that the P_i are subsets of 2^ω or ω^ω or $[\omega]^\omega$, such that being Borel is defined. Then one can ask whether some Borel ψ_e exists on a suitable domain D. The domain D may be replaced by any $\leq_{P_i}$-dense set of representatives of $\mathbf{P}_i/\approx_i$. Moreover, there are various natural choices to describe $\approx_i$-classes in $\mathbf{P}_i$, and they are often Borel computable from each other. In contrast to this, Theorem 2.1 gives inconstructive dense embeddings, which exist in some but possibly not in all models of ZFC.

3. Embeddings

In [3] it is shown that $\mathbf{B}_1$ and $\mathbf{B}_3$ can be not isomorphic. Motivated by this result we work towards showing differences between the notions of forcing. If they are isomorphic in some ZFC-models but are not isomorphic in others this shows that the isomorphisms existing only in some models cannot be absolute. So we try to exclude some possible candidates for isomorphisms who just come from a Borel lifting ψ (in Diagram 1).

Definition 3.1. *The mapping id "identity" is the following*

$$\mathrm{id}\colon P_1/{\approx_1} \to P_2/{\approx_2},$$

$$f/{\approx_1} \mapsto \left(\left\{\omega \setminus X \mid \sum_{n\in X} f(n) < \infty\right\}\right)/{\approx_2} = F_f/{\approx_2}.$$

It is well defined by Vojtáš [11].

Now we prove Theorem 1.2. We use results of Mazur [9]: There are F_σ-filters that are not contained in any summable filter. This idea, that there are many more F_σ-filters than summable filters is used to build an F_σ-filter I^* that does not have a reduction (as in Definition 1.1 - 3) in the summable filters.

We take an arbitrary unbounded $h\colon \omega \to \mathbb{N}$ and iteratively choose finite intervals K_n in ω (whose lengths are written as $|K_n|$) such that

$$K_n = \left[\sum_{i<n} |K_i|, \sum_{i<n} |K_i| + (n \cdot h(n))^n\right); \tag{3.1}$$

Then we take a function $k\colon \omega \to \mathbb{R}_+$ such that

$$\sum_{n\in\omega} k(n) = \infty; \tag{3.2}$$

$$\sum_{n\in\omega} \frac{k(n)}{h(n)} < \infty; \tag{3.3}$$

and finally we choose a function $f\colon \omega \to \mathbb{R}_+$ such that

$$\sum_{i\in K_n} f(i) = k(n). \tag{3.4}$$

For example, we may choose $h(n) = n$ and $k(n) = n^{-\alpha}$ for some $0 < \alpha < 1$ and any f such that equation (3.4) holds.

Let $f/{\approx_1}$ be the equivalence class of f in the separative quotient of P_1. We show that there is some $I^* \leq_{\mathbf{P}_2} F_f$ that is $\mathbf{P}_2$-incompatible with any summable filter below $\mathrm{id}(f/{\approx_1})$. This shows that $I^*/{\approx_2}$ does not have any reduction, and hence the embedding is not complete. We cite two results by Mazur [9, Lemma 1.8, Theorem 1.9]:

Lemma 3.1. *For any $n \in \omega \setminus \{0\}$ and $\varepsilon \in \mathbb{R}_+$ there is a set K_n and a family $\mathcal{R}_n$ of subsets of K_n such that*

a. $\forall v_1, \dots, v_n \in \mathcal{R}_n\; v_1 \cap \dots \cap v_n \neq \emptyset$.

b. *If P is any probability distribution on K_n then there is some $v \in \mathcal{R}_n$ such that $P(v) < \varepsilon$.*

c. $|K_n| = \left(\frac{n}{\varepsilon}\right)^n$.

We set $\mathcal{R}^*_n = \{K_n \setminus v \mid v \in \mathcal{R}_n\}$.

Theorem 3.1. *For all sequences $\langle \varepsilon_n \mid n \in \omega\rangle$ of positive reals there is some F_σ-ideal I and there is a partition $\langle K_n \mid n \in \omega\rangle$ of ω into finite sets K_n such that on $\mathcal{P}(K_n)$ there are $\mathcal{R}_n$ as in the previous lemma and*

1. $|K_n| = \left(\frac{n}{\varepsilon_n}\right)^n$.

2. $I =$ *the ideal generated in* $\mathcal{P}(\omega)$ *by* $\{u \,|\, (\forall^\infty n \in \omega)\ u \cap K_n \in \mathcal{R}_n^*\}$.

3. *If* P_n *is any probability distribution on* K_n *then there is some* $u \in \mathcal{R}_n^*$ *such that* $P(u) \geq 1 - \varepsilon_n$.

Mazur shows that such an I is not contained in any summable ideal. Our aim is to strengthen this by showing that under an appropriate choice of ε_n we have that I^* is $\mathbf{P}_2$-stronger than $\mathrm{id}(f/{\approx_1})$ and $\mathbf{P}_2$-incompatible with any summable filter below $\mathrm{id}(f/{\approx_1})$.

Proof – We fix f, h, k and $\langle K_n \,|\, n \in \omega\rangle$ as in equations (3.1) to (3.4). We take $\varepsilon_n = \frac{1}{h(n)}$. Then we have for the ideal I constructed according to Lemma 3.1 and Theorem 3.1 and the probability distributions P_n on K_n with

$$P_n(\{i\}) = \frac{f(i)}{\sum\limits_{i \in K_n} f(i)}$$

that

$$\forall n\ \exists u_n \in \mathcal{R}_n^*\ \sum_{i \in u_n} f(i) \geq \left(1 - \frac{1}{h(n)}\right) \sum_{i \in K_n} f(i).$$

By definition of I, we have that $u = \bigcup_{n \in \omega} u_n \in I$. Now we have for every $f' \leq^* f$, say $f'(n) \leq f(n)$ for $n \geq k$, that $\omega \setminus u \in I_{f'} := \{X \subseteq \omega \,|\, \sum_{n \in X} f'(n) < \infty\}$, because

$$\sum_{n \in \omega} \sum_{i \in K_n \setminus u} f'(i) \leq \sum_{n \in \omega} \sum_{i \in K_n \setminus u} f(i) + \sum_{i \leq k} f'(i)$$

$$\leq \sum_{n \in \omega} \frac{1}{h(n)} \sum_{i \in K_n} f(i) + \sum_{i \leq k} f'(i) < \infty.$$

Hence $I^* = \{\omega \setminus X \,|\, X \in I\}$ is $\mathbf{P}_2$-incompatible with any summable filter $F_{f'} = I_{f'}^*$, for $f' \leq^* f$. □

Acknowledgements. We thank Andreas Blass for having us asked whether the first two notions of forcing are equivalent.

References

[1] Balcar, B. and Simon, P.: Disjoint refinement, in J. Donald Monk and R. Bonnet, editors, *Handbook of Boolean Algebras*, vol. **II**, North-Holland, Amsterdam (1989), 35–67.

[2] Farah, I.: Completely additive liftings, *Bull. Symbolic Logic* **4** (1998), 37–54.

[3] Fuchino, S., Mildenberger, H., Shelah, S., and Vojtáš, P.: On absolutely divergent series, *Fund. Math.* **160** (1999), 255–268.

[4] Jech, T.: *Set Theory*, Pure and Applied Mathematics, Academic Press, New York-London (1978).

[5] Just, W. and Krawczyk, A.: On Certain Boolean Algebras $\mathscr{P}(\omega)/I$, *Trans. Amer. Math. Soc.* **285** (1984), 411–429.

[6] Kunen, K.: *Set Theory, An Introduction to Independence Proofs*, North-Holland, Amsterdam (1980).

[7] Laflamme, C.: Forcing with filters and complete combinatorics, *Ann. Pure Appl. Logic* **42** (1989), 125–163.

[8] Mathias, A.: Happy families, *Ann. Math. Logic* **12** (1977), 59–111.

[9] Mazur, K.: F_σ-ideals and ω_1, ω_1^*-gaps in $\mathcal{P}(\omega)/I$, *Fund. Math.* **138** (1991), 103–111.

[10] Vojtáš, P.: Boolean isomorphism between partial orderings of convergent and divergent series and infinite subsets of N, *Proc. Amer. Math. Soc.* **117** (1993), 235–242.

[11] Vojtáš, P.: On ω^* and absolutely divergent series, *Top. Proceedings* **19** (1994), 335–348.

Reasonably Complete Forcing Notions

Andrzej Rosłanowski and Saharon Shelah

Contents

1. Introduction

There are serious ZFC obstacles to easy generalizations of properness to the case of iterations with uncountable supports (see, e.g., Shelah [8, Appendix 3.6 - 2]). This paper belongs to the series of works aiming at localizing "good properness conditions" for such iterations and including Shelah [9], [10], Rosłanowski and Shelah [7], [6] and Eisworth [2]. Our results continue Rosłanowski and Shelah [6, Section A.2], but no familiarity with the previous paper is assumed and the current work is fully self-contained.

In Section 2 we introduce 3 bounding–type properties (A, B, C) and we essentially show that the first two are almost preserved in λ–support iterations (Theorems 3.1, 3.2). "Almost" as the limit of the iteration occurs to have a somewhat weaker property, but equally applicable. In the following section we show that reasonably A–bounding forcing notions are exactly the ones introduced in [6, Section A.2], thus showing that Theorem 3.2 improves [6, Thm A.2.4]. In the fourth section of the paper, we give an example of an interesting reasonably B–bounding forcing notion and we use it to show that it is consistent that dominating numbers associated with two normal filters on λ are distinct (Corollary 5.1). Finally, in the last section we present two forcing notions that are not yet covered by existing iteration theorems. We hope that the further development of the theory will include also them.

Like in [6], we assume here that our cardinal λ is inaccessible. We do not know at the moment if any parallel work can be done for a successor cardinal, though some progress will be presented in a subsequent paper [4].

Notation. Our notation is rather standard and compatible with that of classical textbooks (like Jech [3]). In forcing we keep the older convention that a *stronger condition is the larger one.*

1. Ordinal numbers will be denoted by the lower case initial letters of the Greek alphabet ($\alpha, \beta, \gamma, \delta \ldots$) and also by i, j (with possible sub- and superscripts).

*Both authors acknowledge support from the United States-Israel Binational Science Foundation (Grant no. 2002323). This is publication 860 of the second author.

Cardinal numbers will be called κ, λ, μ; λ *will be always assumed to be inaccessible* (we may forget to mention it).

By χ we will denote a *sufficiently large* regular cardinal; $\mathcal{H}(\chi)$ is the family of all sets hereditarily of size less than χ. Moreover, we fix a well ordering $<^*_\chi$ of $\mathcal{H}(\chi)$.

2. For two sequences η, ν we write $\nu \vartriangleleft \eta$ whenever ν is a proper initial segment of η, and $\nu \trianglelefteq \eta$ when either $\nu \vartriangleleft \eta$ or $\nu = \eta$. The length of a sequence η is denoted by $\mathrm{lh}(\eta)$.

3. We will consider several games of two players. One player will be called Generic or Complete or just COM, and we will refer to this player as "she". Her opponent will be called Antigeneric or Incomplete or just INC and will be referred to as "he".

4. For a forcing notion $\mathbb{P}$, $\Gamma_{\mathbb{P}}$ stands for the canonical $\mathbb{P}$–name for the generic filter in $\mathbb{P}$. With this one exception, all $\mathbb{P}$–names for objects in the extension via $\mathbb{P}$ will be denoted with a tilde below (e.g., $\underset{\sim}{\tau}$, $\underset{\sim}{X}$). The weakest element of $\mathbb{P}$ will be denoted by $\emptyset_{\mathbb{P}}$ (and we will always assume that there is one, and that there is no other condition equivalent to it). We will also assume that all forcing notions under consideration are atomless.

   By "λ–support iterations" we mean iterations in which domains of conditions are of size $\leq \lambda$. However, we will pretend that conditions in a λ–support iteration $\bar{\mathbb{Q}} = \langle \mathbb{P}_\zeta, \underset{\sim}{\mathbb{Q}}_\zeta : \zeta < \zeta^* \rangle$ are total functions on ζ^* and for a condition p in the limit $\lim(\bar{\mathbb{Q}})$ of the iteration $\bar{\mathbb{Q}}$ and $\alpha \in \zeta^* \setminus \mathrm{Dom}(p)$ we will let $p(\alpha) = \underset{\sim}{\emptyset}_{\underset{\sim}{\mathbb{Q}}_\alpha}$.

5. For a filter D on λ, the family of all D–positive subsets of λ is called D^+. (So $A \in D^+$ if and only if $A \subseteq \lambda$ and $A \cap B \neq \emptyset$ for all $B \in D$.)

In this paper we assume the following.

Context 1.1 -

a. λ is a strongly inaccessible cardinal,

b. $\bar{\mu} = \langle \mu_\alpha : \alpha < \lambda \rangle$, each μ_α is a regular cardinal satisfying (for $\alpha < \lambda$)

$$\aleph_0 \leq \mu_\alpha \leq \lambda \qquad \text{and} \qquad (\forall f \in {}^\alpha\mu_\alpha)(|\prod_{\xi<\alpha} f(\xi)| < \mu_\alpha),$$

c. $\mathcal{U}$ is a normal filter on λ.

2. Preliminaries on λ–support iterations

Definition 2.1. *Let $\mathbb{P}$ be a forcing notion.*

1. *For a condition $r \in \mathbb{P}$ let $\Game_0^\lambda(\mathbb{P}, r)$ be the following game of two players,* Complete *and* Incomplete*:*

   > *the game lasts at most λ moves and during a play the players construct a sequence $\langle (p_i, q_i) : i < \lambda \rangle$ of pairs of conditions from $\mathbb{P}$ in such a way that $(\forall j < i < \lambda)(r \leq p_j \leq q_j \leq p_i)$ and at the stage $i < \lambda$ of the game, first Incomplete chooses p_i and then Complete chooses q_i.*

   Complete wins if and only if for every $i < \lambda$ there are legal moves for both players.

2. *We say that the forcing notion $\mathbb{P}$ is* strategically $(<\lambda)$–complete *if Complete has a winning strategy in the game $\Game_0^\lambda(\mathbb{P}, r)$ for each condition $r \in \mathbb{P}$.*

3. *Let $N \prec (\mathcal{H}(\chi), \in, <^*_\chi)$ be a model such that ${}^{<\lambda}N \subseteq N$, $|N| = \lambda$ and $\mathbb{P} \in N$. We say that a condition $p \in \mathbb{P}$ is* $(N, \mathbb{P})$–generic in the standard sense *(or just:* $(N, \mathbb{P})$–generic*) if for every $\mathbb{P}$–name $\underset{\sim}{\tau} \in N$ for an ordinal we have $p \Vdash$ "$\underset{\sim}{\tau} \in N$ ".*

4. *$\mathbb{P}$ is* λ–proper in the standard sense *(or just:* λ–proper*) if there is $x \in \mathcal{H}(\chi)$ such that for every model $N \prec (\mathcal{H}(\chi), \in, <^*_\chi)$ satisfying*

   $${}^{<\lambda}N \subseteq N, \quad |N| = \lambda \quad \text{and} \quad \mathbb{P}, x \in N,$$

   and every condition $q \in N \cap \mathbb{P}$ there is an $(N, \mathbb{P})$–generic condition $p \in \mathbb{P}$ stronger than q.

The following two propositions and the definition are essentially quoted from [6, Section A.1].

Proposition 2.1. *Suppose that* $\mathbb{P}$ *is a* $(<\lambda)$*–strategically complete (atomless) forcing notion,* $\alpha^* < \lambda$ *and* $p_\alpha \in \mathbb{P}$ *(for* $\alpha < \alpha^*$*). Then there are conditions* $q_\alpha \in \mathbb{P}$ *(for* $\alpha < \alpha^*$*) such that* $p_\alpha \leq q_\alpha$ *and for distinct* $\alpha, \alpha' < \alpha^*$ *the conditions* $q_\alpha, q_{\alpha'}$ *are incompatible.*

Proposition 2.2. *Suppose* $\bar{\mathbb{Q}} = \langle \mathbb{P}_i, \underset{\sim}{\mathbb{Q}}_i : i < \gamma \rangle$ *is a* λ*–support iteration and, for each* $i < \gamma$,

$$\Vdash_{\mathbb{P}_i} \text{“ } \underset{\sim}{\mathbb{Q}}_i \text{ is strategically } (<\lambda)\text{–complete ”}.$$

Then, for each $\varepsilon \leq \gamma$ *and* $r \in \mathbb{P}_\varepsilon$*, there is a winning strategy* $\mathbf{st}(\varepsilon, r)$ *of Complete in the game* $\partial_0^\lambda(\mathbb{P}_\varepsilon, r)$ *such that, whenever* $\varepsilon_0 < \varepsilon_1 \leq \gamma$ *and* $r \in \mathbb{P}_{\varepsilon_1}$*, we have:*

i. *if* $\langle (p_i, q_i) : i < \lambda \rangle$ *is a play of* $\partial_0^\lambda(\mathbb{P}_{\varepsilon_0}, r{\restriction}\varepsilon_0)$ *in which Complete follows the strategy* $\mathbf{st}(\varepsilon_0, r{\restriction}\varepsilon_0)$*, then* $\langle (p_i{}^\frown r{\restriction}[\varepsilon_0, \varepsilon_1), q_i{}^\frown r{\restriction}[\varepsilon_0, \varepsilon_1)) : i < \lambda \rangle$ *is a play of* $\partial_0^\lambda(\mathbb{P}_{\varepsilon_1}, r)$ *in which Complete uses* $\mathbf{st}(\varepsilon_1, r)$*;*

ii. *if* $\langle (p_i, q_i) : i < \lambda \rangle$ *is a play of* $\partial_0^\lambda(\mathbb{P}_{\varepsilon_1}, r)$ *in which Complete plays according to the strategy* $\mathbf{st}(\varepsilon_1, r)$*, then* $\langle (p_i{\restriction}\varepsilon_0, q_i{\restriction}\varepsilon_0) : i < \lambda \rangle$ *is a play of* $\partial_0^\lambda(\mathbb{P}_{\varepsilon_0}, r{\restriction}\varepsilon_0)$ *in which Complete uses* $\mathbf{st}(\varepsilon_0, r{\restriction}\varepsilon_0)$*;*

iii. *if* ε_1 *is limit and a sequence* $\langle (p_i, q_i) : i < \lambda \rangle \subseteq \mathbb{P}_{\varepsilon_1}$ *is such that for each* $\xi < \varepsilon_1$, $\langle (p_i{\restriction}\xi, q_i{\restriction}\xi) : i < \lambda \rangle$ *is a play of* $\partial_0^\lambda(\mathbb{P}_\xi, r{\restriction}\xi)$ *in which Complete uses the strategy* $\mathbf{st}(\xi, r{\restriction}\xi)$*, then* $\langle (p_i, q_i) : i < \lambda \rangle$ *is a play of* $\partial_0^\lambda(\mathbb{P}_{\varepsilon_1}, r)$ *in which Complete plays according to* $\mathbf{st}(\varepsilon_1, r)$*;*

iv. *if* $\langle (p_i, q_i) : i < i^* \rangle$ *is a partial play of* $\partial_0^\lambda(\mathbb{P}_{\varepsilon_1}, r)$ *in which Complete uses* $\mathbf{st}(\varepsilon_1, r)$ *and* $p' \in \mathbb{P}_{\varepsilon_0}$ *is stronger than all* $p_i{\restriction}\varepsilon_0$ *(for* $i < i^*$*), then there is* $p^* \in \mathbb{P}_{\varepsilon_1}$ *such that* $p' = p^*{\restriction}\varepsilon_0$ *and* $p^* \geq p_i$ *for* $i < i^*$*.*

Definition 2.2.

1. *Let* γ *be an ordinal,* $\emptyset \neq w \subseteq \gamma$*. A* standard $(w, 1)^\gamma$–tree *is a pair* $\mathcal{T} = (T, \mathrm{rk})$ *such that*

   – $\mathrm{rk} : T \longrightarrow w \cup \{\gamma\}$,

- *if* $t \in T$ *and* $\mathrm{rk}(t) = \varepsilon$, *then* t *is a sequence* $\langle (t)_\zeta : \zeta \in w \cap \varepsilon \rangle$,
- $(T, \lhd)$ *is a tree with root* $\langle\rangle$ *and such that every chain in* T *has a* $\lhd$*–upper bound it* T,
- *if* $t \in T$, *then there is* $t' \in T$ *such that* $t \unlhd t'$ *and* $\mathrm{rk}(t') = \gamma$.

We will keep the convention that $\mathcal{T}^x_y$ *is* (T^x_y, rk^x_y).

2. *Let* $\bar{\mathbb{Q}} = \langle \mathbb{P}_i, \underset{\sim}{\mathbb{Q}}_i : i < \gamma \rangle$ *be a* λ*–support iteration. A* standard tree of conditions in $\bar{\mathbb{Q}}$ *is a system* $\bar{p} = \langle p_t : t \in T \rangle$ *such that*
   - (T, rk) *is a standard* $(w,1)^\gamma$*–tree for some* $w \subseteq \gamma$,
   - $p_t \in \mathbb{P}_{\mathrm{rk}(t)}$ *for* $t \in T$, *and*
   - *if* $s, t \in T$, $s \lhd t$, *then* $p_s = p_t {\restriction} \mathrm{rk}(s)$.
3. *Let* $\bar{p}^0, \bar{p}^1$ *be standard trees of conditions in* $\bar{\mathbb{Q}}$, $\bar{p}^i = \langle p^i_t : t \in T \rangle$. *We write* $\bar{p}^0 \le \bar{p}^1$ *whenever for each* $t \in T$ *we have* $p^0_t \le p^1_t$.

Note that our standard trees and trees of conditions are a special case of $(w, \alpha)^\gamma$–trees introduced in [6, Def. A.1.7] (for $\alpha = 1$). Our notation preserves the redundant "1" to keep the compatibility with the established terminology. For the same reason we use $(t)_\zeta$ instead of $t(\zeta)$.

Proposition 2.3. *Assume that* $\bar{\mathbb{Q}} = \langle \mathbb{P}_i, \underset{\sim}{\mathbb{Q}}_i : i < \gamma \rangle$ *is a* λ*–support iteration such that for all* $i < \gamma$ *we have*

$$\Vdash_{\mathbb{P}_i} \text{“ } \underset{\sim}{\mathbb{Q}}_i \text{ is strategically } ({<}\lambda)\text{–complete ”}.$$

1. *[6, Prop. A.1.9] Suppose that* $\bar{p} = \langle p_t : t \in T \rangle$ *is a standard tree of conditions in* $\bar{\mathbb{Q}}$, $|T| < \lambda$, *and* $\mathcal{I} \subseteq \mathbb{P}_\gamma$ *is open dense. Then there is a standard tree of conditions* $\bar{q} = \langle q_t : t \in T \rangle$ *such that* $\bar{p} \le \bar{q}$ *and* $(\forall t \in T)(\mathrm{rk}(t) = \gamma \Rightarrow q_t \in \mathcal{I})$.
2. *If* $\bar{p} = \langle p_t : t \in T \rangle$ *is a standard tree of conditions in* $\bar{\mathbb{Q}}$ *and* $|T| < \lambda$, *then there is a standard tree of conditions* $\bar{q} = \langle q_t : t \in T \rangle$ *such that* $\bar{p} \le \bar{q}$ *and*
   - *if* $t_0, t_1 \in T$, $\mathrm{rk}(t_0) = \mathrm{rk}(t_1)$, $\xi \in \mathrm{Dom}(t_0)$ *and* $(t_0)_\xi \ne (t_1)_\xi$, $t_0 {\restriction} \xi = t_1 {\restriction} \xi$ *then*

$$q_{t_0}{\restriction}\xi \Vdash_{\mathbb{P}_\xi} \text{“ the conditions } q_{t_0}(\xi), q_{t_1}(\xi) \text{ are incompatible in } \underset{\sim}{\mathbb{Q}}_\xi \text{ ”}.$$

3. *Suppose that*

   - $w \subseteq \gamma$, $|w| < \lambda$, $1 < \mu^*_\zeta \leq \lambda$ *for* $\zeta \in w$, *and* $T = \bigcup\limits_{\xi \leq \gamma} \prod\limits_{\zeta \in w \cap \xi} \mu^*_\zeta$ *(so* $\mathcal{T} = (T, \mathrm{rk})$ *is a standard* $(w,1)^\gamma$*–tree),*
   - $\bar{p} = \langle p_t : t \in T \rangle$ *is a standard tree of conditions in* $\bar{\mathbb{Q}}$*,*
   - *for* $\xi \in w$*,* $\underset{\sim}{\varepsilon}_\xi$ *is a* $\mathbb{P}_\xi$*–name for a non-zero ordinal below* μ^*_ξ*.*

   Then there are a standard $(w,1)^\gamma$*–tree* $\mathcal{T}' = (T', \mathrm{rk}')$ *and a tree of conditions* $\bar{q} = \langle q_t : t \in T' \rangle$ *such that*

   - $T' \subseteq T$*,* $\mathrm{rk}' = \mathrm{rk}{\restriction}T'$*, and for every* $t \in T'$ *such that* $\mathrm{rk}'(t) = \xi \in w$*, the condition* q_t *decides the value of* $\underset{\sim}{\varepsilon}_\xi$*, say* $q_t \Vdash \underset{\sim}{\varepsilon}_\xi = \varepsilon^t_\xi$*, and*
   - $p_t \leq q_t$ *for* $t \in T'$*, and*
   - *if* $t \in T'$*,* $\mathrm{rk}(t) = \xi \in w$*, then*

$$\{\alpha < \mu^*_\xi : t \cup \{\langle \xi, \alpha \rangle\} \in T'\} = \varepsilon^t_\xi.$$

PROOF – 2. Straightforward application of Proposition 2.1.

3. Note that we cannot apply the first part directly, as the tree T may be of size λ. So we will proceed inductively constructing initial levels of T' of size $< \lambda$ and applying 1 to them.

For $\varepsilon \leq \gamma$ and $r \in \mathbb{P}_\varepsilon$ let $\mathbf{st}(\varepsilon, r)$ be the winning strategy of Complete in $\Game^\lambda_0(\mathbb{P}_\varepsilon, r)$ given by Proposition 2.2 (so these strategies have the coherence properties listed there). Let $\langle \xi_\beta : \beta \leq \beta^* \rangle$ be the increasing enumeration of $w \cup \{\gamma\}$, $\beta^* < \lambda$. By induction on $\beta \leq \beta^*$ we will pick $\mathcal{T}_\beta, \bar{q}^\beta, \bar{r}^\beta$ and $\bar{\varepsilon}^\beta$ such that

a. $\mathcal{T}_\beta = (T_\beta, \mathrm{rk}_\beta)$ is a standard $(w \cap \xi_\beta, 1)^\gamma$–tree, $T_\beta \subseteq T$, $|T_\beta| < \lambda$, and $\bar{q}^\beta = \langle q^\beta_t : t \in T_\beta \rangle$, $\bar{r}^\beta = \langle r^\beta_t : t \in T_\beta \rangle$ are trees of conditions, $\bar{q}^\beta \leq \bar{r}^\beta$ and $r^\beta_t \in \mathbb{P}_{\mathrm{rk}(t)}$ for $t \in T_\beta$ (note: $\mathrm{rk}(t)$, not $\mathrm{rk}_\beta(t)$);

b. if $\beta_0 < \beta_1 \leq \beta^*$, then $T_{\beta_0} = \{t{\restriction}\xi_{\beta_0} : t \in T_{\beta_1}\}$ and $r^{\beta_0}_{t{\restriction}\xi_{\beta_0}} \leq q^{\beta_1}_t{\restriction}\xi_{\beta_0}$ for $t \in T_{\beta_1}$;

c. if $\beta < \beta^*$, $t \in T_\beta$ and $\mathrm{rk}_\beta(t) = \gamma$ (so $\mathrm{rk}(t) = \xi_\beta$), then

$$\langle (q^\alpha_{t\restriction\xi_\alpha} {}^\frown p_t\restriction[\xi_\alpha,\xi_\beta), r^\alpha_{t\restriction\xi_\alpha} {}^\frown p_t\restriction[\xi_\alpha,\xi_\beta)) : \alpha < \beta\rangle {}^\frown \langle (q^\alpha_t, r^\alpha_t) : \beta \le \alpha < \beta^*\rangle$$

is a partial play of $\partial^\lambda_0(\mathbb{P}_{\xi_\beta}, p_t)$ in which Complete uses her winning strategy $\mathbf{st}(\xi_\beta, p_t)$;

d. $\bar{\varepsilon}^\beta = \langle \varepsilon^\beta_t : t \in T_\beta,\ \mathrm{rk}_\beta(t) = \gamma\rangle \subseteq \mu^*_{\xi_\beta}$;

e. if $\beta < \beta^*$, $t \in T_\beta$ and $\mathrm{rk}_\beta(t) = \gamma$ (so $\mathrm{rk}(t) = \xi_\beta$), then $p_t \le q^\beta_t \in \mathbb{P}_{\xi_\beta}$ and $q^\beta_t \Vdash_{\mathbb{P}_{\xi_\beta}} \underset{\sim}{\varepsilon}_{\xi_\beta} = \varepsilon^\beta_t$;

f. if $\beta < \beta^*$, $t \in T_\beta$ and $\mathrm{rk}_\beta(t) = \gamma$, then $\{\alpha < \mu^*_{\xi_\beta} : t \cup \{\langle \xi_\beta, \alpha\rangle\} \in T_{\beta+1}\} = \varepsilon^\beta_t$.

We let $T_0 = \{\langle\rangle\}$ and we choose $q^0_{\langle\rangle} \in \mathbb{P}_{\xi_0}$ and $\varepsilon^0_{\langle\rangle}$ so that $p_{\langle\rangle} \le q^0_{\langle\rangle}$ and $q^0_{\langle\rangle} \Vdash_{\mathbb{P}_{\xi_0}} \underset{\sim}{\varepsilon}_{\xi_0} = \varepsilon^0_{\langle\rangle}$. Then we let $r^0_{\langle\rangle}$ be the answer given by $\mathbf{st}(\xi_0, p_{\langle\rangle})$ in $\partial^\lambda_0(\mathbb{P}_{\xi_0}, p_{\langle\rangle})$ to $q^0_{\langle\rangle}$. Now suppose that we have defined $\mathcal{T}_\alpha, \bar{q}^\alpha, \bar{r}^\alpha$ and $\bar{\varepsilon}^\alpha$ for $\alpha < \beta \le \beta^*$.

If β is a limit ordinal then the demands *a* and *b* uniquely define the standard tree $\mathcal{T}_\beta$. Note that $|T_\beta| < \lambda$ as λ is inaccessible; remember also clause *f.* It follows from the choice of $\mathbf{st}(\varepsilon, r)$ (see clause Proposition 2.2 - *iii*) and demand *c* at previous stages that

$(\oplus)_\beta$ if $t \in T_\beta$, $\mathrm{rk}_\beta(t) = \gamma$ (so $\mathrm{rk}(t) = \xi_\beta$), then the sequence

$$\langle (q^\alpha_{t\restriction\xi_\alpha} {}^\frown p_t\restriction[\xi_\alpha,\xi_\beta), r^\alpha_{t\restriction\xi_\alpha} {}^\frown p_t\restriction[\xi_\alpha,\xi_\beta)) : \alpha < \beta\rangle$$

is a partial play of $\partial^\lambda_0(\mathbb{P}_{\xi_\beta}, p_t)$ in which Complete uses her winning strategy $\mathbf{st}(\xi_\beta, p_t)$.

For $t \in T_\beta$ we define a condition $q_t \in \mathbb{P}_{\xi_\beta}$ as follows:

- $\mathrm{Dom}(q_t) = \bigcup\limits_{\alpha<\beta} \mathrm{Dom}(r^\alpha_{t\restriction\xi_\alpha}) \cup \mathrm{Dom}(p_t) \subseteq \mathrm{rk}(t)$,
- if $\zeta \in \mathrm{Dom}(q_t)$, then $q_t(\zeta)$ is the $<^*_\chi$–first $\mathbb{P}_\zeta$–name for an element of $\underset{\sim}{\mathbb{Q}}_\zeta$ such that

$$q_t\restriction\zeta \Vdash_{\mathbb{P}_\zeta} \text{“ if the set } \{r^\alpha_{t\restriction\xi_\alpha}(\zeta) : \zeta < \xi_\alpha \ \&\ \alpha < \beta\} \cup \{p_t(\zeta)\} \text{ has an upper bound, then } q_t(\zeta) \text{ is such an upper bound ”.}$$

It follows from $(\oplus)_\beta$ (and Proposition 2.2 - *iv*) that $p_t \le q_t$ and $r^\alpha_{t\restriction\xi_\alpha} \le q_t\restriction\xi_{\alpha+1}$ for $\alpha < \beta$. Now, by "the $<^*_\chi$–first", clearly $\bar{q} = \langle q_t : t \in T_\beta\rangle$ is a tree of conditions. Applying Proposition 2.3 - 1 we may choose a tree of conditions $\bar{q}^\beta = \langle q^\beta_t : t \in T_\beta\rangle$ such that $\bar{q} \le \bar{q}^\beta$ and

– if $\beta < \beta^*$, $t \in T_\beta$ and $\mathrm{rk}_\beta(t) = \gamma$, then the condition q^β_t decides the value of $\underset{\sim}{\varepsilon}_{\xi_\beta}$ (and let $q^\beta_t \Vdash \underset{\sim}{\varepsilon}_{\xi_\beta} = \varepsilon^\beta_t$) and $q^\beta_t \in \mathbb{P}_{\xi_\beta}$.

Then, for $t \in T_\beta$, we let r^β_t be the answer given to Complete by $\mathbf{st}(\mathrm{rk}(t), p_t)$ in the appropriate partial play of $\partial^\lambda_0(\mathbb{P}_{\mathrm{rk}(t)}, p_t)$, where at stage β Incomplete put q^β_t (see *c*, $(\oplus)_\beta$). It follows from Proposition 2.2 - *ii* that $\bar{r}^\beta = \langle r^\beta_t : t \in T_\beta\rangle$ is a tree of conditions. Plainly, $\mathcal{T}_\beta, \bar{q}^\beta, \bar{r}^\beta$ and $\bar{\varepsilon}^\beta$ satisfy all relevant (restrictions of the) demands *a*–*f*.

Now suppose that β is a successor ordinal, say $\beta = \beta_0 + 1$. Let

$$T_\beta = T_{\beta_0} \cup \big\{t \cup \{\langle \xi_{\beta_0}, \varepsilon\rangle\} : t \in T_{\beta_0} \ \&\ \mathrm{rk}_{\beta_0}(t) = \gamma \ \&\ \varepsilon < \varepsilon^{\beta_0}_t\big\}$$

and for $t \in T_\beta$ define q_t as follows:

– if $t \in T_{\beta_0}$, then $q_t = r^{\beta_0}_t$,

– if $t \in T_\beta \setminus T_{\beta_0}$, then $q_t = r^{\beta_0}_{t\restriction\xi_{\beta_0}} {}^\frown p_t\restriction[\xi_{\beta_0}, \xi_\beta)$.

Then $\bar{q} = \langle q_t : t \in T_\beta\rangle$ is a tree of conditions, $r^{\beta_0}_t \le q_t$ for $t \in T_{\beta_0}$. It follows from Proposition 2.3 - 1 that we may choose a tree of conditions $\bar{q}^\beta = \langle q^\beta_t : t \in T_\beta\rangle$ such that $\bar{q} \le \bar{q}^\beta$ and

– if $\beta < \beta^*$, $t \in T_\beta$ and $\mathrm{rk}_\beta(t) = \gamma$, then the condition q^β_t decides $\underset{\sim}{\varepsilon}_{\xi_\beta}$ and, say, $q^\beta_t \Vdash \underset{\sim}{\varepsilon}_{\xi_\beta} = \varepsilon^\beta_t$.

Next, like in the limit case, $\bar{r}^\beta = \langle r^\beta_t : t \in T_\beta\rangle$ is obtained by applying the strategies $\mathbf{st}(\mathrm{rk}(t), p_t)$ suitably. Easily, $\mathcal{T}_\beta, \bar{q}^\beta, \bar{r}^\beta$ and $\bar{\varepsilon}^\beta$ satisfy the demands *a*–*f*.

After the inductive construction is carried out look at T_{β^*}, $\bar{q}^{\beta^*}$ and $\langle \bar{\varepsilon}^\beta : \beta < \beta^*\rangle$. □

3. ABC of reasonable completeness

Remark 3.1 - Note that if a forcing notion $\mathbb{Q}$ is strategically $(<\lambda)$–complete and $\mathcal{U}$ is a normal filter on λ, then the normal filter generated by $\mathcal{U}$ in $\mathbf{V}^{\mathbb{Q}}$ is proper. Abusing notation, we may denote the normal filter generated by $\mathcal{U}$ in $\mathbf{V}^{\mathbb{Q}}$ also by $\mathcal{U}$ or by $\mathcal{U}^{\mathbb{Q}}$. Thus if $\underset{\sim}{A}$ is a $\mathbb{Q}$–name for a subset of λ, then $p \Vdash_{\mathbb{Q}} \underset{\sim}{A} \in \mathcal{U}^{\mathbb{Q}}$ if and only if for some $\mathbb{Q}$–names $\underset{\sim}{A}_\alpha$ for elements of $\mathcal{U}^{\mathbf{V}}$ we have that $p \Vdash_{\mathbb{Q}} \mathop{\triangle}\limits_{\alpha<\lambda} \underset{\sim}{A}_\alpha \subseteq \underset{\sim}{A}$ (where $\triangle$ denotes the operation of diagonal intersection).

Let us note that many of the arguments in this section would be much simpler if we restricted ourselves to $(<\lambda)$–complete forcing notions. Unfortunately, the forcing notions that we would like to cover tend to be only strategically $(<\lambda)$–complete, see [6, Section B.6].

Definition 3.1. *Let $\mathbb{Q}$ be a strategically $(<\lambda)$–complete forcing notion.*

1. *For a condition $p \in \mathbb{Q}$ we define a game[1] $\Game^{\rm rcA}_{\bar{\mu}}(p,\mathbb{Q})$ between two players, Generic and Antigeneric, as follows. A play of $\Game^{\rm rcA}_{\bar{\mu}}(p,\mathbb{Q})$ lasts λ steps and during a play a sequence*

$$\Big\langle I_\alpha, \langle p^\alpha_t, q^\alpha_t : t \in I_\alpha\rangle : \alpha < \lambda \Big\rangle$$

is constructed. Suppose that the players have arrived to a stage $\alpha < \lambda$ of the game. Now,

$(\aleph)_\alpha$ *first Generic chooses a non-empty set I_α of cardinality $< \mu_\alpha$ and a system $\langle p^\alpha_t : t \in I_\alpha\rangle$ of conditions from $\mathbb{Q}$,*

$(\beth)_\alpha$ *then Antigeneric answers by picking a system $\langle q^\alpha_t : t \in I_\alpha\rangle$ of conditions from $\mathbb{Q}$ such that $(\forall t \in I_\alpha)(p^\alpha_t \le q^\alpha_t)$.*

At the end, Generic wins the play

$$\Big\langle I_\alpha, \langle p^\alpha_t, q^\alpha_t : t \in I_\alpha\rangle : \alpha < \lambda \Big\rangle$$

of $\Game^{\rm rcA}_{\bar{\mu}}(p,\mathbb{Q})$ if and only if

[1] rc stands for **r**easonable **c**ompleteness

$(\circledast)^{\rm rc}_{\rm A}$ *there is a condition* $p^*\in\mathbb{Q}$ *stronger than* p *and such that*[2]

$$p^*\Vdash_{\mathbb{Q}}\text{``}\ \big\{\alpha<\lambda:\big(\exists t\in I_\alpha\big)\big(q^\alpha_t\in\Gamma_{\mathbb{Q}}\big)\big\}=\lambda\ \text{''}.$$

2. *Games* $\Game^{\rm rcB}_{\mathcal{U},\bar\mu}(p,\mathbb{Q}),\Game^{\rm rcC}_{\mathcal{U},\bar\mu}(p,\mathbb{Q})$ *are defined similarly, except that the winning criterion* $(\circledast)^{\rm rc}_{\rm A}$ *is replaced by*

$(\circledast)^{\rm rc}_{\rm B}$ *there is a condition* $p^*\in\mathbb{Q}$ *stronger than* p *and such that*

$$p^*\Vdash_{\mathbb{Q}}\text{``}\ \big\{\alpha<\lambda:\big(\exists t\in I_\alpha\big)\big(q^\alpha_t\in\Gamma_{\mathbb{Q}}\big)\big\}\in\mathcal{U}^{\mathbb{Q}}\ \text{''},$$

$(\circledast)^{\rm rc}_{\rm C}$ *there is a condition* $p^*\in\mathbb{Q}$ *stronger than* p *and such that*

$$p^*\Vdash_{\mathbb{Q}}\text{``}\ \big\{\alpha<\lambda:\big(\exists t\in I_\alpha\big)\big(q^\alpha_t\in\Gamma_{\mathbb{Q}}\big)\big\}\in\big(\mathcal{U}^{\mathbb{Q}}\big)^+\ \text{''},$$

respectively.

3. *For a condition* $p\in\mathbb{Q}$ *we define a game* $\Game^{\rm rc\mathbf{b}}_{\mathcal{U},\bar\mu}(p,\mathbb{Q})$ *between Generic and Antigeneric as follows. A play of* $\Game^{\rm rc\mathbf{b}}_{\mathcal{U},\bar\mu}(p,\mathbb{Q})$ *lasts* λ *steps and during a play a sequence*

$$\Big\langle\zeta_\alpha,\langle p^\alpha_\xi,q^\alpha_\xi:\xi<\zeta_\alpha\rangle:\alpha<\lambda\Big\rangle$$

is constructed. Suppose that the players have arrived to a stage $\alpha<\lambda$ *of the game. Now, Generic chooses a non-zero ordinal* $\zeta_\alpha<\mu_\alpha$ *and then the two players play a subgame of length* ζ_α *alternately choosing successive terms of a sequence* $\langle p^\alpha_\xi,q^\alpha_\xi:\xi<\zeta_\alpha\rangle$. *At a stage* $\xi<\zeta_\alpha$ *of the subgame, first Generic picks a condition* $p^\alpha_\xi\in\mathbb{Q}$ *and then Antigeneric answers with a condition* q^α_ξ *stronger than* p^α_ξ.

At the end, Generic wins the play

$$\Big\langle\zeta_\alpha,\langle p^\alpha_\xi,q^\alpha_\xi:\xi<\zeta_\alpha\rangle:\alpha<\lambda\Big\rangle$$

of $\Game^{\rm rc\mathbf{b}}_{\mathcal{U},\bar\mu}(p,\mathbb{Q})$ *if and only if*

$(\circledast)^{\rm rc}_{\mathbf{b}}$ *there is a condition* $p^*\in\mathbb{Q}$ *stronger than* p *and such that*

$$p^*\Vdash_{\mathbb{Q}}\text{``}\ \big\{\alpha<\lambda:\big(\exists\xi<\zeta_\alpha\big)\big(q^\alpha_\xi\in\Gamma_{\mathbb{Q}}\big)\big\}\in\mathcal{U}^{\mathbb{Q}}\ \text{''}.$$

[2]equivalently, for every $\alpha<\lambda$ the set $\big\{q^\alpha_t:t\in I_\alpha\big\}$ is pre-dense above p^*

4. *Games* $\Game^{\mathrm{rc}\mathbf{a}}_{\bar{\mu}}(p,\mathbb{Q})$ *and* $\Game^{\mathrm{rc}\mathbf{c}}_{\mathcal{U},\bar{\mu}}(p,\mathbb{Q})$ *are defined similarly except that the winning criterion* $(\circledast)^{\mathrm{rc}}_{\mathbf{b}}$ *is changed so that "*$\in \mathcal{U}^{\mathbb{Q}}$*" is replaced by "*$= \lambda$*" or "*$\in \left(\mathcal{U}^{\mathbb{Q}}\right)^{+}$*", respectively.*

5. *We say that a forcing notion* $\mathbb{Q}$ *is* reasonably A–bounding over $\bar{\mu}$ *if*

   a. $\mathbb{Q}$ *is strategically* $({<}\lambda)$*–complete, and*

   b. *for any* $p \in \mathbb{Q}$*, Generic has a winning strategy in the game* $\Game^{\mathrm{rcA}}_{\bar{\mu}}(p,\mathbb{Q})$*.*

   In an analogous manner we define when the forcing notion $\mathbb{Q}$ *is* reasonably X–bounding over $\mathcal{U},\bar{\mu}$ *(for* $\mathrm{X} \in \{\mathrm{B},\mathrm{C},\mathbf{a},\mathbf{b},\mathbf{c}\}$*) — just using the game* $\Game^{\mathrm{rcX}}_{\mathcal{U},\bar{\mu}}(p,\mathbb{Q})$ *appropriately.*

   If $\mu_\alpha = \lambda$ *for each* $\alpha < \lambda$*, then we may omit* $\bar{\mu}$ *and say* reasonably B–bounding over $\mathcal{U}$ *etc. If* $\mathcal{U}$ *is the filter generated by club subsets of* λ*, we may omit it as well.*

6. *Let* $\mathbf{st}$ *be a strategy for Generic in the game* $\Game^{\mathrm{rcB}}_{\mathcal{U},\bar{\mu}}(p,\mathbb{Q})$*. We will say that a sequence* $\left\langle I_\alpha, \langle p^\alpha_t, q^\alpha_t : t \in I_\alpha\rangle : \delta < \alpha < \lambda\right\rangle$ *is a* δ–delayed play according to $\mathbf{st}$ *if it has an extension* $\left\langle I_\alpha, \langle p^\alpha_t, q^\alpha_t : t \in I_\alpha\rangle : \alpha < \lambda\right\rangle$ *which is a play agreeing with* $\mathbf{st}$ *and such that* $p^\alpha_t = q^\alpha_t$ *for* $\alpha \le \delta$*,* $t \in I_\alpha$*.*

Remark 3.2 - If $\mathbf{st}$ is a winning strategy for Generic in the game $\Game^{\mathrm{rcB}}_{\mathcal{U},\bar{\mu}}(p,\mathbb{Q})$, and $\bar{\sigma} = \left\langle I_\alpha, \langle p^\alpha_t, q^\alpha_t : t \in I_\alpha\rangle : \delta \le \alpha < \lambda\right\rangle$ is a δ–delayed play according to $\mathbf{st}$, then $\bar{\sigma}$ satisfies the condition $(\circledast)^{\mathrm{rc}}_{\mathrm{B}}$.

Remark 3.3 - For $\mathcal{U},\bar{\mu}$ as in Context 1.1, $X \in \{A,B,C,\mathbf{a},\mathbf{b},\mathbf{c}\}$ and a forcing notion $\mathbb{Q}$, let $\Phi(\mathbb{Q},X,\mathcal{U},\bar{\mu})$ be the statement

"$\mathbb{Q}$ is reasonably X–bounding over $\mathcal{U},\bar{\mu}$".

Then the following implications hold

$$\begin{array}{ccccccc}
\Phi(\mathbb{Q},A,\bar{\mu}) & \Rightarrow & \Phi(\mathbb{Q},B,\mathcal{U},\bar{\mu}) & \Rightarrow & \Phi(\mathbb{Q},C,\mathcal{U},\bar{\mu}) & & \\
\Downarrow & & \Downarrow & & \Downarrow & & \\
\Phi(\mathbb{Q},\mathbf{a},\bar{\mu}) & \Rightarrow & \Phi(\mathbb{Q},\mathbf{b},\mathcal{U},\bar{\mu}) & \Rightarrow & \Phi(\mathbb{Q},\mathbf{c},\mathcal{U},\bar{\mu}) & \Rightarrow & \mathbb{Q} \text{ is } \lambda\text{–proper.}
\end{array}$$

Theorem 3.1. *Assume that* $\lambda, \mathcal{U}, \bar{\mu}$ *are as in Context 1.1 and* $\bar{\mathbb{Q}} = \langle \mathbb{P}_\xi, \mathop{\mathbb{Q}_\xi}\limits_{\sim} : \xi < \gamma \rangle$ *is a* λ*–support iteration such that for every* $\xi < \gamma$,

$$\Vdash_{\mathbb{P}_\xi} \text{“ } \mathop{\mathbb{Q}_\xi}\limits_{\sim} \text{ is reasonably B–bounding over } \mathcal{U}, \bar{\mu} \text{ ”.}$$

Then $\mathbb{P}_\gamma = \lim(\bar{\mathbb{Q}})$ *is reasonably* **b***–bounding over* $\mathcal{U}, \bar{\mu}$ *(and so also* λ*–proper).*

Proof – For each $\xi < \gamma$ pick a $\mathbb{P}_\xi$–name $\mathop{\mathbf{st}^0_\xi}\limits_{\sim}$ such that

$$\Vdash_{\mathbb{P}_\xi} \text{“ } \mathop{\mathbf{st}^0_\xi}\limits_{\sim} \text{ is a winning strategy for Complete in } \partial^\lambda_0(\mathop{\mathbb{Q}_\xi}\limits_{\sim}, \mathop{\emptyset_{\mathbb{Q}_\xi}}\limits_{\sim}) \text{ such that}$$
$$\text{if Incomplete plays } \mathop{\emptyset_{\mathbb{Q}_\xi}}\limits_{\sim} \text{ then Complete answers with } \mathop{\emptyset_{\mathbb{Q}_\xi}}\limits_{\sim} \text{ as well ”.}$$

Also, for $\xi \leq \gamma$ and $r \in \mathbb{P}_\xi$, let $\mathbf{st}(\xi, r)$ be a winning strategy of Complete in $\partial^\lambda_0(\mathbb{P}_\xi, r)$ with the coherence properties given in Proposition 2.2.

We are going to describe a strategy $\mathbf{st}$ for Generic in the game $\partial^{\mathbf{rcb}}_{\mathcal{U},\bar{\mu}}(p, \mathbb{P}_\gamma)$. In the course of the play, at a stage $\delta < \lambda$, Generic will be instructed to construct aside

$(\otimes)_\delta$ $\quad \mathcal{T}_\delta, \bar{p}^\delta_*, \bar{q}^\delta_*, r^-_\delta, r_\delta, w_\delta, \langle \varepsilon_{\delta,\xi}, \bar{p}_{\delta,\xi}, \mathop{\bar{q}_{\delta,\xi}}\limits_{\sim} : \xi \in w_\delta \rangle$, and $\mathop{\mathbf{st}_\xi}\limits_{\sim}$ for $\xi \in w_{\delta+1} \setminus w_\delta$.

These objects will be chosen so that if

$$\langle \zeta_\delta, \langle p^\delta_\zeta, q^\delta_\zeta : \zeta < \zeta_\delta \rangle : \delta < \lambda \rangle$$

is a play of $\partial^{\mathbf{rcb}}_{\mathcal{U},\bar{\mu}}(p, \mathbb{P}_\gamma)$ in which Generic follows $\mathbf{st}$, and the side objects constructed at stage $\delta < \lambda$ are listed in $(\otimes)_\delta$, then the following conditions are satisfied (for each $\delta < \lambda$).

$(*)_1$ $r^-_\delta, r_\delta \in \mathbb{P}_\gamma$, $r_0(0) = p(0)$, $w_\delta \subseteq \gamma$, $|w_\delta| = |\delta| + 1$, $\bigcup\limits_{\alpha<\lambda} \mathrm{Dom}(r_\alpha) = \bigcup\limits_{\alpha<\lambda} w_\alpha$, $w_0 = \{0\}$, $w_\delta \subseteq w_{\delta+1}$ and if δ is limit then $w_\delta = \bigcup\limits_{\alpha<\delta} w_\alpha$.

$(*)_2$ For each $\alpha < \delta < \lambda$ we have $(\forall \xi \in w_{\alpha+1})(r_\alpha(\xi) = r_\delta(\xi))$ and $p \leq r^-_\alpha \leq r_\alpha \leq r^-_\delta \leq r_\delta$.

$(*)_3$ If $\xi \in \gamma \setminus w_\delta$, then

$$r_\delta{\restriction}\xi \Vdash \text{“ the sequence } \langle r^-_\alpha(\xi), r_\alpha(\xi) : \alpha \leq \delta \rangle \text{ is a legal partial play of}$$
$$\partial^\lambda_0(\mathop{\mathbb{Q}_\xi}\limits_{\sim}, \mathop{\emptyset_{\mathbb{Q}_\xi}}\limits_{\sim}) \text{ in which Complete follows } \mathop{\mathbf{st}^0_\xi}\limits_{\sim} \text{ ”}$$

and if $\xi \in w_{\delta+1} \setminus w_\delta$, then $\underset{\sim}{\mathbf{st}}_\xi$ is a $\mathbb{P}_\xi$–name for a winning strategy of Generic in $\Game^{\mathrm{rcB}}_{\mathcal{U},\bar{\mu}}(r_\delta(\xi), \underset{\sim}{\mathbb{Q}}_\xi)$ such that if $\langle p^\alpha_t : t \in I_\alpha \rangle$ is given by that strategy to Generic at stage α, then I_α is an ordinal below μ_α. (And $\mathbf{st}_0$ is a suitable winning strategy of Generic in $\Game^{\mathrm{rcB}}_{\mathcal{U},\bar{\mu}}(p(0), \mathbb{Q}_0)$.)

$(*)_4$ $\mathcal{T}_\delta = (T_\delta, \mathrm{rk}_\delta)$ is a standard $(w_\delta, 1)^\gamma$–tree, $|T_\delta| < \mu_\delta$.

$(*)_5$ $\bar{p}^\delta_* = \langle p^\delta_{*,t} : t \in T_\delta \rangle$ and $\bar{q}^\delta_* = \langle q^\delta_{*,t} : t \in T_\delta \rangle$ are standard trees of conditions, $\bar{p}^\delta_* \leq \bar{q}^\delta_*$.

$(*)_6$ For $t \in T_\delta$ we have $\big(\mathrm{Dom}(p) \cup \bigcup\limits_{\alpha<\delta} \mathrm{Dom}(r_\alpha) \cup w_\delta\big) \cap \mathrm{rk}_\delta(t) \subseteq \mathrm{Dom}(p^\delta_{*,t})$ and for each $\xi \in \mathrm{Dom}(p^\delta_{*,t}) \setminus w_\delta$:

$$p^\delta_{*,t}{\restriction}\xi \Vdash_{\mathbb{P}_\xi} \text{“ if the set } \{r_\alpha(\xi) : \alpha < \delta\} \cup \{p(\xi)\} \text{ has an upper bound in } \underset{\sim}{\mathbb{Q}}_\xi, \text{ then } p^\delta_{*,t}(\xi) \text{ is such an upper bound ”.}$$

$(*)_7$ $\zeta_\delta = |\{t \in T_\delta : \mathrm{rk}_\delta(t) = \gamma\}|$ and for some enumeration $\{t \in T_\delta : \mathrm{rk}_\delta(t) = \gamma\} = \{t_\zeta : \zeta < \zeta_\delta\}$, for each $\zeta < \zeta_\delta$ we have

$$p^\delta_{*,t_\zeta} \leq p^\delta_\zeta \leq q^\delta_\zeta \leq q^\delta_{*,t_\zeta}.$$

$(*)_8$ If $\xi \in w_\delta$, then $\underset{\sim}{\varepsilon}_{\delta,\xi}$ is a $\mathbb{P}_\xi$–name for an ordinal below μ_δ, $\underset{\sim}{\bar{p}}_{\delta,\xi}, \underset{\sim}{\bar{q}}_{\delta,\xi}$ are $\mathbb{P}_\xi$–names for sequences of conditions in $\underset{\sim}{\mathbb{Q}}_\xi$ of length $\underset{\sim}{\varepsilon}_{\delta,\xi}$.

$(*)_9$ If $\xi \in w_{\beta+1} \setminus w_\beta$, $\beta < \lambda$, then

$$\Vdash_{\mathbb{P}_\xi} \text{“ } \langle \underset{\sim}{\varepsilon}_{\alpha,\xi}, \underset{\sim}{\bar{p}}_{\alpha,\xi}, \underset{\sim}{\bar{q}}_{\alpha,\xi} : \beta < \alpha < \lambda \rangle \text{ is a delayed play of } \Game^{\mathrm{rcB}}_{\mathcal{U},\bar{\mu}}(r_\beta(\xi), \underset{\sim}{\mathbb{Q}}_\xi) \text{ in which Generic uses } \underset{\sim}{\mathbf{st}}_\xi \text{ ”.}$$

$(*)_{10}$ If $t \in T_\delta$, $\mathrm{rk}_\delta(t) = \xi < \gamma$, then the condition $p^\delta_{*,t}$ decides the value of $\underset{\sim}{\varepsilon}_{\delta,\xi}$, say $p^\delta_{*,t} \Vdash$“$\underset{\sim}{\varepsilon}_{\delta,\xi} = \varepsilon^t_{\delta,\xi}$”, and $\{(s)_\xi : t \vartriangleleft s \in T_\delta\} = \varepsilon^t_{\delta,\xi}$ and

$$q^\delta_{*,t} \Vdash_{\mathbb{P}_\xi} \text{“ } \underset{\sim}{\bar{p}}_{\delta,\xi}(\varepsilon) \leq p^\delta_{*,t^\frown\langle\varepsilon\rangle}(\xi) \text{ for } \varepsilon < \varepsilon^t_{\delta,\xi} \text{ and } \underset{\sim}{\bar{q}}_{\delta,\xi} = \langle q^\delta_{*,s}(\xi) : t \vartriangleleft s \in T_\delta \rangle \text{ ”.}$$

$(*)_{11}$ If $t_0, t_1 \in T_\delta$, $\mathrm{rk}_\delta(t_0) = \mathrm{rk}_\delta(t_1)$ and $\xi \in w_\delta \cap \mathrm{rk}_\delta(t_0)$, $t_0{\restriction}\xi = t_1{\restriction}\xi$ but $(t_0)_\xi \neq (t_1)_\xi$, then

$$q^\delta_{*,t_0\restriction\xi} \Vdash_{\mathbb{P}_\xi} \text{“ the conditions } q^\delta_{*,t_0}(\xi), q^\delta_{*,t_1}(\xi) \text{ are incompatible ”.}$$

$(*)_{12}$ $\mathrm{Dom}(r_\delta) = \bigcup\limits_{t\in T_\delta} \mathrm{Dom}(q^\delta_{*,t}) \cup \mathrm{Dom}(p)$ and if $t \in T_\delta$, $\xi \in \mathrm{Dom}(r_\delta) \cap \mathrm{rk}_\delta(t) \setminus w_\delta$, and $q^\delta_{*,t}{\restriction}\xi \leq q \in \mathbb{P}_\xi$, $r_\delta{\restriction}\xi \leq q$, then

$$q \Vdash_{\mathbb{P}_\xi} \text{“ if the set } \{r_\alpha(\xi) : \alpha < \delta\} \cup \{q^\delta_{*,t}(\xi), p(\xi)\} \text{ has an upper bound in } \underset{\sim}{\mathbb{Q}}_\xi, \text{ then } r_\delta(\xi) \text{ is such an upper bound ”.}$$

To describe the instructions given by $\mathbf{st}$ at stage $\delta < \lambda$ of a play of $\partial^{\mathbf{rcb}}_{\mathcal{U},\bar{\mu}}(p, \mathbb{P}_\gamma)$ let us assume that

$$\langle \zeta_\alpha, \langle p^\alpha_\zeta, q^\alpha_\zeta : \zeta < \zeta_\alpha\rangle : \alpha < \delta\rangle$$

is the result of the play so far and that Generic constructed objects listed in $(\otimes)_\alpha$ (for $\alpha < \delta$) with properties $(*)_1$–$(*)_{12}$.

First, Generic uses her favourite bookkeeping device to determine w_δ such that the demands in $(*)_1$ are satisfied (and that at the end we will have $\bigcup\limits_{\alpha<\lambda} \mathrm{Dom}(r_\alpha) = \bigcup\limits_{\alpha<\lambda} w_\alpha$). Now Generic lets $\mathcal{T}'_\delta$ be a standard $(w_\delta, 1)^\gamma$–tree such that for each $\xi \in w_\delta \cup \{\gamma\}$ we have $\{t \in T'_\delta : \mathrm{rk}'_\delta(t) = \xi\} = \prod\limits_{\varepsilon\in w_\delta\cap\xi} \mu_\delta$. Then for $\xi \in w_\delta$ she chooses $\mathbb{P}_\xi$–names $\underset{\sim}{\varepsilon}_{\delta,\xi}, \underset{\sim}{\bar{p}}_{\delta,\xi}$ such that $\underset{\sim}{\varepsilon}_{\delta,\xi}$ is a name for an ordinal below μ_δ and $\underset{\sim}{\bar{p}}_{\delta,\xi}$ is a name for a sequence of conditions in $\underset{\sim}{\mathbb{Q}}_\xi$ of length $\underset{\sim}{\varepsilon}_{\delta,\xi}$ and

$$\Vdash_{\mathbb{P}_\xi} \text{“ } \underset{\sim}{\varepsilon}_{\delta,\xi}, \underset{\sim}{\bar{p}}_{\delta,\xi} \text{ is the answer to the delayed play } \langle \underset{\sim}{\varepsilon}_{\alpha,\xi}, \underset{\sim}{\bar{p}}_{\alpha,\xi}, \underset{\sim}{\bar{q}}_{\alpha,\xi} : \xi \in w_\alpha \;\&\; \alpha < \delta\rangle \text{ given to Generic by } \underset{\sim}{\mathbf{st}}_\xi \text{ ”.}$$

She lets $\bar{p}^{\delta,0}_* = \langle p^{\delta,0}_{*,t} : t \in T'_\delta\rangle$ be a tree of conditions defined so that $\mathrm{Dom}(p^{\delta,0}_{*,t}) = \left(\mathrm{Dom}(p) \cup \bigcup\limits_{\alpha<\delta} \mathrm{Dom}(r_\alpha) \cup w_\delta\right) \cap \mathrm{rk}'_\delta(t)$ and for each $\xi \in \mathrm{Dom}(p^{\delta,0}_{*,t})$

$(*)_{13}$ $p^{\delta,0}_{*,t}(\xi)$ is the $<^*_\chi$–first $\mathbb{P}_\xi$–name for an element of $\underset{\sim}{\mathbb{Q}}_\xi$ such that

– if $\xi \in w_\delta$, then

$$\Vdash_{\mathbb{P}_\xi} \text{“ if } (t)_\xi < \underset{\sim}{\varepsilon}_{\delta,\xi} \text{ then } p^{\delta,0}_{*,t}(\xi) = \underset{\sim}{\bar{p}}_{\delta,\xi}\big((t)_\xi\big), \text{ otherwise } p^{\delta,0}_{*,t}(\xi) = \underset{\sim}{\emptyset}_{\mathbb{Q}_\xi} \text{ ”},$$

– if $\xi \notin w_\delta$, then

$$\Vdash_{\mathbb{P}_\xi} \text{“ if the set } \{r_\alpha(\xi) : \alpha < \delta\} \cup \{p(\xi)\} \text{ has an upper bound in } \underset{\sim}{\mathbb{Q}}_\xi, \text{ then } p^{\delta,0}_{*,t}(\xi) \text{ is such an upper bound ”}.$$

Now Generic uses Proposition 2.3 - 3 and then Proposition 2.3 - 2 to choose a standard tree $(w_\delta, 1)^\gamma$–tree $\mathcal{T}_\delta = (T_\delta, \mathrm{rk}_\delta)$ and a tree of conditions $\bar{p}^\delta_* = \langle p^\delta_{*,t} : t \in T_\delta\rangle$ such that

$(*)^{\rm a}_{14}$ $T_\delta \subseteq T'_\delta$ and for every $t \in T_\delta$ such that $\mathrm{rk}_\delta(t) = \xi \in w_\delta$ the condition $p^\delta_{*,t}$ decides the value of $\underset{\sim}{\varepsilon}_{\delta,\xi}$, say $p^\delta_{*,t} \Vdash \underset{\sim}{\varepsilon}_{\delta,\xi} = \varepsilon^t_{\delta,\xi}$, and

$(*)^{\rm b}_{14}$ if $t \in T_\delta$, $\mathrm{rk}_\delta(t) = \xi \in w_\delta$, then $\{\alpha < \lambda : t \cup \{\langle \xi, \alpha\rangle\} \in T_\delta\} = \varepsilon^t_{\delta,\xi}$, and

$(*)^{\rm c}_{14}$ $p^{\delta,0}_{*,t} \leq p^\delta_{*,t}$ for all $t \in T_\delta$, and if $t_0, t_1 \in T_\delta$, $\mathrm{rk}_\delta(t_0) = \mathrm{rk}_\delta(t_1)$, $\xi \in \mathrm{Dom}(t_0)$, and $t_0{\restriction}\xi = t_1{\restriction}\xi$ but $(t_0)_\xi \neq (t_1)_\xi$, then

$$p^\delta_{*,t_0{\restriction}\xi} \Vdash_{\mathbb{P}_\xi} \text{“ the conditions } p^\delta_{*,t_0}(\xi), p^\delta_{*,t_1}(\xi) \text{ are incompatible in } \underset{\sim}{\mathbb{Q}}_\xi \text{ ”},$$

Thus Generic has written aside $\mathcal{T}_\delta$, $\bar{p}^\delta_*$, w_δ and $\langle \underset{\sim}{\varepsilon}_{\delta,\xi}, \underset{\sim}{\bar{p}}_{\delta,\xi} : \xi \in w_\delta\rangle$. (It should be clear that they satisfy the relevant demands in $(*)_1$, $(*)_4$–$(*)_6$, $(*)_8$ and $(*)_9, (*)_{10}$.) Now she turns to the play of $\partial^{\mathbf{rcb}}_{\mathcal{U},\bar{\mu}}(p, \mathbb{P}_\gamma)$ and she puts

$$\zeta_\delta = |\{t \in T_\delta : \mathrm{rk}_\delta(t) = \gamma\}|$$

and she also picks an enumeration $\langle t_\zeta : \zeta < \zeta_\delta\rangle$ of $\{t \in T_\delta : \mathrm{rk}_\delta(t) = \gamma\}$. The two players start playing the subgame of level δ of length ζ_δ. During the subgame Generic constructs partial plays $\langle (r^\zeta_i, s^\zeta_i) : i \leq \zeta_\delta\rangle$ of $\partial^\lambda_0(\mathbb{P}_\gamma, p^\delta_{*,t_\zeta})$ (for $\zeta < \zeta_\delta$) in which Complete uses the strategy $\mathbf{st}(\gamma, p^\delta_{*,t_\zeta})$ and such that

$(*)^{\rm a}_{15}$ if $\zeta, \xi < \zeta_\delta$, $t \in T_\delta$, $t \vartriangleleft t_\zeta$, $t \vartriangleleft t_\xi$, $i \leq \zeta_\delta$, then $r^\zeta_i{\restriction}\mathrm{rk}_\delta(t) = r^\xi_i{\restriction}\mathrm{rk}_\delta(t)$ and $s^\zeta_i{\restriction}\mathrm{rk}_\delta(t) = s^\xi_i{\restriction}\mathrm{rk}_\delta(t)$;

$(*)^{\rm b}_{15}$ if $p^\delta_\zeta, q^\delta_\zeta$ are the conditions played at stage ζ of the subgame, then $p^\delta_{*,t_\zeta} \leq r^\zeta_i \leq p^\delta_\zeta \leq q^\delta_\zeta = r^\zeta_\zeta$ for all $i < \zeta$.

So suppose that the two players have arrived at a stage $\zeta < \zeta_\delta$ of the subgame and $\langle\langle(r^\xi_i, s^\xi_i) : i < \zeta\rangle : \xi < \zeta_\delta\rangle$ has been defined. Generic looks at $\langle(r^\zeta_i, s^\zeta_i) : i < \zeta\rangle$ – it is a play of $\partial^\lambda_0(\mathbb{P}_\gamma, p^\delta_{*,t_\zeta})$ in which Complete uses $\mathbf{st}(\gamma, p^\delta_{*,t_\zeta})$, so we may find a condition $p^\delta_\zeta \in \mathbb{P}_\gamma$ stronger than all r^ζ_i, s^ζ_i for $i < \zeta$ (and $p^\delta_\zeta \geq p^\delta_{*,t_\zeta}$). She plays this condition as her move at stage ζ of the subgame and Antigeneric answers with $q^\delta_\zeta \geq p^\delta_\zeta$. Generic lets $r^\zeta_\zeta = q^\delta_\zeta$ and she defines r^ξ_ζ for $\xi < \zeta_\delta$, $\xi \neq \zeta$, as follows. Let $t \in T_\delta$ be such that $t \vartriangleleft t_\zeta$, $t \vartriangleleft t_\xi$ and $\mathrm{rk}_\delta(t)$ is the largest possible. Generic declares that

$$\mathrm{Dom}(r^\xi_\zeta) = \big(\mathrm{Dom}(r^\zeta_\zeta) \cap \mathrm{rk}_\delta(t)\big) \cup \bigcup_{i<\zeta} \mathrm{Dom}(s^\xi_i) \cup \mathrm{Dom}(p^\delta_{*,t_\xi}),$$

and $r^\xi_\zeta {\restriction} \mathrm{rk}_\delta(t) = r^\zeta_\zeta {\restriction} \mathrm{rk}_\delta(t)$, and for $\varepsilon \in \mathrm{Dom}(r^\xi_\zeta) \setminus \mathrm{rk}_\delta(t)$ she lets $r^\xi_\zeta(\varepsilon)$ be the $<^*_\chi$–first $\mathbb{P}_\varepsilon$–name for a member of $\underset{\sim}{\mathbb{Q}}_\varepsilon$ such that

$$r^\xi_\zeta {\restriction} \varepsilon \Vdash_{\mathbb{P}_\varepsilon} \text{“ } r^\xi_\zeta(\varepsilon) \text{ is an upper bound to } \{p^\delta_{*,t_\xi}(\varepsilon)\} \cup \{s^\xi_i(\varepsilon) : i < \zeta\} \text{ ”}$$

(remember Proposition 2.2 - *iv*). Finally, s^ξ_ζ (for $\xi < \zeta_\delta$) is defined as the condition given to Complete by $\mathbf{st}(\gamma, p^\delta_{*,t_\xi})$ in answer to $\langle(r^\xi_i, s^\xi_i) : i < \zeta\rangle^\frown\langle r^\xi_\zeta\rangle$. It follows from Proposition 2.2 - *ii* that $(*)^{\rm a}_{15}$ is still satisfied for the s^ξ_ζ.

After the subgame is completed and both $p^\delta_\zeta, q^\delta_\zeta$ and $\langle\langle(r^\xi_i, s^\xi_i) : i < \zeta_\delta\rangle : \xi < \zeta_\delta\rangle$ have been determined, Generic chooses $r^0_{\zeta_\delta}$ as any upper bound to $\langle s^0_i : i < \zeta_\delta\rangle$ and then defines $r^\xi_{\zeta_\delta}$ for $\xi \in \zeta_\delta \setminus 1$ like r^ξ_ζ for $\xi \neq \zeta$ above. Also $s^\xi_{\zeta_\delta}$ (for $\xi < \zeta_\delta$) are chosen like earlier (as results of applying $\mathbf{st}(\gamma, p^\delta_{*,t_\xi})$). Finally, Generic picks a standard tree of conditions $\bar{q}^\delta_* = \langle q^\delta_{*,t} : t \in T_\delta\rangle$ such that $(\forall\zeta < \zeta_\delta)(q^\delta_{*,t_\zeta} = s^\zeta_{\zeta_\delta})$. (Note that $(*)_5$, $(*)_7$ hold.)

Now Generic defines $r^-_\delta, r_\delta \in \mathbb{P}_\gamma$ so that

$$\mathrm{Dom}(r^-_\delta) = \mathrm{Dom}(r_\delta) = \bigcup_{t\in T_\delta} \mathrm{Dom}(q^\delta_{*,t}) \cup \mathrm{Dom}(p)$$

and

$(*)^{\rm a}_{16}$ if $\xi \in \mathrm{Dom}(r^-_\delta) \setminus w_\delta$, then:

$r^-_\delta(\xi)$ is the $<^*_\chi$–first $\mathbb{P}_\xi$–name for an element of $\mathbb{Q}_\xi$ such that

$$r^-_\delta{\restriction}\xi \Vdash_{\mathbb{P}_\xi} \text{“ } r^-_\delta(\xi) \text{ is an upper bound of } \{r_\alpha(\xi) : \alpha < \delta\} \cup \{p(\xi)\} \text{ and if } t \in T_\delta,\ \mathrm{rk}_\delta(t) > \xi, \text{ and } q^\delta_{*,t}{\restriction}\xi \in \Gamma_{\mathbb{P}_\xi} \text{ and the set } \{r_\alpha(\xi) : \alpha < \delta\} \cup \{q^\delta_{*,t}(\xi), p(\xi)\} \text{ has an upper bound in } \mathbb{Q}_\xi, \text{ then } r^-_\delta(\xi) \text{ is such an upper bound ”},$$

and $r_\delta(\xi)$ is the $<^*_\chi$–first $\mathbb{P}_\xi$–name for an element of $\mathbb{Q}_\xi$ such that

$$r_\delta{\restriction}\xi \Vdash_{\mathbb{P}_\xi} \text{“ } r_\delta(\xi) \text{ is given to Complete by } \mathbf{st}^0_\xi \text{ as the answer to } \langle r^-_\alpha(\xi), r_\alpha(\xi) : \alpha < \delta\rangle^\frown\langle r^-_\delta(\xi)\rangle \text{ ”}$$

$(*)^{\mathrm{b}}_{16}$ if $\xi \in w_{\alpha+1}$, $\alpha < \delta$, then $r^-_\delta(\xi) = r_\delta(\xi) = r_\alpha(\xi)$.

(Note that by a straightforward induction on $\xi \in \mathrm{Dom}(r_\delta)$ one easily applies $(*)_3$ from previous stages to show that r^-_δ, r_δ are well defined and $r_\delta \geq r^-_\delta \geq r_\alpha, p$ for $\alpha < \delta$. Remember also $(*)_{11}$ and/or $(*)^{\mathrm{c}}_{14}$.) If $\delta = 0$ we also stipulate $r^-_0(0) = r_0(0) = p(0)$.

Finally, for each $\xi \in w_\delta$, Generic chooses a $\mathbb{P}_\xi$–name $\bar{q}_{\delta,\xi}$ for a sequence of conditions in $\mathbb{Q}_\xi$ of length $\varepsilon_{\delta,\xi}$ such that

$$\Vdash_{\mathbb{P}_\xi} \text{“ } (\forall \varepsilon < \varepsilon_{\delta,\xi})(\bar{p}_{\delta,\xi}(\varepsilon) \leq \bar{q}_{\delta,\xi}(\varepsilon)) \text{ and if } t \in T_\delta,\ \mathrm{rk}_\delta(t) > \xi, \text{ and } q^\delta_{*,t}{\restriction}\xi \in \Gamma_{\mathbb{P}_\xi} \text{ then } \bar{q}_{\delta,\xi}\big((t)_\xi\big) = q^\delta_{*,t}(\xi) \text{ ”}.$$

Generic also picks $w_{\delta+1}$ by the bookkeeping device mentioned at the beginning and for $\xi \in w_{\delta+1} \setminus w_\delta$ she fixes $\mathbf{st}_\xi$ as in $(*)_3$.

This completes the description of the side objects constructed by Generic at stage δ. Verification that they satisfy our demands $(*)_1$–$(*)_{12}$ is straightforward, and thus the description of the strategy **st** is complete.

We are going to argue now that **st** is a winning strategy for Generic. To this end suppose that

$$\big\langle \zeta_\delta, \langle p^\delta_\zeta, q^\delta_\zeta : \zeta < \zeta_\delta\rangle : \delta < \lambda\big\rangle$$

is the result of a play of $\partial^{\mathrm{rc}\mathbf{b}}_{\mathcal{U},\bar{\mu}}(p, \mathbb{P}_\gamma)$ in which Generic followed **st** and constructed aside objects listed in $(\otimes)_\delta$ (for $\delta < \lambda$) so that $(*)_1$–$(*)_{12}$ hold.

We define a condition $r \in \mathbb{P}_\gamma$ as follows. Let $\mathrm{Dom}(r) = \bigcup\limits_{\delta<\lambda} \mathrm{Dom}(r_\delta)$ and for $\xi \in \mathrm{Dom}(r)$ let $r(\xi)$ be a $\mathbb{P}_\xi$–name for a condition in $\underset{\sim}{\mathbb{Q}}_\xi$ such that if $\xi \in w_{\alpha+1} \setminus w_\alpha$, $\alpha < \lambda$ (or $\xi = 0 = \alpha$), then

$$\Vdash_{\mathbb{P}_\xi} \text{“} r(\xi) \geq r_\alpha(\xi) \text{ and } r(\xi) \Vdash_{\underset{\sim}{\mathbb{Q}}_\xi} \big\{\delta<\lambda\colon \big(\exists \varepsilon < \underset{\sim}{\varepsilon}_{\delta,\xi}\big)\big(\underset{\sim}{\bar{q}}_{\delta,\xi}(\varepsilon) \in \Gamma_{\underset{\sim}{\mathbb{Q}}_\xi}\big)\big\} \in (\mathcal{U}^{\mathbb{P}_\xi})^{\underset{\sim}{\mathbb{Q}}_\xi}\text{”}.$$

Clearly r is well defined (remember $(*)_9$) and $(\forall \delta < \lambda)(r_\delta \leq r)$ and $r \geq p$. For each $\xi \in \mathrm{Dom}(r)$ choose a sequence $\langle \underset{\sim}{A}^\xi_i : i < \lambda\rangle$ of $\mathbb{P}_{\xi+1}$–names for elements of $\mathcal{U} \cap \mathbf{V}$ such that

$(*)^\xi_{17}$ $\quad r{\restriction}(\xi+1) \Vdash_{\mathbb{P}_{\xi+1}} \big(\forall \delta \in \mathop{\triangle}\limits_{i<\lambda} \underset{\sim}{A}^\xi_i\big)\big(\exists \varepsilon < \underset{\sim}{\varepsilon}_{\delta,\xi}\big)\big(\underset{\sim}{\bar{q}}_{\delta,\xi}(\varepsilon) \in \Gamma_{\underset{\sim}{\mathbb{Q}}_\xi}\big)$.

Claim 3.1 - For each limit ordinal $\delta < \lambda$,

$$r \Vdash_{\mathbb{P}_\gamma} \text{“ } \big(\forall \xi \in w_\delta\big)\big(\delta \in \mathop{\triangle}\limits_{i<\lambda} \underset{\sim}{A}^\xi_i\big) \Rightarrow \big(\exists t \in T_\delta\big)\big(\mathrm{rk}_\delta(t) = \gamma \;\&\; q^\delta_{*,t} \in \Gamma_{\mathbb{P}_\gamma}\big)\text{ ”}.$$

Proof of the Claim – Suppose that $r' \geq r$ and a limit ordinal $\delta < \lambda$ are such that

$(*)_{18}$ $\quad r' \Vdash_{\mathbb{P}_\gamma}$“ $\big(\forall \xi \in w_\delta\big)\big(\delta \in \mathop{\triangle}\limits_{i<\lambda} \underset{\sim}{A}^\xi_i\big)$ ”.

We are going to show that there is $t \in T_\delta$ such that $\mathrm{rk}_\delta(t) = \gamma$ and the conditions $q^\delta_{*,t}$ and r' are compatible (and then the claim will readily follow). To this end let $\langle \varepsilon_\alpha : \alpha \leq \alpha^*\rangle = w_\delta \cup \{\gamma\}$ be the increasing enumeration. By induction on $\alpha \leq \alpha^*$ we will choose conditions $r^*_\alpha, r^{**}_\alpha \in \mathbb{P}_{\varepsilon_\alpha}$ and $t = \langle (t)_{\varepsilon_\alpha} : \alpha < \alpha^*\rangle \in T_\delta$ such that letting $t^\alpha_\circ = \langle (t)_{\varepsilon_\beta} : \beta < \alpha\rangle \in T_\delta$ we have

$(*)^\alpha_{19}$ $q^\delta_{*,t^\alpha_\circ} \leq r^*_\alpha$ and $r'{\restriction}\varepsilon_\alpha \leq r^*_\alpha$,

$(*)^\alpha_{20}$ $\langle r^*_\beta{}^\frown r'{\restriction}[\varepsilon_\beta,\gamma), r^{**}_\beta{}^\frown r'{\restriction}[\varepsilon_\beta,\gamma) : \beta < \alpha\rangle$ is a partial legal play of $\partial^\lambda_0(\mathbb{P}_\gamma, r')$ in which Complete uses her winning strategy $\mathbf{st}(\gamma, r')$.

Suppose that $\alpha \leq \alpha^*$ is a limit ordinal and we have already defined $t^\alpha_\circ = \langle (t)_{\varepsilon_\beta} : \beta < \alpha\rangle$ and $\langle r^*_\beta, r^{**}_\beta : \beta < \alpha\rangle$. Let $\xi = \sup(\varepsilon_\beta : \beta < \alpha)$. It follows from $(*)^\beta_{20}$ (for $\beta < \alpha$) that we may find a condition $s \in \mathbb{P}_\xi$ stronger than all r^{**}_β (for $\beta < \alpha$). Let $r^*_\alpha \in \mathbb{P}_{\varepsilon_\alpha}$ be such that $r^*_\alpha{\restriction}\xi = s$ and $r^*_\alpha{\restriction}[\xi, \varepsilon_\alpha) = r'{\restriction}[\xi, \varepsilon_\alpha)$. It follows from $(*)^\beta_{19}$ that $q^\delta_{*,t^\alpha_\circ}{\restriction}\xi \leq s = r^*_\alpha{\restriction}\xi$ and $r'{\restriction}\xi \leq s = r^*_\alpha{\restriction}\xi$. Note also that

$(\forall\beta < \alpha)(r^{**}_\beta \le s{\restriction}\varepsilon_\beta = r^*_\alpha{\restriction}\varepsilon_\beta)$, so $(\forall\beta < \alpha)(r^{**}_\beta{}^\frown r'{\restriction}[\varepsilon_\beta,\gamma) \le r^*_\alpha{}^\frown r'{\restriction}[\varepsilon_\alpha,\gamma))$. Now by induction on $\zeta \le \varepsilon_\alpha$ we show that $q^\delta_{*,t^\alpha_\circ}{\restriction}\zeta \le r^*_\alpha{\restriction}\zeta$ and $r'{\restriction}\zeta \le r^*_\alpha{\restriction}\zeta$. For $\zeta \le \xi$ we are already done, so assume that $\zeta \in [\xi,\varepsilon_\alpha)$ and we have shown $q^\delta_{*,t^\alpha_\circ}{\restriction}\zeta \le r^*_\alpha{\restriction}\zeta$ and $r'{\restriction}\zeta \le r^*_\alpha{\restriction}\zeta$. It follows from $(*)_6 + (*)_3$ that $r^*_\alpha{\restriction}\zeta \Vdash (\forall i < \delta)(r_i(\zeta) \le p^\delta_{*,t^\alpha_\circ}(\zeta))$ and therefore we may use $(*)_{12}$ to conclude that

$$r^*_\alpha{\restriction}\zeta \Vdash_{\mathbb{P}_\zeta} q^\delta_{*,t^\alpha_\circ}(\zeta) \le r_\delta(\zeta) \le r(\zeta) \le r'(\zeta) = r^*_\alpha(\zeta).$$

The limit stages are trivial and we see that $(*)^\alpha_{19}$ and (a part of) $(*)^\alpha_{20}$ hold. Finally we let $r^{**}_\alpha \in \mathbb{P}_{\varepsilon_\alpha}$ be the condition given to Complete by $\mathbf{st}(\gamma, r')$ as the response to $\langle r^*_\beta{}^\frown r'{\restriction}[\varepsilon_\beta,\gamma), r^{**}_\beta{}^\frown r'{\restriction}[\varepsilon_\beta,\gamma) : \beta < \alpha\rangle^\frown\langle r^*_\alpha\rangle$.

Now suppose that $\alpha = \beta+1 \le \alpha^*$ and we have already defined $r^*_\beta, r^{**}_\beta \in \mathbb{P}_{\varepsilon_\beta}$ and $t^\beta_\circ \in T_\delta$. It follows from $(*)^{\varepsilon_\beta}_{17} + (*)_{18} + (*)^\beta_{19} + (*)_{10}$ that

$$r^{**}_\beta \Vdash_{\mathbb{P}_{\varepsilon_\beta}} \text{“ } r'(\varepsilon_\beta) \Vdash_{\mathop{\mathbb{Q}}\limits_{\sim}{}_{\varepsilon_\beta}} \big(\exists\varepsilon < \varepsilon^{t^\beta_\circ}_{\delta,\varepsilon_\beta}\big)\big(q^\delta_{*,t^\beta_\circ{}^\frown\langle\varepsilon\rangle}(\varepsilon_\beta) \in \Gamma_{\mathop{\mathbb{Q}}\limits_{\sim}{}_{\varepsilon_\beta}}\big)\text{ ”}.$$

Therefore we may choose $\varepsilon = (t)_{\varepsilon_\beta} < \varepsilon^{t^\beta_\circ}_{\delta,\varepsilon_\beta}$ (thus defining $t^\alpha_\circ$) and a condition $s \in \mathbb{P}_{\varepsilon_\beta+1}$ such that $s{\restriction}\varepsilon_\beta \ge r^{**}_\beta \ge q^\delta_{*,t^\beta_\circ}$ and

$$s{\restriction}\varepsilon_\beta \Vdash \text{“ } s(\varepsilon_\beta) \ge r'(\varepsilon_\beta) \ \&\ s(\varepsilon_\beta) \ge q^\delta_{*,t^\alpha_\circ}(\varepsilon_\beta)\text{ ”}.$$

We let $r^*_\alpha \in \mathbb{P}_{\varepsilon_\alpha}$ be such that $r^*_\alpha{\restriction}(\varepsilon_\beta+1) = s$ and $r^*_\alpha{\restriction}(\varepsilon_\beta,\varepsilon_\alpha) = r'{\restriction}(\varepsilon_\beta,\varepsilon_\alpha)$. Exactly like in the limit case we argue that $(*)^\alpha_{19}$ and (a part of) $(*)^\alpha_{20}$ hold and then in the same manner as there we define r^{**}_α.

Finally note that $t \in T_\delta$, $\mathrm{rk}_\delta(t) = \gamma$, and the condition $r^*_{\alpha^*}$ witnesses that r' and $q^\delta_{*,t}$ are compatible. □

Now note that

$$\Vdash_{\mathbb{P}_\gamma} \text{“ } \big\{\delta < \lambda : \big(\forall\xi \in w_\delta\big)\big(\delta \in \mathop{\triangle}\limits_{i<\lambda} \mathop{A}\limits_{\sim}{}^\xi_i\big)\big\} \in \mathcal{U}^{\mathbb{P}_\gamma}\text{ ”},$$

and hence by Claim 3.1 we have

$$r \Vdash_{\mathbb{P}_\gamma} \text{“ } \big\{\delta < \lambda : \big(\exists t \in T_\delta\big)\big(\mathrm{rk}_\delta(t) = \gamma \ \&\ q^\delta_{*,t} \in \Gamma_{\mathbb{P}_\gamma}\big)\big\} \in \mathcal{U}^{\mathbb{P}_\gamma}\text{ ”}.$$

Therefore, by $(*)_7$,

$$r \Vdash_{\mathbb{P}_\gamma} \text{“ } \big\{\delta < \lambda : \big(\exists\zeta < \zeta_\delta\big)\big(q^\delta_\zeta \in \Gamma_{\mathbb{P}_\gamma}\big)\big\} \in \mathcal{U}^{\mathbb{P}_\gamma}\text{ ”}$$

and the proof of the theorem is complete. □

Remark 3.4 - The reason for the weaker "**b**–bounding" in the conclusion of Theorem 3.1 (and not "B–bounding") is that in our description of the strategy **st**, we would have to make sure that the conditions played by Antigeneric form a tree of conditions. Playing a subgame and keeping the demands of $(*)_{15}$ are a convenient way to deal with this issue.

Similar work and arguments may be carried out for A/**a** –bounding. However, in a subsequent paper [5] we find out that getting reasonably **a**–bounding for the limit of the iteration is not sufficient for the applications there. With these applications in mind we introduce a stronger property which more precisely captures what can be claimed on iterations of reasonably A–bounding forcing notions.

Definition 3.2. *Let $\bar{\mathbb{Q}} = \langle \mathbb{P}_\xi, \underset{\sim}{\mathbb{Q}}_\xi : \xi < \gamma \rangle$ be a λ–support iteration.*

1. *For a condition $p \in \mathbb{P}_\gamma = \lim(\bar{\mathbb{Q}})$ we define a game $\Game^{\mathrm{tree}\mathbf{A}}_{\bar{\mu}}(p, \bar{\mathbb{Q}})$ between two players, Generic and Antigeneric, as follows. A play of $\Game^{\mathrm{tree}\mathbf{A}}_{\bar{\mu}}(p, \bar{\mathbb{Q}})$ lasts λ steps and in the course of the play a sequence $\langle \mathcal{T}_\alpha, \bar{p}^\alpha, \bar{q}^\alpha : \alpha < \lambda \rangle$ is constructed. Suppose that the players have arrived to a stage $\alpha < \lambda$ of the game. Now,*

   $(\aleph)_\alpha$ *first Generic chooses a standard $(w, 1)^\gamma$–tree $\mathcal{T}_\alpha$ such that $|\mathcal{T}_\alpha| < \mu_\alpha$ and a tree of conditions $\bar{p}^\alpha = \langle p^\alpha_t : t \in \mathcal{T}_\alpha \rangle \subseteq \mathbb{P}_\gamma$,*

   $(\beth)_\alpha$ *then Antigeneric answers by picking a tree of conditions $\bar{q}^\alpha = \langle q^\alpha_t : t \in \mathcal{T}_\alpha \rangle \subseteq \mathbb{P}_\gamma$ such that $\bar{p}^\alpha \leq \bar{q}^\alpha$.*

   At the end, Generic wins the play $\langle \mathcal{T}_\alpha, \bar{p}^\alpha, \bar{q}^\alpha : \alpha < \lambda \rangle$ of $\Game^{\mathrm{tree}\mathbf{A}}_{\bar{\mu}}(p, \bar{\mathbb{Q}})$ if and only if

   $(\circledast)^{\mathrm{tree}}_{\mathbf{A}}$ *there is a condition $p^* \in \mathbb{P}_\gamma$ stronger than p and such that*

   $$p^* \Vdash_{\mathbb{P}_\gamma} \text{“ } (\forall \alpha < \lambda)\big(\exists t \in \mathcal{T}_\alpha\big)\big(\mathrm{rk}_\alpha(t) = \gamma \ \& \ q^\alpha_t \in \Gamma_{\mathbb{P}_\gamma}\big) \text{ ”}$$

2. *We say that $\mathbb{P}_\gamma = \lim(\bar{\mathbb{Q}})$ is reasonably* $A(\bar{\mathbb{Q}})$–bounding over $\bar{\mu}$ if Generic has a winning strategy in the game $\Game^{\mathrm{tree}\mathbf{A}}_{\bar{\mu}}(p, \bar{\mathbb{Q}})$ for every $p \in \mathbb{P}_\gamma$.*

Theorem 3.2. *Assume that $\lambda, \bar{\mu}$ are as in Context 1.1 and $\bar{\mathbb{Q}} = \langle \mathbb{P}_\xi, \underset{\sim}{\mathbb{Q}}_\xi : \xi < \gamma \rangle$ is a λ–support iteration such that for every $\xi < \gamma$,*

$$\Vdash_{\mathbb{P}_\xi} \text{“ } \underset{\sim}{\mathbb{Q}}_\xi \text{ is reasonably } A\text{–bounding over } \bar{\mu} \text{ ”.}$$

Then $\mathbb{P}_\gamma = \lim(\bar{\mathbb{Q}})$ is reasonably $A(\bar{\mathbb{Q}})$–bounding over $\bar{\mu}$.*

Proof – This is a variation on the proof of Theorem 3.1, but let us sketch the proof of our present version. For each $\xi < \gamma$ pick a $\mathbb{P}_\xi$–name $\underset{\sim}{\mathbf{st}}^0_\xi$ such that

$$\Vdash_{\mathbb{P}_\xi} \text{“ } \underset{\sim}{\mathbf{st}}^0_\xi \text{ is a winning strategy for Complete in } \Game^\lambda_0(\underset{\sim}{\mathbb{Q}}_\xi, \underset{\sim}{\emptyset}_{\mathbb{Q}_\xi}) \text{ such that}$$
$$\text{if Incomplete plays } \underset{\sim}{\emptyset}_{\mathbb{Q}_\xi} \text{ then Complete answers with } \underset{\sim}{\emptyset}_{\mathbb{Q}_\xi} \text{ as well ”.}$$

Let $p \in \mathbb{P}_\gamma$. We are going to describe a strategy $\mathbf{st}$ for Generic in the game $\Game^{\mathrm{tree}\mathbf{A}}_{\bar{\mu}}(p, \bar{\mathbb{Q}})$. In the course of the play, at a stage $\delta < \lambda$, Generic is instructed to construct aside

$(\otimes)_\delta$ $\quad r^-_\delta, r_\delta, w_\delta, \langle \underset{\sim}{\varepsilon}_{\delta,\xi}, \underset{\sim}{\bar{p}}_{\delta,\xi}, \underset{\sim}{\bar{q}}_{\delta,\xi} : \xi \in w_\delta \rangle$, and $\underset{\sim}{\mathbf{st}}_\xi$ for $\xi \in w_{\delta+1} \setminus w_\delta$.

These objects are to be chosen so that if $\langle \mathcal{T}_\delta, \bar{p}^\delta, \bar{q}^\delta : \delta < \lambda \rangle$ is a play of $\Game^{\mathrm{tree}\mathbf{A}}_{\bar{\mu}}(p, \bar{\mathbb{Q}})$ in which Generic follows $\mathbf{st}$, and the additional objects constructed at stage $\delta < \lambda$ are listed in $(\otimes)_\delta$, then the following conditions are satisfied (for each $\delta < \lambda$).

$(*)_1$ $r^-_\delta, r_\delta \in \mathbb{P}_\gamma$, $r_0(0) = p(0)$, $w_\delta \subseteq \gamma$, $|w_\delta| = |\delta + 1|$, $\bigcup\limits_{\alpha<\lambda} \mathrm{Dom}(r_\alpha) = \bigcup\limits_{\alpha<\lambda} w_\alpha$, $w_0 = \{0\}$, $w_\delta \subseteq w_{\delta+1}$ and if δ is limit then $w_\delta = \bigcup\limits_{\alpha<\delta} w_\alpha$.

$(*)_2$ For each $\alpha < \delta < \lambda$ we have $(\forall \xi \in w_{\alpha+1})(r_\alpha(\xi) = r_\delta(\xi))$ and $p \le r^-_\alpha \le r_\alpha \le r^-_\delta \le r_\delta$.

$(*)_3$ If $\xi \in \gamma \setminus w_\delta$, then

$$r_\delta {\restriction} \xi \Vdash \text{ “ the sequence } \langle r^-_\alpha(\xi), r_\alpha(\xi) : \alpha \le \delta \rangle \text{ is a legal partial play of}$$
$$\Game^\lambda_0(\underset{\sim}{\mathbb{Q}}_\xi, \underset{\sim}{\emptyset}_{\mathbb{Q}_\xi}) \text{ in which Complete follows } \underset{\sim}{\mathbf{st}}^0_\xi \text{ ”}$$

and if $\xi \in w_{\delta+1} \setminus w_\delta$, then $\underset{\sim}{\mathbf{st}}_\xi$ is a $\mathbb{P}_\xi$–name for a winning strategy of Generic in $\Game^{\mathrm{rcA}}_{\bar{\mu}}(r_\delta(\xi), \underset{\sim}{\mathbb{Q}}_\xi)$ such that if $\langle p^\alpha_t : t \in I_\alpha \rangle$ is given by that strategy to Generic at stage α, then I_α is an ordinal below μ_α. Also $\mathbf{st}_0$ is a suitable winning strategy of Generic in $\Game^{\mathrm{rcA}}_{\bar{\mu}}(p(0), \mathbb{Q}_0)$.

$(*)_4$ $\mathcal{T}_\delta = (T_\delta, \mathrm{rk}_\delta)$ is a standard $(w_\delta, 1)^\gamma$–tree, $|T_\delta| < \mu_\delta$.

$(*)_5$ $\bar{p}^\delta = \langle p^\delta_t : t \in T_\delta \rangle$ and $\bar{q}^\delta = \langle q^\delta_t : t \in T_\delta \rangle$ are standard trees of conditions in $\bar{\mathbb{Q}}$, $\bar{p}^\delta \le \bar{q}^\delta$ and $\mathcal{T}_\delta, \bar{p}^\delta$ and $\bar{q}^\delta$ are the innings of the two players at stage δ.

$(*)_6$ For $t \in T_\delta$ we have $\big(\mathrm{Dom}(p) \cup \bigcup\limits_{\alpha<\delta} \mathrm{Dom}(r_\alpha) \cup w_\delta\big) \cap \mathrm{rk}_\delta(t) \subseteq \mathrm{Dom}(p^\delta_t)$ and for each $\xi \in \mathrm{Dom}(p^\delta_t) \setminus w_\delta$:

$$p^\delta_t{\restriction}\xi \Vdash_{\mathbb{P}_\xi} \text{" if the set } \{r_\alpha(\xi) : \alpha < \delta\} \cup \{p(\xi)\} \text{ has an upper bound in } \underset{\sim}{\mathbb{Q}}_\xi, \text{ then } p^\delta_t(\xi) \text{ is such an upper bound ".}$$

$(*)_7$ If $\xi \in w_\delta$, then $\underset{\sim}{\varepsilon}_{\delta,\xi}$ is a $\mathbb{P}_\xi$–name for an ordinal below μ_δ, $\underset{\sim}{\bar{p}}_{\delta,\xi}, \underset{\sim}{\bar{q}}_{\delta,\xi}$ are $\mathbb{P}_\xi$–names for $\underset{\sim}{\varepsilon}_{\delta,\xi}$–sequences of conditions in $\underset{\sim}{\mathbb{Q}}_\xi$.

$(*)_8$ If $\xi \in w_{\beta+1} \setminus w_\beta$, $\beta < \lambda$, then

$$\Vdash_{\mathbb{P}_\xi} \text{" } \langle \underset{\sim}{\varepsilon}_{\alpha,\xi}, \underset{\sim}{\bar{p}}_{\alpha,\xi}, \underset{\sim}{\bar{q}}_{\alpha,\xi} : \beta < \alpha < \lambda \rangle \text{ is a delayed play of } \Game^{\mathrm{rcA}}_{\bar{\mu}}(r_\beta(\xi), \underset{\sim}{\mathbb{Q}}_\xi) \text{ in which Generic uses } \underset{\sim}{\mathbf{st}}_\xi \text{ ".}$$

$(*)_9$ If $t \in T_\delta$, $\mathrm{rk}_\delta(t) = \xi < \gamma$, then the condition p^δ_t decides the value of $\underset{\sim}{\varepsilon}_{\delta,\xi}$, say $p^\delta_t \Vdash$"$\underset{\sim}{\varepsilon}_{\delta,\xi} = \varepsilon^t_{\delta,\xi}$", and $\{(s)_\xi : t \lhd s \in T_\delta\} = \varepsilon^t_{\delta,\xi}$ and

$$q^\delta_t \Vdash_{\mathbb{P}_\xi} \text{" } \underset{\sim}{\bar{p}}_{\delta,\xi}(\varepsilon) \le p^\delta_{t\frown\langle\varepsilon\rangle}(\xi) \text{ and } \underset{\sim}{\bar{q}}_{\delta,\xi}(\varepsilon) = q^\delta_{t\frown\langle\varepsilon\rangle}(\xi) \text{ for } \varepsilon < \varepsilon^t_{\delta,\xi} \text{ ".}$$

$(*)_{10}$ If $t_0, t_1 \in T_\delta$, $\mathrm{rk}_\delta(t_0) = \mathrm{rk}_\delta(t_1)$ and $\xi \in w_\delta \cap \mathrm{rk}_\delta(t_0)$, $t_0{\restriction}\xi = t_1{\restriction}\xi$ but $(t_0)_\xi \ne (t_1)_\xi$, then

$$p^\delta_{t_0{\restriction}\xi} \Vdash_{\mathbb{P}_\xi} \text{" the conditions } p^\delta_{t_0}(\xi), p^\delta_{t_1}(\xi) \text{ are incompatible ".}$$

$(*)_{11}$ $\mathrm{Dom}(r_\delta) = \bigcup\limits_{t\in T_\delta} \mathrm{Dom}(q^\delta_t) \cup \mathrm{Dom}(p)$ and if $t \in T_\delta$, $\xi \in \mathrm{Dom}(r_\delta) \cap \mathrm{rk}_\delta(t) \setminus w_\delta$, and $q^\delta_t{\restriction}\xi \le q \in \mathbb{P}_\xi$, $r_\delta{\restriction}\xi \le q$, then

$$q \Vdash_{\mathbb{P}_\xi} \text{" if the set } \{r_\alpha(\xi) : \alpha < \delta\} \cup \{q^\delta_t(\xi), p(\xi)\} \text{ has an upper bound in } \underset{\sim}{\mathbb{Q}}_\xi, \text{ then } r^-_\delta(\xi) \text{ is such an upper bound ".}$$

The detailed description of the strategy **st** closely follows the description of the strategy **st** in the proof of Theorem 3.1 (after the formulation of $(*)_1$–$(*)_{12}$ there). To argue that **st** is a winning strategy for Generic, suppose that $\langle \mathcal{T}_\delta, \bar{p}^\delta, \bar{q}^\delta : \delta < \lambda \rangle$ is the result of a play of $\partial^{\rm tree\mathbf{A}}_{\bar{\mu}}(p, \bar{\mathbb{Q}})$ in which Generic followed **st** and constructed objects listed in $(\otimes)_\delta$ (for $\delta < \lambda$) so that $(*)_1$–$(*)_{11}$ hold. Define a condition $r \in \mathbb{P}_\gamma$ as follows. Let $\mathrm{Dom}(r) = \bigcup\limits_{\delta<\lambda} \mathrm{Dom}(r_\delta)$ and for $\xi \in \mathrm{Dom}(r)$ let $r(\xi)$ be a $\mathbb{P}_\xi$–name for a condition in $\mathbb{Q}_\xi$ such that if $\xi \in w_{\alpha+1} \setminus w_\alpha$, $\alpha < \lambda$ (or $\xi = 0 = \alpha$), then

$$\Vdash_{\mathbb{P}_\xi} \text{“ } r(\xi) \geq r_\alpha(\xi) \text{ and } r(\xi) \Vdash_{\mathbb{Q}_\xi} (\forall \delta < \lambda)(\exists \varepsilon < \varepsilon_{\delta,\xi})(\bar{q}_{\delta,\xi}(\varepsilon) \in \Gamma_{\mathbb{Q}_\xi}) \text{ ”}.$$

Clearly r is well defined (remember $(*)_8$) and $(\forall \delta < \lambda)(r_\delta \leq r)$ and $r \geq p$. An argument following the lines of the proof of Claim 3.1 shows that for each $\delta < \lambda$ the family $\{q^\delta_t : t \in T_\delta \ \&\ \mathrm{rk}_\delta(t) = \gamma\}$ is pre-dense above r. □

We do not know if iterations of reasonably x–bounding forcing notions are reasonably x–bounding or even λ–proper (for $x \in \{\mathbf{a}, \mathbf{b}\}$). In a subsequent paper [5] we introduce a property called nice double x–bounding and we show that it is preserved in λ–support iterations (see [5, 2.9, 2.10]). This property is in some cases stronger than being reasonably x–bounding, but it puts some restrictions on $\bar{\mu}$. In this context the following problem is very natural.

Problem 3.1.

1. *Do we have a result parallel to Theorem 3.1 for reasonably C–bounding forcings?*
2. *Let $x \in \{\mathbf{a}, \mathbf{b}\}$. Are λ–support iterations of reasonably x–bounding forcing notions still reasonably x–bounding? At least λ–proper?*

4. Consequences of reasonable ABC

Let us note that Theorem 3.2 improves [6, Theorem A.2.4]. Before we explain why, we should recall the following definition (see [6, Def. A.2.1]).

Definition 4.1. *Let $\mathbb{P}$ be a forcing notion.*

1. *A complete λ–tree of height $\alpha < \lambda$ is a set of sequences $s \subseteq {}^{\leq\alpha}\lambda$ such that*

   – *s has the $\lhd$–smallest element denoted* $\mathrm{root}(s)$,

   – *s is closed under initial segments longer than* $\mathrm{lh}(\mathrm{root}(s))$, *and*

   – *the union of any $\lhd$–increasing sequence of members of s is in s, and*

   – $(\forall \eta \in s)(\exists \nu \in s)(\eta \lhd \nu \ \&\ \mathrm{lh}(\nu) = \alpha)$.

2. *For a condition $p \in \mathbb{P}$ and an ordinal $i_0 < \lambda$ we define a game $\Game^{\rm Sacks}_{\bar{\mu}}(i_0, p, \mathbb{P})$ of two players,* Generic *and* Antigeneric. *A play lasts at most λ moves indexed by ordinals from the interval $[i_0, \lambda)$, and during it the players construct a sequence $\langle (s_i, \bar{p}^i, \bar{q}^i) : i_0 \leq i < \lambda \rangle$ as follows. At stage i of the play (where $i_0 \leq i < \lambda$), first Generic chooses $s_i \subseteq {}^{\leq i+1}\lambda$ and a system $\bar{p}^i = \langle p^i_\eta : \eta \in s_i \cap {}^{i+1}\lambda \rangle$ such that*

   α. *s_i is a complete λ–tree of height $i+1$ and* $\mathrm{lh}(\mathrm{root}(s_i)) = i_0$,

   β. *for all j such that $i_0 \leq j < i$ we have $s_j = s_i \cap {}^{\leq j+1}\lambda$,*

   γ. *$p^i_\eta \in \mathbb{P}$ for all $\eta \in s_i \cap {}^{i+1}\lambda$, and*

   δ. *if $i_0 \leq j < i$, $\nu \in s_i \cap {}^{j+1}\lambda$ and $\nu \lhd \eta \in s_i \cap {}^{i+1}\lambda$, then $q^j_\nu \leq p^i_\eta$ and $p \leq p^i_\eta$,*

   ε. $|s_i \cap {}^{i+1}\lambda| < \mu_i$.

   Then Antigeneric answers choosing a system $\bar{q}^i = \langle q^i_\eta : \eta \in s_i \cap {}^{i+1}\lambda \rangle$ of conditions in $\mathbb{P}$ such that $p^i_\eta \leq q^i_\eta$ for each $\eta \in s_i \cap {}^{i+1}\lambda$.

   Generic wins a play if she always has legal moves (so the play really lasts λ steps) and there are a condition $q \geq p$ and a $\mathbb{P}$–name $\underset{\sim}{\rho}$ such that

   $$(\circledast) \qquad q \Vdash_{\mathbb{P}} \text{“ } \underset{\sim}{\rho} \in {}^{\lambda}\lambda \ \&\ (\forall i \in [i_0, \lambda))\big(\underset{\sim}{\rho}{\restriction}(i+1) \in s_i \ \&\ q^i_{\underset{\sim}{\rho}\restriction(i+1)} \in \Gamma_{\mathbb{P}}\big) \text{ ”.}$$

3. *We say that $\mathbb{P}$ has the* strong $\bar{\mu}$–Sacks property *whenever*

   a. *$\mathbb{P}$ is strategically $(<\lambda)$–complete, and*

   b. *Generic has a winning strategy in the game $\Game^{\rm Sacks}_{\bar{\mu}}(i_0, p, \mathbb{P})$ for any $i_0 < \lambda$ and $p \in \mathbb{P}$.*

The following proposition explains why Theorem 3.2 is stronger than [6, Theorem A.2.4].

Proposition 4.1. *Assume that $\lambda, \bar{\mu}$ are as in Context 1.1 and that additionally $(\forall i < j < \lambda)(\mu_i \leq \mu_j)$. Let $\mathbb{Q}$ be a forcing notion. Then*

$$\mathbb{Q} \text{ is reasonably A–bounding over } \bar{\mu}$$

if and only if

$$\mathbb{Q} \text{ has the strong } \bar{\mu}\text{–Sacks property.}$$

PROOF – Suppose that $\mathbb{Q}$ is reasonably A–bounding over $\bar{\mu}$. Since the sequence $\bar{\mu}$ is non-decreasing, it is enough to show that Generic has a winning strategy in $\Game^{\rm Sacks}_{\bar{\mu}}(0, p, \mathbb{Q})$ for each $p \in \mathbb{Q}$ (as then almost the same strategy will be good in $\Game^{\rm Sacks}_{\bar{\mu}}(i, p, \mathbb{Q})$ for any $i < \lambda$).

Let $p \in \mathbb{Q}$. We are going to define a strategy $\mathbf{st}$ for Generic in the game $\Game^{\rm Sacks}_{\bar{\mu}}(0, p, \mathbb{Q})$. To describe it, let us fix a winning strategy $\mathbf{st}_0$ of Complete in $\Game^{\lambda}_0(\mathbb{Q}, p)$ and a winning strategy $\mathbf{st}_1$ of Generic in $\Game^{\rm rcA}_{\bar{\mu}}(p, \mathbb{Q})$. Now, at a stage $\delta < \lambda$ of the play the strategy $\mathbf{st}$ will tell Generic to write aside

$(\boxtimes)_\delta$ $\quad I_\delta$ and $\langle r^{0,\delta}_t, r^{1,\delta}_t : t \in I_\delta\rangle$ and $\langle r^\delta_\eta : \eta \in s_\delta \cap {}^{\delta+1}\lambda\rangle$

so that if $\langle (s_\delta, \bar{p}^\delta, \bar{q}^\delta) : \delta < \lambda\rangle$ is a play of $\Game^{\rm Sacks}_{\bar{\mu}}(0, p, \mathbb{Q})$ in which Generic follows $\mathbf{st}$, then the following conditions $(\odot)_1$–$(\odot)_4$ are satisfied (for each $\delta < \lambda$).

$(\odot)_1$ $\langle I_\alpha, \langle r^{0,\alpha}_t, r^{1,\alpha}_t : t \in I_\alpha\rangle : \alpha \leq \delta\rangle$ is a partial legal play of $\Game^{\rm rcA}_{\bar{\mu}}(p, \mathbb{Q})$ in which Generic uses $\mathbf{st}_1$.

$(\odot)_2$ For each $\eta \in s_\delta \cap {}^{\delta+1}\lambda$ the sequence $\langle q^\alpha_{\eta\restriction(\alpha+1)}, r^\alpha_{\eta\restriction(\alpha+1)} : \alpha \leq \delta\rangle$ is a partial legal play of $\Game^{\lambda}_0(\mathbb{Q}, p)$ in which Complete uses $\mathbf{st}_0$.

$(\odot)_3$ If $t \in I_\delta$, $\alpha < \delta$, $\nu \in s_\alpha \cap {}^{\alpha+1}\lambda$, then either $r^\alpha_\nu, r^{1,\delta}_t$ are incompatible or $r^\alpha_\nu \leq r^{1,\delta}_t$.

$(\odot)_4$ $\langle p^\delta_\nu : \nu \in s_\delta \cap {}^{\delta+1}\lambda\rangle$ is an antichain in $\mathbb{Q}$.

So suppose that the two players arrived to a stage $\delta < \lambda$ of the game $\Game^{\rm Sacks}_{\bar{\mu}}(0, p, \mathbb{Q})$ and the objects listed in $(\boxtimes)_\alpha$ (for $\alpha < \delta$) as well as $\langle (s_\alpha, \bar{p}^\alpha, \bar{q}^\alpha) :$

$\alpha < \delta\rangle$ have been constructed. First Generic uses $\mathbf{st}_1$ to pick the answer $\big(I_\delta, \langle r^{0,\delta}_t : t \in I_\delta\rangle\big)$ to $\big\langle I_\alpha, \langle r^{0,\alpha}_t, r^{1,\alpha}_t : t \in I_\alpha\rangle : \alpha < \delta\big\rangle$ in $\eth^{\rm rcA}_{\bar\mu}(p, \mathbb{Q})$. Then she uses the strategic completeness of $\mathbb{Q}$ and Proposition 2.1 to choose a system $\langle r^*_t : t \in I_\delta\rangle$ of conditions in $\mathbb{Q}$ such that

$(\odot)_5$ if $t \in I_\delta$, then $r^{0,\delta}_t \le r^*_t$ and for every $\alpha < \delta$ and $\nu \in s_\alpha \cap {}^{\alpha+1}\lambda$, either r^α_ν, r^*_t are incompatible or $r^\alpha_\nu \le r^*_t$, and also either p, r^*_t are incompatible or $p \le r^*_t$,

$(\odot)_6$ if $t_0, t_1 \in I_\delta$, $t_0 \ne t_1$, then the conditions $r^*_{t_0}, r^*_{t_1}$ are incompatible in $\mathbb{Q}$.

Now she lets $s^* = \{\eta \in {}^\delta\lambda : (\forall \alpha < \delta)(\eta{\restriction}(\alpha+1) \in s_\alpha)\}$ and

$$s^- = \{\eta \in s^* : (\exists t \in I_\delta)(\forall \alpha < \delta)(r^\alpha_{\eta\restriction(\alpha+1)} \le r^*_t \ \&\ p \le r^*_t)\},$$

and for each $\eta \in s^-$ she fixes an enumeration $\langle t^\eta_\xi : \xi < \xi_\eta\rangle$ of the set

$$\big\{t \in I_\delta : (\forall \alpha < \delta)\big(r^\alpha_{\eta\restriction(\alpha+1)} \le r^*_t \ \&\ p \le r^*_t\big)\big\}.$$

Now Generic defines

$$s^+_\delta = \big\{\nu \in {}^{\delta+1}\lambda : \big(\nu{\restriction}\delta \in s^* \setminus s^- \ \&\ \nu(\delta) = 0\big) \text{ or } \big(\nu{\restriction}\delta \in s^- \ \&\ \nu(\delta) < \xi_{\nu\restriction\delta}\big)\big\}$$

and she lets s_δ be a λ–tree of height $\delta+1$ such that $s_\delta \cap {}^{\delta+1}\lambda = s^+_\delta$. For $\nu \in s^+_\delta$ she also chooses p^δ_ν so that

- if $\nu{\restriction}\delta \notin s^-$, then $p^\delta_\nu \in \mathbb{Q}$ is an upper bound to $\{r^\alpha_{\nu\restriction(\alpha+1)} : \alpha < \delta\} \cup \{p\}$ (remember $(\odot)_2$),
- if $\nu{\restriction}\delta \in s^-$, then $p^\delta_\nu = r^*_{t^{\nu\restriction\delta}_{\nu(\delta)}}$.

And now, in the play of $\eth^{\rm Sacks}_{\bar\mu}(0, p, \mathbb{Q})$, Generic puts

$$s_\delta \quad \text{and} \quad \langle p^\delta_\nu : \nu \in s^+_\delta\rangle$$

and Antigeneric answers with $\langle q^\delta_\nu : \nu \in s^+_\delta\rangle$ (so that $q^\delta_\nu \ge p^\delta_\nu$). Conditions r^δ_ν (for $\nu \in s^+_\delta$) are determined using $\mathbf{st}_0$ (so that the demand in $(\odot)_2$ is satisfied). Finally, Generic defines also $r^{1,\delta}_t$ for $t \in I_\delta$ so that

- if $t = t^\eta_\xi$ for some $\eta \in s^-$ and $\xi < \xi_\eta$, then $r^{1,\delta}_t = r^\delta_{\eta^\frown\langle\xi\rangle}$,

– otherwise $r^{1,\delta}_t = r^*_t$.

This completes the description of what Generic plays and what she writes aside — it should be clear that the requirements of $(\odot)_1$–$(\odot)_4$ are satisfied. Now, why is **st** a winning strategy? So suppose that $\langle (s_\delta, \bar{p}^\delta, \bar{q}^\delta) : \delta < \lambda\rangle$ is a play of $\Game^{\rm Sacks}_{\bar{\mu}}(0, p, \mathbb{Q})$ in which Generic follows **st**, and I_δ, $\langle r^{0,\delta}_t, r^{1,\delta}_t : t \in I_\delta\rangle$ and $\langle r^\delta_\eta : \eta \in s_\delta \cap {}^{\delta+1}\lambda\rangle$ (for $\delta < \lambda$) are the objects constructed by Generic aside, so they satisfy $(\odot)_1$–$(\odot)_4$. It follows from $(\odot)_1$ and the choice of $\mathbf{st}_1$ that there is a condition $p^* \geq p$ such that

$(\odot)_7$ for every $\delta < \lambda$ the set $\{r^{1,\delta}_t : t \in I_\delta\}$ is pre-dense above p^*.

We claim that then also

$(\odot)_8$ for every $\delta < \lambda$ the set $\{r^\delta_\eta : \eta \in s_\delta \cap {}^{\delta+1}\lambda\}$ is pre-dense above p^*

(and this clearly implies that Generic won the play, remember $(\odot)_4$). Assume towards contradiction that $(\odot)_8$ fails and let $\delta < \lambda$ be the smallest ordinal for which we may find a condition $q \geq p^*$ such that q is incompatible with every r^δ_η for $\eta \in s_\delta \cap {}^{\delta+1}\lambda$. It follows from $(\odot)_7$ that we may pick $t \in I_\delta$ such that the conditions $r^{1,\delta}_t, q$ are compatible. By the previous sentence and by the definition of $r^{1,\delta}_t$ we get that $t \neq t^\eta_\xi$ for all $\xi < \xi_\eta$, $\eta \in s^-$ and thus $r^{1,\delta}_t = r^*_t$. Look at the condition r^*_t (satisfying $(\odot)_5 + (\odot)_6$) — it must be stronger than p and by the minimality of δ we have that $(\forall \alpha < \delta)(\exists \nu \in s_\alpha \cap {}^{\alpha+1}\lambda)(r^\alpha_\nu \leq r^*_t)$. It follows from $(\odot)_4$ from stages $\alpha < \delta$ that there is $\eta \in s^*$ such that $(\forall \alpha < \delta)(r^\alpha_{\eta\restriction(\alpha+1)} \leq r^*_t)$. Then $t \in s^-$ and hence $t = t^\eta_\xi$ for some $\xi < \xi_\eta$, contradicting what we already got.

The converse implication should be clear. □

The following easy proposition explains why the names of the properties defined in Definition 3.1 include the adjective "bounding".

Proposition 4.2. *Let $\lambda, \mathcal{U}$ and $\bar{\mu}$ be as in Context 1.1. Assume that $\mathbb{Q}$ is a forcing notion, $p \in \mathbb{Q}$ and $\underset{\sim}{\tau}$ is a $\mathbb{Q}$–name for an element of ${}^\lambda\lambda$.*

1. *If $\mathbb{Q}$ is reasonably* **a**–*bounding over $\bar{\mu}$, then there are a condition $q \geq p$ and a sequence $\bar{a} = \langle a_\alpha : \alpha < \lambda\rangle$ such that*

- $a_\alpha \subseteq \lambda$, $|a_\alpha| < \mu_\alpha$ for all $\alpha < \lambda$,
- $q \Vdash_{\mathbb{Q}}$ " $(\forall \alpha < \lambda)(\underaccent{\sim}{\tau}(\alpha) \in a_\alpha)$ ".

2. *If $\mathbb{Q}$ is reasonably* **b**–*bounding over $\mathcal{U}, \bar{\mu}$, then there are a condition $q \geq p$ and a sequence $\bar{a} = \langle a_\alpha : \alpha < \lambda \rangle$ such that*

   - $a_\alpha \subseteq \lambda$, $|a_\alpha| < \mu_\alpha$ for all $\alpha < \lambda$,
   - $q \Vdash_{\mathbb{Q}}$ " $\{\alpha < \lambda : \underaccent{\sim}{\tau}(\alpha) \in a_\alpha\} \in \mathcal{U}^{\mathbb{Q}}$ ".

3. *If $\mathbb{Q}$ is reasonably* **c**–*bounding over $\mathcal{U}, \bar{\mu}$, then there are a condition $q \geq p$ and a sequence $\bar{a} = \langle a_\alpha : \alpha < \lambda \rangle$ such that*

   - $a_\alpha \subseteq \lambda$, $|a_\alpha| < \mu_\alpha$ for all $\alpha < \lambda$,
   - $q \Vdash_{\mathbb{Q}}$ " $\{\alpha < \lambda : \underaccent{\sim}{\tau}(\alpha) \in a_\alpha\} \in \left(\mathcal{U}^{\mathbb{Q}}\right)^+$ ".

5. A model

In this section, in addition to the assumptions stated in Context 1.1 we will also assume that

Context 5.1 -

d. $S \subseteq \lambda$ is stationary and co-stationary, $S \in \mathcal{U}$,

e. $\mathcal{V}$ is a normal filter on λ, $\lambda \setminus S \in \mathcal{V}$.

Definition 5.1.

1. *Let $\alpha < \beta < \lambda$. An (α, β)*–extending function *is a mapping $c : \mathcal{P}(\alpha) \longrightarrow \mathcal{P}(\beta) \setminus \mathcal{P}(\alpha)$ such that $c(u) \cap \alpha = u$ for all $u \in \mathcal{P}(\alpha)$.*

2. *Let C be an unbounded subset of λ. A C*–extending sequence *is a sequence $\mathfrak{c} = \langle c_\alpha : \alpha \in C \rangle$ such that each c_α is an $(\alpha, \min(C \setminus (\alpha + 1)))$–extending function.*

3. *Let $C \subseteq \lambda$, $|C| = \lambda$, $\beta \in C$, $w \subseteq \beta$ and let $\mathfrak{c} = \langle c_\alpha : \alpha \in C \rangle$ be a C–extending sequence. We define $\mathrm{pos}^+(w, \mathfrak{c}, \beta)$ as the family of all subsets u of β such that*

i. *if* $\alpha_0 = \min\left(\{\alpha \in C : (\forall \xi \in w)(\xi < \alpha)\}\right)$, *then* $u \cap \alpha_0 = w$ *(so if* $\alpha_0 = \beta$, *then* $u = w$*), and*

ii. *if* $\alpha_0, \alpha_1 \in C$, $w \subseteq \alpha_0 < \alpha_1 = \min(C \setminus (\alpha_0 + 1)) \leq \beta$, *then either* $c_{\alpha_0}(u \cap \alpha_0) = u \cap \alpha_1$ or $u \cap \alpha_0 = u \cap \alpha_1$,

iii. *if* $\sup(w) < \alpha_0 = \sup(C \cap \alpha_0) \notin C$, $\alpha_1 = \min\left(C \setminus (\alpha_0 + 1)\right) \leq \beta$, *then* $u \cap \alpha_1 = u \cap \alpha_0$.

For $\alpha_0 \in \beta \cap C$ *such that* $w \subseteq \alpha_0$, *the family* $\mathrm{pos}(w, \mathfrak{c}, \alpha_0, \beta)$ *consists of all elements* u *of* $\mathrm{pos}^+(w, \mathfrak{c}, \beta)$ *which satisfy also the following condition:*

iv. *if* $\alpha_1 = \min\left(C \setminus (\alpha_0 + 1)\right) \leq \beta$, *then* $u \cap \alpha_1 = c_{\alpha_0}(w)$.

4. *A* C*–extending sequence* $\mathfrak{c} = \langle c_\alpha : \alpha \in C \rangle$ *is* S–closed *provided that*

   i. C *is a club of* λ*, and*

   ii. *if* $\alpha \in C$ *and* $u \subseteq \alpha$, *then* $\alpha \in c_\alpha(u)$*, and*

   iii. *if* $\xi \in S \setminus C$, $\alpha \in C \cap \xi$, $u \subseteq \alpha$ *and* $\xi = \sup\left(c_\alpha(u) \cap \xi\right)$, *then* $\xi \in c_\alpha(u)$.

5. *A set* $w \subseteq \lambda$ *is* S*–closed if* $\xi = \sup\left(w \cap \xi\right) \in S$ *implies* $\xi \in w$.

6. *Let* $\mathfrak{c} = \langle c_\alpha : \alpha \in C \rangle$ *be an* S*–closed* C*–extending sequence,* $\alpha_0, \beta \in C$, $w \subseteq \alpha_0 < \beta$. *Assume also that* $w \cup \{\alpha_0\}$ *is* S*–closed. Then we let*

$$\mathrm{pos}^+_S(w, \mathfrak{c}, \beta) = \{u \in \mathrm{pos}^+(w, \mathfrak{c}, \beta) : u \cup \{\beta\} \text{ is } S\text{–closed}\}$$

$$\mathrm{pos}_S(w, \mathfrak{c}, \alpha_0, \beta) = \{u \in \mathrm{pos}(w, \mathfrak{c}, \alpha_0, \beta) : u \cup \{\beta\} \text{ is } S\text{–closed}\}$$

Note that if C is a club (e.g. $\mathfrak{c}$ is S–closed), then clause Definition 5.1 - 3 *iii* is satisfied vacuously.

Remark 5.1 -

1. Assume that $\mathfrak{c}$ is a C–extending sequence, $\alpha_0, \alpha_1, \beta \in C$, $\alpha_0 < \alpha_1 \leq \beta$ and $w \subseteq \alpha_0$.

   a. If $u \in \mathrm{pos}(w, \mathfrak{c}, \alpha_0, \alpha_1)$ and $v \in \mathrm{pos}^+(u, \mathfrak{c}, \beta)$, then $v \in \mathrm{pos}(w, \mathfrak{c}, \alpha_0, \beta)$.

   b. If $v \in \mathrm{pos}(w, \mathfrak{c}, \alpha_0, \beta)$, then $v \cap \alpha_1 \in \mathrm{pos}(w, \mathfrak{c}, \alpha_0, \alpha_1)$ and $v \in \mathrm{pos}^+(v \cap \alpha_1, \mathfrak{c}, \beta)$.

c. Similarly for pos^+.

2. Assume that $\mathfrak{c}$ is an S–closed C–extending sequence, $\alpha_0, \alpha_1, \beta \in C$, $\alpha_0 < \alpha_1 \leq \beta$, $w \subseteq \alpha_0$ and $w \cup \{\alpha_0\}$ is S–closed.

   a. If $u \in \mathrm{pos}_S(w, \mathfrak{c}, \alpha_0, \alpha_1)$ and $v \in \mathrm{pos}^+_S(u, \mathfrak{c}, \beta)$, then $v \in \mathrm{pos}_S(w, \mathfrak{c}, \alpha_0, \beta)$.

   b. If $v \in \mathrm{pos}_S(w, \mathfrak{c}, \alpha_0, \beta)$, then $v \cap \alpha_1 \in \mathrm{pos}_S(w, \mathfrak{c}, \alpha_0, \alpha_1)$ and $v \in \mathrm{pos}^+_S(v \cap \alpha_1, \mathfrak{c}, \beta)$.

   c. Similarly for pos^+_S.

   d. $\emptyset \neq \mathrm{pos}_S(w, \mathfrak{c}, \alpha_0, \beta) = \{u \in \mathrm{pos}^+_S(w, \mathfrak{c}, \beta) : u \cap \alpha_0 = w \ \&\ \alpha_0 \in u\}$.

Definition 5.2. *We define a forcing notion $\mathbb{Q}^1_S$ as follows.*
A condition in $\mathbb{Q}^1_S$ *is a triple $p = (w^p, C^p, \mathfrak{c}^p)$ such that*

i. *$C^p \subseteq \lambda$ is a club of λ and $w^p \subseteq \min(C^p)$ is such that the set $w^p \cup \{\min(C^p)\}$ is S–closed,*

ii. *$\mathfrak{c}^p = \langle c^p_\alpha : \alpha \in C^p \rangle$ is an S–closed C^p–extending sequence.*

The order $\leq_{\mathbb{Q}^1_S} = \leq$ **of** $\mathbb{Q}^1_S$ *is given by*
$p \leq_{\mathbb{Q}^1_S} q$ *if and only if*

a. *$C^q \subseteq C^p$ and $w^q \in \mathrm{pos}^+_S(w^p, \mathfrak{c}^p, \min(C^q))$ and*

b. *if $\alpha_0 < \alpha_1$ are two successive members of C^q, $u \in \mathrm{pos}^+_S(w^q, \mathfrak{c}^q, \alpha_0)$, then $c^q_{\alpha_0}(u) \in \mathrm{pos}_S(u, \mathfrak{c}^p, \alpha_0, \alpha_1)$.*

For $p \in \mathbb{Q}^1_S$, $\alpha \in C^p$ and $u \in \mathrm{pos}^+_S(w^p, \mathfrak{c}^p, \alpha)$ we let $p\restriction_\alpha u \stackrel{\mathrm{def}}{=} (u, C^p \setminus \alpha, \mathfrak{c}^p \restriction (C^p \setminus \alpha))$.

Remark 5.2 - Note that in Definition 5.2 - *b* we may replace $\mathrm{pos}_S(u, \mathfrak{c}^p, \alpha_0, \alpha_1)$ by $\mathrm{pos}^+_S(u, c^p, \alpha_1)$ (remember Definition 5.1 - 4 *ii* and Remark 5.1 - 2 *d*).

Proposition 5.1.

1. *$\mathbb{Q}^1_S$ is a $(<\lambda)$–complete forcing notion of cardinality 2^λ.*

2. *If $p \in \mathbb{Q}^1_S$ and $\alpha \in C^p$, then*

- *for each* $u \in \mathrm{pos}^+_S(w^p, \mathfrak{c}^p, \alpha)$, $p{\restriction}_\alpha u \in \mathbb{Q}^1_S$ *is a condition stronger than* p, *and*
- *the family* $\{p{\restriction}_\alpha u : u \in \mathrm{pos}^+_S(w^p, \mathfrak{c}^p, \alpha)\}$ *is pre-dense above* p.

3. *Let* $p \in \mathbb{Q}^1_S$ *and* $\alpha < \beta$ *be two successive members of* C^p. *Suppose that for each* $u \in \mathrm{pos}^+_S(w^p, \mathfrak{c}^p, \alpha)$ *we are given a condition* $q_u \in \mathbb{Q}^1_S$ *such that* $p{\restriction}_\beta c^p_\alpha(u) \leq q_u$. *Then there is a condition* $q \in \mathbb{Q}^1_S$ *such that letting* $\alpha' = \min(C^q \setminus \beta)$ *we have*

   a. $p \leq q$, $w^q = w^p$, $C^q \cap \beta = C^p \cap \beta$ *and* $c^q_\delta = c^p_\delta$ *for* $\delta \in C^q \cap \alpha$, *and*

   b. $\bigcup\{w^{q_u} : u \in \mathrm{pos}^+_S(w^p, \mathfrak{c}^p, \alpha)\} \subseteq \alpha'$, *and*

   c. $q_u \leq q{\restriction}_{\alpha'} c^q_\alpha(u)$ *for every* $u \in \mathrm{pos}^+_S(w^p, \mathfrak{c}^p, \alpha)$.

4. *Assume that* $p \in \mathbb{Q}^1_S$, $\alpha \in C^p$ *and* $\underset{\sim}{\tau}$ *is a* $\mathbb{Q}^1_S$*–name such that* $p \Vdash$"$\underset{\sim}{\tau} \in \mathbf{V}$". *Then there is a condition* $q \in \mathbb{Q}^1_S$ *stronger than* p *and such that*

   a. $w^q = w^p$, $\alpha \in C^q$ *and* $C^q \cap \alpha = C^p \cap \alpha$, *and*

   b. *if* $u \in \mathrm{pos}^+_S(w^q, \mathfrak{c}^q, \alpha)$ *and* $\gamma = \min(C^q \setminus (\alpha+1))$, *then the condition* $q{\restriction}_\gamma c^q_\alpha(u)$ *forces a value to* $\underset{\sim}{\tau}$.

PROOF – 1. It should be clear that $\mathbb{Q}^1_S$ is a forcing notion of size 2^λ. To show that it is $(<\lambda)$–complete suppose that $\gamma < \lambda$ is a limit ordinal and $\bar{p} = \langle p_\xi : \xi < \gamma\rangle \subseteq \mathbb{Q}^1_S$ is $\leq_{\mathbb{Q}^1_S}$–increasing. We put $w^q = \bigcup\limits_{\xi<\gamma} w^{p_\xi}$, $C^q = \bigcap\limits_{\xi<\gamma} C^{p_\xi}$ and for $\delta \in C^q$ we define $c^q_\delta : \mathcal{P}(\delta) \longrightarrow \mathcal{P}\big(\min(C^q \setminus (\delta+1))\big)$ so that

- if $u \in \bigcap\limits_{\xi<\gamma} \mathrm{pos}^+_S(w^{p_\xi}, \mathfrak{c}^{p_\xi}, \delta)$, then $c^q_\delta(u) = \bigcup\limits_{\xi<\gamma} c^{p_\xi}_\delta(u)$,
- if $u \subseteq \delta$ but it is not in $\bigcap\limits_{\xi<\gamma} \mathrm{pos}^+_S(w^{p_\xi}, \mathfrak{c}^{p_\xi}, \delta)$, then $c^q_\delta(u) = u \cup \{\delta\}$.

Finally we put $\mathfrak{c}^q = \langle c^q_\delta : \delta \in C^q\rangle$ and $q = (w^q, C^q, \mathfrak{c}^q)$. One easily checks that $q \in \mathbb{Q}^1_S$ is a condition stronger than all p_ξ's.

2. Straightforward (remember Remark 5.1 - 2).

3. Note that if $u \in \mathrm{pos}^+_S(w^p, \mathfrak{c}^p, \alpha)$, $\alpha_u = \min(C^{q_u})$, then $w^{q_u} \in \mathrm{pos}(u, c^p, \alpha, \alpha_u)$. We let $w^q = w^p$ and $C^q = (C^p \cap \beta) \cup \bigcap\{C^{q_u} : u \in \mathrm{pos}^+_S(w^p, \mathfrak{c}^p, \alpha)\}$ (plainly, C^q

is a club of λ). Let $\alpha' = \min(C^q \setminus (\alpha+1)) = \min(C^q \setminus \beta)$. For $\delta \in C^q \cap \alpha = C^p \cap \alpha$ put $c^q_\delta = c^p_\delta$. Next, choose an (α, α')–extending function $c^q_\alpha : \mathcal{P}(\alpha) \longrightarrow \mathcal{P}(\alpha')$ such that $(\forall u \in \mathrm{pos}^+_S(w^p, \mathfrak{c}^p, \alpha))(c^q_\alpha(u) \in \mathrm{pos}^+_S(w^{q_u}, \mathfrak{c}^{q_u}, \alpha'))$ and $(c^q_\alpha(u) \setminus \alpha) \cup \{\alpha'\}$ is S–closed for each $u \subseteq \alpha$. (Remember Remark 5.1 - 2 d; note that, by the definition of C^q, $w^{q_u} \subseteq \alpha'$ for each $u \in \mathrm{pos}^+_S(w^p, \mathfrak{c}^p, \alpha)$.) Finally, if $\delta_0 < \delta_1$ are two successive members of $C^q \setminus \alpha'$, then choose a (δ_0, δ_1)–extending function $c^q_{\delta_0} : \mathcal{P}(\delta_0) \longrightarrow \mathcal{P}(\delta_1)$ so that

i. if $v \subseteq \delta_0$, $u = v \cap \alpha \in \mathrm{pos}^+_S(w^p, \mathfrak{c}^p, \alpha)$ and $v \in \mathrm{pos}^+_S(w^{q_u}, \mathfrak{c}^{q_u}, \delta_0)$, then $c^q_{\delta_0}(v) \in \mathrm{pos}_S(v, \mathfrak{c}^{q_u}, \delta_0, \delta_1)$;

ii. if $v \subseteq \delta_0$, $v \in \mathrm{pos}^+_S(w^p, \mathfrak{c}^p, \delta_0)$ but we are not in a case covered by *i*, then $c^q_{\delta_0}(v) \in \mathrm{pos}_S(v, \mathfrak{c}^p, \delta_0, \delta_1)$;

iii. in all other cases we let $c^q_{\delta_0}(v) = v \cup \{\delta_0\}$.

Let $\mathfrak{c}^q = \langle c^q_\delta : \delta \in C^q \rangle$ and $q = (w^q, C^q, \mathfrak{c}^q)$. It should be clear that $q \in \mathbb{Q}^1_S$ is a condition as required.

4. Easily follows from 3. □

Definition 5.3. *Suppose that $\gamma < \lambda$ is a limit ordinal and $\bar{p} = \langle p_\xi : \xi < \gamma \rangle \subseteq \mathbb{Q}^1_S$ is $\leq_{\mathbb{Q}^1_S}$–increasing. The condition q constructed as in the proof of Proposition 5.1 - 1 for $\bar{p}$ will be called* the natural limit of $\bar{p}$.

Proposition 5.2.

1. *Suppose $\bar{p} = \langle p_\xi : \xi < \lambda \rangle$ is a $\leq_{\mathbb{Q}^1_S}$–increasing sequence of conditions from $\mathbb{Q}^1_S$ such that*

   a. *$w^{p_\xi} = w^{p_0}$ for all $\xi < \lambda$, and*

   b. *if $\gamma < \lambda$ is limit, then p_γ is the natural limit of $\bar{p} \restriction \gamma$, and*

   c. *for each $\xi < \lambda$, if $\delta \in C^{p_\xi}$, $\mathrm{otp}(C^{p_\xi} \cap \delta) = \xi$, then $C^{p_{\xi+1}} \cap (\delta+1) = C^{p_\xi} \cap (\delta+1)$ and for every $\alpha \in C^{p_{\xi+1}} \cap \delta$ we have $c^{p_{\xi+1}}_\alpha = c^{p_\xi}_\alpha$.*

   Then the sequence $\bar{p}$ has an upper bound in $\mathbb{Q}^1_S$.

2. *Suppose that $p \in \mathbb{Q}^1_S$ and $\underset{\sim}{h}$ is a $\mathbb{Q}^1_S$–name such that $p \Vdash$"$\underset{\sim}{h} : \lambda \longrightarrow \mathbf{V}$". Then there is a condition $q \in \mathbb{Q}^1_S$ stronger than p and such that*

($\otimes$) *if* $\delta < \delta'$ *are two successive points of* C^q, $u \in \mathrm{pos}^+_S(w^q, \mathfrak{c}^q, \delta)$, *then the condition* $q{\restriction}_{\delta'} c^q_\delta(u)$ *decides the value of* $\underset{\sim}{h}{\restriction}(\delta+1)$.

PROOF – 1. First let us note that if $\delta \in \underset{\xi<\lambda}{\triangle} C^{p_\xi}$ is a limit ordinal, then $\delta \in \bigcap\limits_{\xi<\lambda} C^{p_\xi}$ and $c^{p_{\delta+1}}_\delta = c^{p_\xi}_\delta$ for all $\xi \geq \delta + 2$ (by assumptions b and c). Now, we put $w^q = w^{p_0}$ and $C^q = \{\delta \in \underset{\xi<\lambda}{\triangle} C^{p_\xi} : \delta \text{ is limit }\}$, and for $\delta \in C^q$ we let $c^q_\delta = c^{p_{\delta+1}}_\delta$ (thus defining $\mathfrak{c}^q = \langle c^q_\delta : \delta \in C^q\rangle$). It should be clear that $q = (w^q, C^q, \mathfrak{c}^q) \in \mathbb{Q}^1_S$ is an upper bound to $\bar{p}$.

2. Follows from 1 above and Proposition 5.1 - 4. □

Definition 5.4. *We let* $\underset{\sim}{W}$ *and* $\underset{\sim}{\eta}, \underset{\sim}{\nu}$ *be* $\mathbb{Q}^1_S$*–names such that*

$$\Vdash_{\mathbb{Q}^1_S} \underset{\sim}{W} = \bigcup \{w^p : p \in \Gamma_{\mathbb{Q}^1_S}\}$$

and

$$\Vdash_{\mathbb{Q}^1_S} \text{“ } \underset{\sim}{\eta}, \underset{\sim}{\nu} \in {}^\lambda\lambda \text{ and if } \langle \delta_\xi : \xi < \lambda\rangle \text{ is the increasing enumeration of } \mathrm{cl}(\underset{\sim}{W}), \text{ and } \delta_\xi \leq \alpha < \delta_{\xi+1},\ \xi < \lambda, \text{ then } \underset{\sim}{\eta}(\alpha) = \xi \text{ and } \underset{\sim}{\nu}(\alpha) = \delta_{\xi+4} \text{ ”.}$$

Proposition 5.3.

1. $\Vdash_{\mathbb{Q}^1_S}$ *“* $\underset{\sim}{W}$ *is an unbounded* S*–closed subset of* λ *”. Consequently* $\Vdash_{\mathbb{Q}^1_S}$ *“* $\underset{\sim}{W} \in \mathcal{U}^{\mathbb{Q}^1_S}$ *”.*

2. $\Vdash_{\mathbb{Q}^1_S}$ *“* $\underset{\sim}{W}, \lambda \setminus \underset{\sim}{W} \in \big(\mathcal{V}^{\mathbb{Q}^1_S}\big)^+$ *”.*

3. $\Vdash_{\mathbb{Q}^1_S} \big(\forall f \in {}^\lambda\lambda \cap \mathbf{V}\big)\big(\forall A \in \mathcal{V}^{\mathbb{Q}^1_S}\big)\big(\exists \alpha \in A\big)\big(f(\alpha) < \underset{\sim}{\nu}(\alpha)\big)$.

PROOF – 2. Suppose that $p \in \mathbb{Q}^1_S$ and $\underset{\sim}{A}_i$ (for $i < \lambda$) are $\mathbb{Q}^1_S$–names for elements of $\mathcal{V} \cap \mathbf{V}$. Build inductively sequences $\langle p_i : i < \lambda\rangle \subseteq \mathbb{Q}^1_S$ and $\langle A_i : i < \lambda\rangle \subseteq \mathcal{V}$ such that

a. $\big(\forall i < j < \lambda\big)\big(p \leq p_i \leq p_j\big)$,

b. $p_{i+1} \Vdash_{\mathbb{Q}^1_S} \underset{\sim}{A}_i = A_i$ and $i \leq \sup(w^{p_i})$ for all $i < \lambda$,

c. if $\gamma < \lambda$ is limit, then p_γ is the natural limit of $\langle p_i : i < \gamma\rangle$.

Pick a limit ordinal $\delta \in \mathop{\triangle}\limits_{i<\lambda} A_i \setminus S$ such that $\delta = \sup\big(\bigcup\limits_{i<\delta} w^{p_i}\big) \in C^{p_\delta}$ (possible by the normality of $\mathcal{V}$; remember (b,c) above). Then $p_\delta \Vdash \delta \in \mathop{\triangle}\limits_{i<\lambda} \underaccent{\tilde}{A}_i$. Put $\beta = \min\big(C^{p_\delta} \setminus (\delta+1)\big)$.

Let $w = c^{p_\delta}_\delta(w^{p_\delta})$ and $p^* = p_\delta{\restriction}_\beta w$. Then $p^* \geq p_\delta$ and $p^* \Vdash \delta \in \underaccent{\tilde}{W}$.

On the other hand, since $\delta = \sup(w^{p_\delta}) \notin S$, we have $w^{p_\delta} \in \mathrm{pos}^+_S(w^{p_\delta}, \mathfrak{c}^{p_\delta}, \beta)$ so we may let $p^{**} = p_\delta{\restriction}_\beta w^{p_\delta}$. Then $p^{**} \geq p_\delta$ and $p^{**} \Vdash \delta \notin \underaccent{\tilde}{W}$.

3. Suppose that $p \in \mathbb{Q}^1_S$, $f \in {}^\lambda\lambda$ and $\langle \underaccent{\tilde}{A}_\alpha : \alpha < \lambda\rangle$ is a sequence of $\mathbb{Q}^1_S$–names for members of $\mathcal{V} \cap \mathbf{V}$. By induction on $\alpha < \lambda$ construct a sequence $\langle p_\alpha, A_\alpha : \alpha < \lambda\rangle$ such that for each α:

i. $p_\alpha \in \mathbb{Q}^1_S$, $A_\alpha \subseteq \lambda$, $A_\alpha \in \mathcal{V}$, $p_0 = p$, $p_\alpha \leq_{\mathbb{Q}^1_S} p_{\alpha+1}$, $\min(C^{p_{\alpha+1}}) > \alpha$ and

ii. if α is a limit ordinal, then p_α is the natural limit of $\langle p_\beta : \beta < \alpha\rangle$, and

iii. $p_{\alpha+1} \Vdash_{\mathbb{Q}^1_S} \underaccent{\tilde}{A}_\alpha \cap (\lambda \setminus S) = A_\alpha$.

Next pick a limit ordinal $\delta \in \mathop{\triangle}\limits_{\alpha<\lambda} A_\alpha \cap (\lambda \setminus S)$ such that $(\forall \alpha < \delta)(\min(C^{p_\alpha}) < \delta)$. Then $p_\delta \Vdash \delta \in \mathop{\triangle}\limits_{\alpha<\lambda} \underaccent{\tilde}{A}_\alpha$ and $\delta = \min(C^{p_\delta})$ and $w^{p_\delta} \subseteq \delta$ is S–closed, so we may let $w^q = w^{p_\delta}$, $C^q = C^{p_\delta} \setminus \big(f(\delta)+1\big)$ and $\mathfrak{c}^q = \mathfrak{c}^{p_\delta}{\restriction} C^q$ to get a condition $q \in \mathbb{Q}^1_S$ stronger than p and such that

$$q \Vdash_{\mathbb{Q}^1_S} \text{“ } \delta \in \mathop{\triangle}\limits_{\alpha<\lambda} \underaccent{\tilde}{A}_\alpha \text{ and } f(\delta) < \underaccent{\tilde}{\nu}(\delta) \text{ ”}.$$

□

Proposition 5.4. *The forcing notion $\mathbb{Q}^1_S$ is reasonably B–bounding over $\mathcal{U}$.*

Proof – By Proposition 5.1 - 1, $\mathbb{Q}^1_S$ is $(<\lambda)$–complete, so we have to verify Definition 3.1 - 5 *b* only. Let $p \in \mathbb{Q}^1_S$ and let $\bar{\mu}' = \langle \mu'_\alpha : \alpha < \lambda\rangle$, $\mu'_\alpha = \lambda$ for each $\alpha < \lambda$. We are going to describe a strategy **st** for Generic in $\Game^{\mathrm{rcB}}_{\mathcal{U},\bar{\mu}'}(p, \mathbb{Q}^1_S)$.

In the course of a play the strategy **st** instructs Generic to build aside an increasing sequence of conditions $\bar{p}^* = \langle p^*_\alpha : \alpha < \lambda\rangle \subseteq \mathbb{Q}^1_S$ such that

a. $p^*_0 = p$ and $w^{p^*_\alpha} = w^p$ for all $\alpha < \lambda$, and

b. if $\gamma < \lambda$ is limit, then p^*_γ is the natural limit of $\bar{p}^*{\restriction}\gamma$, and

c. for each $\alpha < \lambda$, if $\delta \in C^{p^*_\alpha}$, $\mathrm{otp}(C^{p^*_\alpha} \cap \delta) = \alpha$, then $C^{p^*_{\alpha+1}} \cap (\delta+1) = C^{p^*_\alpha} \cap (\delta+1)$ and for every $\xi \in C^{p^*_{\alpha+1}} \cap \delta$ we have $c^{p^*_{\alpha+1}}_\xi = c^{p^*_\alpha}_\xi$, and

d. after stage $\alpha < \lambda$ of the play of $\partial^{\mathrm{rcB}}_{\mathcal{U},\bar{\mu}'}(p, \mathbb{Q}^1_S)$, the condition $p^*_{\alpha+1}$ is determined (conditions p^*_α for non-successor $\alpha < \lambda$ are determined by a,b above).

So suppose that the players arrived to a stage $\alpha < \lambda$ of $\partial^{\mathrm{rcB}}_{\mathcal{U},\bar{\mu}'}(p, \mathbb{Q}^1_S)$, and Generic (playing according to **st** so far) has constructed aside an increasing sequence $\langle p^*_\xi : \xi \leq \alpha\rangle$ of conditions (satisfying a$-d$). Let $\delta \in C^{p^*_\alpha}$ be such that $\mathrm{otp}(C^{p^*_\alpha} \cap \delta) = \alpha$ and let $\gamma = \min(C^{p^*_\alpha} \setminus (\delta+1))$. Now Generic makes her move in $\partial^{\mathrm{rcB}}_{\mathcal{U},\bar{\mu}'}(p, \mathbb{Q}^1_S)$:

- $I_\alpha = \mathrm{pos}^+_S(w^{p^*_\alpha}, \mathfrak{c}^{p^*_\alpha}, \delta)$, and
- $p^\alpha_u = p^*_\alpha \restriction_\gamma c^{p^*_\alpha}_\delta(u)$ for $u \in I_\alpha$.

Let $\langle q^\alpha_u : u \in I_\alpha\rangle \subseteq \mathbb{Q}^1_S$ be the answer of Antigeneric, so $p^*_\alpha \restriction_\gamma c^{p^*_\alpha}_\delta(u) \leq q^\alpha_u$ for each $u \in \mathrm{pos}^+(w^{p^*_\alpha}, \mathfrak{c}^{p^*_\alpha}, \delta)$. Now Generic uses Proposition 5.1 - 3 (with $\delta, \gamma, p^*_\alpha, q^\alpha_u$ here standing for α, β, p, q_u there) to pick a condition $p^*_{\alpha+1}$ such that, letting $\alpha' = \min(C^{p^*_{\alpha+1}} \setminus \gamma)$, we have

e. $p^*_\alpha \leq p^*_{\alpha+1}$, $w^{p^*_{\alpha+1}} = w^p$, $C^{p^*_{\alpha+1}} \cap \gamma = C^{p^*_\alpha} \cap \gamma$ and $c^{p^*_{\alpha+1}}_\xi = c^{p^*_\alpha}_\xi$ for $\xi \in C^{p^*_{\alpha+1}} \cap \delta$, and

f. $\bigcup\{w^{q^\alpha_u} : u \in I_\alpha\} \subseteq \alpha'$, and

g. $q^\alpha_u \leq p^*_{\alpha+1} \restriction_{\alpha'} c^{p^*_{\alpha+1}}_\delta(u)$ for every $u \in I_\alpha$.

We claim that **st** is a winning strategy for Generic in $\partial^{\mathrm{rcB}}_{\mathcal{U},\bar{\mu}'}(p, \mathbb{Q}^1_S)$. So suppose that

$$\Big\langle I_\alpha, \langle p^\alpha_u, q^\alpha_u : u \in I_\alpha\rangle : \alpha < \lambda\Big\rangle$$

is a play of $\partial^{\mathrm{rcB}}_{\mathcal{U},\bar{\mu}'}(p, \mathbb{Q}^1_S)$ in which Generic uses **st**, and let $\bar{p}^* = \langle p^*_\alpha : \alpha < \lambda\rangle \subseteq \mathbb{Q}^1_S$ be the sequence constructed aside by Generic, so it satisfies a$-c$ above, and thus also the assumptions of Proposition 5.2 - 1. Let p^* be an upper bound to $\bar{p}$ (which exists by Proposition 5.2 - 1). Now note that

$$p^* \Vdash_{\mathbb{Q}^1_S} \text{“ if } \alpha \in C^{p^*} \cap \underset{\sim}{W} \text{ and } u = \underset{\sim}{W} \cap \alpha, \text{ then } q^\alpha_u \in \Gamma_{\mathbb{Q}^1_S} \text{ ”}$$

and therefore

$$p^* \Vdash_{\mathbb{Q}^1_S} \text{“ } (\forall \alpha \in C^{p^*} \cap \underset{\sim}{W})(\exists u \in I_\alpha)(q^\alpha_u \in \Gamma_{\mathbb{Q}^1_S}) \text{ ”.}$$

Since $p^* \Vdash C^{p^*} \cap \underset{\sim}{W} \in \mathcal{U}^{\mathbb{Q}^1_S}$ (by Proposition 5.3) we may conclude that the condition p^* witnesses that Generic won the play. □

Definition 5.5. *Let $\mathcal{F}$ be a filter on λ including all co-bounded subsets of λ, $\emptyset \notin \mathcal{F}$.*

1. *We say that a family $F \subseteq {}^\lambda\lambda$ is $\mathcal{F}$–dominating whenever*

$$(\forall g \in {}^\lambda\lambda)(\exists f \in F)(\{\alpha < \lambda : g(\alpha) < f(\alpha)\} \in \mathcal{F}).$$

2. *The $\mathcal{F}$–dominating number $\mathfrak{d}_{\mathcal{F}}$ is the minimal size of an $\mathcal{F}$–dominating family in ${}^\lambda\lambda$.*

3. *If $\mathcal{F}$ is the filter of co-bounded subsets of λ, then the corresponding dominating number is also denoted by $\mathfrak{d}_\lambda$. If $\mathcal{F}$ is the filter generated by club subsets of λ, then the corresponding dominating number is called $\mathfrak{d}_{\rm cl}$.*

It was shown in Cummings and Shelah [1] that $\mathfrak{d}_\lambda = \mathfrak{d}_{\rm cl}$ (whenever $\lambda > \beth_\omega$ is regular). The following corollary is an interesting addition to that result.

Corollary 5.1. *It is consistent that λ is an inaccessible cardinal and there are two normal filters $\mathcal{U}', \mathcal{U}''$ on λ such that $\mathfrak{d}_{\mathcal{U}'} \neq \mathfrak{d}_{\mathcal{U}''}$.*

PROOF – Start with the universe where $\lambda, \mathcal{U}, \mathcal{V}, S$ are as in Context 1.1 + 5.1 and $2^\lambda = \lambda^+$. Let $\bar{\mathbb{Q}} = \langle \mathbb{P}_\xi, \underset{\sim}{\mathbb{Q}}_\xi : \xi < \lambda^{++} \rangle$ be a λ–support iteration such that for every $\xi < \lambda^{++}$, $\Vdash_{\mathbb{P}_\xi}$ “ $\underset{\sim}{\mathbb{Q}}_\xi = \mathbb{Q}^1_S$ ”.

It follows from Theorem 3.1 that $\mathbb{P}_{\lambda^{++}}$ is reasonably **b**–bounding over $\mathcal{U}$, and hence also λ–proper. Therefore using Proposition 5.1 - 1 and [6, Theorem A.1.10] (see also Eisworth [2, Section 3]) one can easily argue that the limit $\mathbb{P}_{\lambda^{++}}$ of the iteration satisfies the λ^{++}–cc, $\Vdash_{\mathbb{P}_{\lambda^{++}}} 2^\lambda = \lambda^{++}$, $\mathbb{P}_{\lambda^{++}}$ is strategically $({<}\lambda)$–complete and λ–proper. Thus, the forcing with $\mathbb{P}_{\lambda^{++}}$ does not collapse cardinals and it follows from Proposition 4.2 that

$$\Vdash_{\mathbb{P}_{\lambda^{++}}} \text{“ } {}^\lambda\lambda \cap \mathbf{V} \text{ is } (\mathcal{U})^{\mathbb{P}_{\lambda^{++}}}\text{–dominating in } {}^\lambda\lambda \text{ ”}$$

and it follows from Proposition 5.3 - 3 that for each $\xi < \lambda^{++}$

$$\Vdash_{\mathbb{P}_{\lambda^{++}}} \text{" } {}^{\lambda}\lambda \cap \mathbf{V}^{\mathbb{P}_\xi} \text{ is not } (\mathcal{V})^{\mathbb{P}_{\lambda^{++}}}\text{–dominating in } {}^{\lambda}\lambda \text{ "}$$

Therefore we may easily conclude that

$$\Vdash_{\mathbb{P}_{\lambda^{++}}} \text{" if } \mathcal{U}' = (\mathcal{U})^{\mathbb{P}_{\lambda^{++}}},\ \mathcal{U}'' = (\mathcal{V})^{\mathbb{P}_{\lambda^{++}}} \text{ then}$$
$$\mathfrak{b}_\lambda = \mathfrak{d}_{\mathcal{U}'} = \lambda^+ < 2^\lambda = \lambda^{++} = \mathfrak{d}_{\mathcal{U}''} = \mathfrak{d}_{\text{cl}} = \mathfrak{d}_\lambda \text{ ".}$$

□

6. Two bad examples of forcing notions

In this section we give two more examples of forcing notions that have some of the properties studied in the paper - but not strong enough to allow us to quote results obtained earlier. They are test cases for our future research.

Definition 6.1. *We define a forcing notion $\mathbb{P}^{\bar{\mu}}$ as follows.*
A condition in $\mathbb{P}^{\bar{\mu}}$ *is a pair $p = (f^p, C^p)$ such that*

$$C^p \subseteq \lambda \text{ is a club of } \lambda \text{ and } f^p \in \prod\{\mu_\iota : \iota \in \lambda \setminus C^p\}.$$

The order $\leq_{\mathbb{P}^{\bar{\mu}}} = \leq$ **of** $\mathbb{P}^{\bar{\mu}}$ *is given by:*
$p \leq_{\mathbb{P}^{\bar{\mu}}} q$ *if and only if* $C^q \subseteq C^p$ *and* $f^p \subseteq f^q$.

Proposition 6.1. *Assume in addition to Context 1.1 that the sequence $\bar{\mu}$ is increasing unbounded in λ. Let $\mathcal{D}_\lambda$ be the club filter on λ. Then the forcing notion $\mathbb{P}^{\bar{\mu}}$ is reasonably* **b**–*bounding over $\mathcal{D}_\lambda$ but it is not reasonably B–bounding over $\mathcal{D}_\lambda$.*

PROOF – It should be clear that $\mathbb{P}^{\bar{\mu}}$ is a $(< \lambda)$–complete forcing notion.

Let $p \in \mathbb{P}^{\bar{\mu}}$. We are going to describe a winning strategy **st** for Generic in the game $\partial^{\text{rcb}}(p, \mathbb{P}^{\bar{\mu}})$. The strategy **st** instructs Generic to construct aside (in addition to the innings in the play)

$(\odot)_1$ a closed increasing sequence $\langle \delta_\alpha : \alpha < \lambda \rangle \subseteq \lambda$ and an increasing sequence $\langle r_\alpha : \alpha < \lambda \rangle$ of conditions in $\mathbb{P}^{\bar{\mu}}$, so that

$(\odot)_2$ $r_0 = p$, $\delta_0 = \min(C^p)$ and $\{\delta_\beta : \beta \le \alpha\} = C^{r_\alpha} \cap (\delta_\alpha + 1)$ for $\alpha < \lambda$,

$(\odot)_3$ $\delta_{\alpha+1}$ and $r_{\alpha+1}$ are known right after the stage α of the play.

Suppose that the players have arrived at a stage $\alpha < \lambda$ of the play, and Generic playing according to **st** constructed aside $\langle \delta_{\beta+1}, r_{\beta+1} : \beta < \alpha \rangle$. If α is limit, then $(\odot)_1$ determines δ_α and Generic lets $r_\alpha \in \mathbb{P}^{\bar{\mu}}$ be such that $C^{r_\alpha} = \bigcap\limits_{\beta<\alpha} C^{r_\beta}$ and $f^{r_\alpha} = \bigcup\limits_{\beta<\alpha} f^{r_\beta}$. (Clearly, relevant parts of $(\odot)_1 + (\odot)_2$ are satisfied.) Now Generic picks an enumeration $\langle \eta_\xi : \xi < \zeta_\alpha \rangle$ of $\prod \{\mu_\iota : \iota \in \{\delta_\beta : \beta \le \alpha\}\}$ (for some limit $\zeta_\alpha < \lambda$) and puts ζ_α as her inning in $\Game^{\mathrm{rc}\mathbf{b}}(p, \mathbb{P}^{\bar{\mu}})$. Now the two players start a subgame of length ζ_α. The innings of Generic in the subgame are essentially determined by the following demands:

$(\odot)_4$ $p^\alpha_0 = (\eta_0 \cup f^{r_\alpha}, C^{r_\alpha} \setminus \{\delta_\beta : \beta \le \alpha\})$,

and for $\xi < \zeta < \zeta_\alpha$

$(\odot)_5$ $\eta_\xi \subseteq f^{p^\alpha_\xi}$ and $(f^{q^\alpha_\xi} {\restriction} (\lambda \setminus \{\delta_\beta : \beta \le \alpha\}), C^{q^\alpha_\xi} \cup \{\delta_\beta : \beta \le \alpha\}) \le_{\mathbb{P}^{\bar{\mu}}} (f^{p^\alpha_\zeta} {\restriction} (\lambda \setminus \{\delta_\beta : \beta \le \alpha\}), C^{p^\alpha_\zeta} \cup \{\delta_\beta : \beta \le \alpha\})$.

After the subgame is over, Generic lets $r_{\alpha+1} \in \mathbb{P}^{\bar{\mu}}$ be such that

$$f^{r_{\alpha+1}} = \bigcup \{ f^{q^\alpha_\zeta} {\restriction} (\lambda \setminus \{\delta_\beta : \beta \le \alpha\}) : \zeta < \zeta_\alpha \} \quad \text{and}$$

$$C^{r_{\alpha+1}} = \bigcap_{\zeta<\zeta_\alpha} C^{q^\alpha_\zeta} \cup \{\delta_\beta : \beta \le \alpha\}$$

and she takes $\delta_{\alpha+1} = \min\big(C^{r_{\alpha+1}} \setminus (\delta_\alpha + 1)\big)$.

This completes the description of the strategy **st**. Suppose that

$$\Big\langle \zeta_\alpha, \langle p^\alpha_\xi, q^\alpha_\xi : \xi < \zeta_\alpha \rangle : \alpha < \lambda \Big\rangle$$

is a play of $\Game^{\mathrm{rc}\mathbf{b}}(p, \mathbb{P}^{\bar{\mu}})$ in which Generic followed **st** and she constructed aside $\langle \delta_\alpha, r_\alpha : \alpha < \lambda \rangle$. Then $(\bigcup\limits_{\alpha<\lambda} f^{r_\alpha}, \bigcap\limits_{\alpha<\lambda} C^{r_\alpha}) \in \mathbb{P}^{\bar{\mu}}$ is a condition witnessing that $(\circledast)^{\mathrm{rc}}_{\mathbf{b}}$ of Definition 3.1 - 3 holds, so Generic wins this play.

To show that $\mathbb{P}^{\bar{\mu}}$ is not reasonably B–bounding over $\mathcal{D}_\lambda$, we will describe a winning strategy $\mathbf{st}^*$ for Antigeneric in $\Game^{\mathrm{rcB}}(\emptyset_{\mathbb{P}^{\bar{\mu}}}, \mathbb{P}^{\bar{\mu}})$. In the course of the play

Antigeneric will construct (in addition to his innings) an increasing sequence $\langle \xi_\delta : \delta < \lambda \rangle$ of ordinals below λ. Suppose that the two players have arrived at the stage $\delta < \lambda$ of the play and Generic has chosen a set I_δ of cardinality less than λ and a system $\langle p^\delta_t : t \in I_\delta \rangle$ of conditions in $\mathbb{P}^{\bar{\mu}}$. Now, Antigeneric lets $C = \bigcap \{C^{p^\delta_t} : t \in I_\delta\}$ (it is a club of λ) and then he picks $\xi_\delta \in C$ such that

$(\boxdot)_1$ $\mu_{\xi_\delta} > |I_\delta|$ and $\xi_\delta > \sup(\xi_\alpha : \alpha < \delta) + \delta$.

Next for every $t \in I_\delta$, Antigeneric chooses a condition $q^\delta_t \in \mathbb{P}^{\bar{\mu}}$ such that

$(\boxdot)_2$ $C^{q^\delta_t} = C \setminus (\xi_\delta + 1)$, $p^\delta_t \leq q^\delta_t$ and

$(\boxdot)_3$ if $t, s \in I_\delta$ are distinct, then $f^{q^\delta_t}(\xi_\delta) \neq f^{q^\delta_s}(\xi_\delta)$.

This completes the description of $\mathbf{st}^*$.

Suppose that $\langle I_\delta, \langle p^\delta_t, q^\delta_t : t \in I_\delta \rangle : \delta < \lambda \rangle$ is a play of $\Game^{\rm rcB}(\emptyset_{\mathbb{P}^{\bar{\mu}}}, \mathbb{P}^{\bar{\mu}})$ in which Antigeneric plays according to $\mathbf{st}^*$ and $\langle \xi_\delta : \delta < \lambda \rangle$ is the sequence constructed aside. Let $p^* \in \mathbb{P}^{\bar{\mu}}$. We claim that for every $\delta \in C^{p^*}$ the family $\{q^\delta_t : t \in I_\delta\}$ is not predense above p^*. So suppose that $\delta \in C^{p^*}$. It follows from $(\boxdot)_1 + (\boxdot)_2$ that $\delta, \xi_\delta \in \bigcap \{\lambda \setminus C^{q^\delta_t} : t \in I_\delta\}$.

Case 1 – $\xi_\delta \in C^{p^*}$.

Then we may find a condition $r \geq p^*$ such that $f^r(\xi_\delta) \notin \{f^{q^\delta_t}(\xi_\delta) : t \in I_\delta\}$ (remember $(\boxdot)_1$). The condition r is incompatible with all q^δ_t (for $t \in I_\delta$).

Case 2 – $\xi_\delta \notin C^{p^*}$.

If $f^{p^*}(\xi_\delta) \notin \{f^{q^\delta_t}(\xi_\delta) : t \in I_\delta\}$, then p^* is incompatible with all q^δ_t for $t \in I_\delta$. Otherwise there is a unique $s \in I_\delta$ such that $f^{p^*}(\xi_\delta) = f^{q^\delta_s}(\xi_\delta)$ and the condition p^* is incompatible with all q^δ_t for $t \in I_\delta \setminus \{s\}$. Since $\delta \in C^{p^*} \setminus C^{q^\delta_s}$, we may pick a condition $r \geq p^*$ such that $\delta \notin C^r$ and $f^r(\delta) \neq f^{q^\delta_s}(\delta)$. Then r is incompatible with all q^δ_t for $t \in I_\delta$.

Now we may easily argue that Generic lost the play. □

An iterable property which will be introduced in the subsequent paper [5] will capture also $\mathbb{P}^{\bar{\mu}}$. The second example is a very close relative of the forcing notion $\mathbb{Q}^1_S$ from the previous section. Yet at the moment we do not know if we can iterate it.

Definition 6.2. *We define a forcing notion $\mathbb{Q}^2_{\mathcal{U}}$ as follows.*
A condition in $\mathbb{Q}^2_{\mathcal{U}}$ *is a triple $p=(w^p,C^p,\mathfrak{c}^p)$ such that*

i. $C^p\in\mathcal{U}$, $w^p\subseteq\min(C^p)$,

ii. $\mathfrak{c}^p=\langle c^p_\alpha:\alpha\in C^p\rangle$ *is a C^p–extending sequence.*

The order $\leq_{\mathbb{Q}^2_{\mathcal{U}}}=\leq$ **of** $\mathbb{Q}^2_{\mathcal{U}}$ *is given by*
$p\leq_{\mathbb{Q}^2_{\mathcal{U}}} q$ *if and only if*

a. $C^q\subseteq C^p$ *and* $w^q\in\mathrm{pos}^+(w^p,\mathfrak{c}^p,\min(C^q))$ *and*

b. *if* $\alpha_0,\alpha_1\in C^q$, $\alpha_0<\alpha_1=\min(C^q\setminus(\alpha_0+1))$ *and* $u\in\mathrm{pos}^+(w^q,\mathfrak{c}^q,\alpha_0)$, *then* $c^q_{\alpha_0}(u)\in\mathrm{pos}(u,\mathfrak{c}^p,\alpha_0,\alpha_1)$.

For $p\in\mathbb{Q}^2_{\mathcal{U}}$, $\alpha\in C^p$ *and* $u\in\mathrm{pos}^+(w^p,\mathfrak{c}^p,\alpha)$ *we let* $p{\restriction}_\alpha u\stackrel{\rm def}{=}(u,C^p\setminus\alpha,\mathfrak{c}^p{\restriction}(C^p\setminus\alpha))$.

Proposition 6.2.

1. $\mathbb{Q}^2_{\mathcal{U}}$ *is a $({<}\lambda)$–complete forcing notion of cardinality 2^λ.*

2. *If* $p\in\mathbb{Q}^2_{\mathcal{U}}$ *and* $\alpha\in C^p$, *then*

   – *for each* $u\in\mathrm{pos}^+(w^p,\mathfrak{c}^p,\alpha)$, $p{\restriction}_\alpha u\in\mathbb{Q}^2_{\mathcal{U}}$ *is a condition stronger than* p, *and*

   – *the family* $\{p{\restriction}_\alpha u:u\in\mathrm{pos}^+(w^p,\mathfrak{c}^p,\alpha)\}$ *is pre-dense above* p.

3. *Let* $p\in\mathbb{Q}^2_{\mathcal{U}}$ *and* $\alpha<\beta$ *be two successive members of* C^p. *Suppose that for each* $u\in\mathrm{pos}^+(w^p,\mathfrak{c}^p,\alpha)$ *we are given a condition* $q_u\in\mathbb{Q}^2_{\mathcal{U}}$ *such that* $p{\restriction}_\beta c^p_\alpha(u)\leq q_u$. *Then there is a condition* $q\in\mathbb{Q}^2_{\mathcal{U}}$ *such that letting* $\alpha'=\min(C^q\setminus\beta)$ *we have*

   a. $p\leq q$, $w^q=w^p$, $C^q\cap\beta=C^p\cap\beta$ *and* $c^q_\delta=c^p_\delta$ *for* $\delta\in C^q\cap\alpha$, *and*

   b. $\bigcup\big\{w^{q_u}:u\in\mathrm{pos}^+(w^p,\mathfrak{c}^p,\alpha)\big\}\subseteq\alpha'$, *and*

   c. $q_u\leq q{\restriction}_{\alpha'}c^q_\alpha(u)$ *for every* $u\in\mathrm{pos}^+(w^p,\mathfrak{c}^p,\alpha)$.

4. *Assume that* $p\in\mathbb{Q}^2_{\mathcal{U}}$, $\alpha\in C^p$ *and* $\underset{\sim}{\tau}$ *is a* $\mathbb{Q}^2_{\mathcal{U}}$*–name such that* $p\Vdash$“$\underset{\sim}{\tau}\in\mathbf{V}$”. *Then there is a condition* $q\in\mathbb{Q}^2_{\mathcal{U}}$ *stronger than* p *and such that*

a. $w^q = w^p$, $\alpha \in C^q$ and $C^q \cap \alpha = C^p \cap \alpha$, and

b. if $u \in \mathrm{pos}^+(w^q, \mathfrak{c}^q, \alpha)$ and $\gamma = \min(C^q \setminus (\alpha+1))$, then the condition $q{\restriction}_\gamma c^q(u)$ forces a value to $\underset{\sim}{\tau}$.

PROOF – Fully parallel to Proposition 5.1. ☐

Definition 6.3. *The* natural limit *of an $\leq_{\mathbb{Q}^2_{\mathcal{U}}}$–increasing sequence $\bar{p} = \langle p_\xi : \xi < \gamma\rangle \subseteq \mathbb{Q}^2_{\mathcal{U}}$ (where $\gamma < \lambda$ is a limit ordinal) is the condition $q = (w^q, C^q, \mathfrak{c}^q)$ defined as follows:*

– $w^q = \bigcup\limits_{\xi<\gamma} w^{p_\xi}$, $C^q = \bigcap\limits_{\xi<\gamma} C^{p_\xi}$ and

– $\mathfrak{c}^q = \langle c^q_\delta : \delta \in C^q\rangle$ is such that for $\delta \in C^q$ and $u \subseteq \delta$ we have $c^q_\delta(u) = \bigcup\limits_{\xi<\gamma} c^{p_\xi}_\delta(u)$.

Proposition 6.3.

1. *Suppose $\bar{p} = \langle p_\xi : \xi < \lambda\rangle$ is a $\leq_{\mathbb{Q}^2_{\mathcal{U}}}$–increasing sequence of conditions from $\mathbb{Q}^2_{\mathcal{U}}$ such that*

   a. $w^{p_\xi} = w^{p_0}$ for all $\xi < \lambda$, and

   b. if $\gamma < \lambda$ is limit, then p_γ is the natural limit of $\bar{p}{\restriction}\gamma$, and

   c. for each $\xi < \lambda$, if $\delta \in C^{p_\xi}$, $\mathrm{otp}(C^{p_\xi} \cap \delta) = \xi$, then $C^{p_{\xi+1}} \cap (\delta+1) = C^{p_\xi} \cap (\delta+1)$ and for every $\alpha \in C^{p_{\xi+1}} \cap \delta$ we have $c^{p_{\xi+1}}_\alpha = c^{p_\xi}_\alpha$.

   Then the sequence $\bar{p}$ has an upper bound in $\mathbb{Q}^2_{\mathcal{U}}$.

2. *Suppose that $p \in \mathbb{Q}^2_{\mathcal{U}}$ and $\underset{\sim}{h}$ is a $\mathbb{Q}^2_{\mathcal{U}}$–name such that $p \Vdash$"$\underset{\sim}{h} : \lambda \longrightarrow \mathbf{V}$". Then there is a condition $q \in \mathbb{Q}^2_{\mathcal{U}}$ stronger than p and such that*

   ($\otimes$) if $\delta < \delta'$ are two successive points of C^q, $u \in \mathrm{pos}(w^q, \mathfrak{c}^q, \delta)$, then the condition $q{\restriction}_{\delta'} c^q_\delta(u)$ decides the value of $\underset{\sim}{h}{\restriction}(\delta+1)$.

PROOF – Fully parallel to Proposition 5.2. ☐

Definition 6.4. *We let $\underset{\sim}{W}$ and $\underset{\sim}{\eta}, \underset{\sim}{\nu}$ be $\mathbb{Q}^2_{\mathcal{U}}$–names such that*

$$\Vdash_{\mathbb{Q}^2_{\mathcal{U}}} \underset{\sim}{W} = \bigcup\{w^p : p \in \Gamma_{\mathbb{Q}^2_{\mathcal{U}}}\}$$

and

$$\Vdash_{\mathbb{Q}^2_{\mathcal{U}}} \text{“ } \underset{\sim}{\eta}, \underset{\sim}{\nu} \in {}^{\lambda}\lambda \text{ and if } \langle \delta_\xi : \xi < \lambda \rangle \text{ is the increasing enumeration of } \mathrm{cl}(\underset{\sim}{W}), \text{ and } \delta_\xi \le \alpha < \delta_{\xi+1},\ \xi < \lambda, \text{ then } \underset{\sim}{\eta}(\alpha) = \xi \text{ and } \underset{\sim}{\nu}(\alpha) = \delta_{\xi+4} \text{ ”.}$$

Note that if $p \in \mathbb{Q}^2_{\mathcal{U}}$, then

$$p \Vdash_{\mathbb{Q}^2_{\mathcal{U}}} \text{“ } \underset{\sim}{W} \subseteq \bigcup \big\{ [\alpha_0, \alpha_1) : \alpha_0, \alpha_1 \in C^p \ \&\ \alpha_1 = \min\big(C^p \setminus (\alpha_0 + 1)\big) \big\} \text{ ”}$$

and

$$p \Vdash_{\mathbb{Q}^2_{\mathcal{U}}} \text{“ } \big\{\alpha \in C^p : \big[\alpha, \min(C^p \setminus (\alpha+1))\big) \cap \underset{\sim}{W} \neq \emptyset\big\}, \big\{\alpha \in C^p : \big[\alpha, \min(C^p \setminus (\alpha+1))\big) \cap \underset{\sim}{W} = \emptyset\big\} \in \big(\mathcal{U}^{\mathbb{Q}^2_{\mathcal{U}}}\big)^+ \text{ ”.}$$

Proposition 6.4. $\Vdash_{\mathbb{Q}^2_{\mathcal{U}}} \big(\forall f \in {}^{\lambda}\lambda \cap \mathbf{V}\big)\big(\forall A \in \mathcal{U}^{\mathbb{Q}^2_{\mathcal{U}}}\big)\big(\exists \alpha \in A\big)\big(f(\alpha) < \underset{\sim}{\nu}(\alpha)\big)$.

PROOF – Fully parallel to Proposition 5.3. □

Proposition 6.5. *The forcing notion $\mathbb{Q}^2_{\mathcal{U}}$ is reasonably C–bounding over $\mathcal{U}$.*

PROOF – Fully parallel to Proposition 5.4. □

The following problem is a particular case of Problem 3.1 - 1.

Problem 6.1. *Are λ–support iterations of $\mathbb{Q}^2_{\mathcal{U}}$ λ–proper?*

Acknowledgements. We would like to thank the anonymous referee for very valuable comments and corrections.

References

[1] CUMMINGS, J. and SHELAH, S.: Cardinal invariants above the continuum, *Annals of Pure and Applied Logic* **75** (1995), 251–268.

[2] EISWORTH, T.: On iterated forcing for successors of regular cardinals, *Fundamenta Mathematicae* **179** (2003), 249–266.

[3] JECH, T.: *Set theory*, Academic Press, New York (1978).

[4] ROSLANOWSKI, A. and SHELAH, S.: Lords of the iteration, in preparation.

[5] ROSLANOWSKI, A. and SHELAH, S.: Reasonable ultrafilters, again, *Notre Dame Journal of Formal Logic*, submitted.

[6] ROSLANOWSKI, A. and SHELAH, S.: Sheva-Sheva-Sheva: Large Creatures, *Israel Journal of Mathematics*, accepted.

[7] ROSLANOWSKI, A. and SHELAH, S.: Iteration of λ-complete forcing notions not collapsing λ^+, *International Journal of Mathematics and Mathematical Sciences* **28** (2001), 63–82.

[8] SHELAH, S.: *Proper and improper forcing*, Perspectives in Mathematical Logic, Springer-Verlag, Berlin (1998).

[9] SHELAH, S.: Not collapsing cardinals $\leq \kappa$ in $(< \kappa)$–support iterations, *Israel Journal of Mathematics* **136** (2003), 29–115.

[10] SHELAH, S.: Successor of singulars: combinatorics and not collapsing cardinals $\leq \kappa$ in $(< \kappa)$-support iterations, *Israel Journal of Mathematics* **134** (2003), 127–155.

Finito di stampare nel mese di gennaio del 2007
dalla «Aracne editrice S.r.l.» di Roma

The Series *Quaderni di Matematica* is planned as a sequence of thematic volumes, each presenting one or more papers that deal with relevant problems in the same branch of mathematical research. The collected papers in each issue, further than new and meaningful results in their field, should also offer an uptodated picture of at least the recent researches about the problems they treat, together—if possible—with a quick look to their origin and relevance. Accordingly, will be published in the volumes of the Series only those papers whose Authors will have been explicitly invited by an Editor appointed by the Editorial Board, composed of the professors of the Department of Mathematics of the Second University of Naples.

Quaderni di Matematica publishes, as a rule, two volumes per year. The price of each volume is 32 euro in Italy and 42 euro abroad. Subscription requests and orders should be sent to

Aracne Editrice
via Raffaele Garofalo, 133
00173 Roma (Italy)
Telefax +39 06726722. 22–33
www.aracne.net

Any correspondence about exchange with other publications or any other matter should be sent to

Paola D'Aquino
Segretario di *Quaderni di Matematica*
Dipartimento di Matematica
Seconda Università di Napoli
Via Vivaldi,43 - 81100 Caserta (Italy)
Tel. +39 0823274716, Fax +39 0823274753
e-mail: quaderni@unina2.it